T0192346

Field Sampling Methods for Remedial Investigations

Originally published in 1994, the first edition of *Field Sampling Methods for Remedial Investigations* soon became a premier resource in this field. The "Princeton Groundwater" course designated it as one of the top books on the market that addresses strategies for groundwater characterization, groundwater well installation, well completion, and groundwater sampling. This long awaited third edition provides most current and most cost-effective environmental media characterization methods and approaches supporting all aspects of remediation activities. This book integrates recommendations from over one hundred of the most current US EPA, State EPA, US Geological Survey, US Army Corps of Engineers, and National Laboratory environmental guidance and/or technical documents.

This book provides guidance, examples, and/or case studies for the following subjects:

- Implementing the EPA's latest Data Quality Objectives process
- Developing cost-effective statistical and non-statistical sampling designs supporting all aspects of environmental remediation activities, and available statistical sample design software
- Aerial photography, surface geophysics, airborne/surface/downhole/building radiological surveys, soil gas surveying, environmental media sampling, DNAPL screening, portable X-ray fluorescence measurements
- Direct push groundwater sampling, well installation, well development, well purging, no-purge/low-flow/standard groundwater sampling, depth-discrete ground sampling, groundwater modeling
- Tracer testing, slug testing, waste container and building material sampling, pipe surveying, defining background conditions
- Documentation, quality control sampling, data verification/validation, data quality assessment, decontamination, health and safety, management of investigation waste

A recognized expert on this subject, author Mark Byrnes provides standard operating procedures and guidance on the proper implementation of these methods, focusing on proven technologies that are acknowledged by EPA and State regulatory agencies as reputable techniques.

Field Sampling Methods for Remedial Investigations

Third Edition

Mark Edward Byrnes

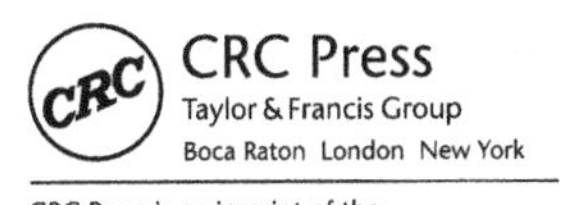

CRC Press is an imprint of the
Taylor & Francis Group, an **informa** business

Third edition published 2023
by CRC Press
6000 Broken Sound Parkway NW, Suite 300, Boca Raton, FL 33487-2742

and by CRC Press
4 Park Square, Milton Park, Abingdon, Oxon, OX14 4RN

CRC Press is an imprint of Taylor & Francis Group, LLC

First edition published by CRC Press 1994

Second edition published by CRC Press 2008

Library of Congress Cataloging-in-Publication Data
Names: Byrnes, Mark E., author.
Title: Field sampling methods for remedial investigations / Mark Edward Byrnes.
Description: Third edition. | Boca Raton : CRC Press, 2022. |
Includes bibliographical references and index.
Identifiers: LCCN 2022003073 (print) | LCCN 2022003074 (ebook) |
ISBN 9781032033013 (hardback) | ISBN 9781032255705 (paperback) |
ISBN 9781003284000 (ebook)
Subjects: LCSH: Soil pollution–Measurement–Technique. |
Water–Pollution–Measurement–Technique. | Soils–Sampling–Technique. |
Water–Sampling–Technique. | Groundwater–Sampling–Technique. |
Soil pollution–Measurement–Fieldwork. |
Water–Pollution–Measurement–Fieldwork. | Soils–Sampling–Fieldwork. |
Water–Sampling–Fieldwork. | Groundwater–Sampling–Fieldwork.
Classification: LCC TD878 .B96 2022 (print) |
LCC TD878 (ebook) | DDC 628.5/5–dc23/eng/20220215
LC record available at https://lccn.loc.gov/2022003073
LC ebook record available at https://lccn.loc.gov/2022003074

ISBN: 978-1-032-03301-3 (hbk)
ISBN: 978-1-032-25570-5 (pbk)
ISBN: 978-1-003-28400-0 (ebk)

DOI: 10.1201/9781003284000

Typeset in Times
by Newgen Publishing UK

Dedication

This book is dedicated to:

- *My father, Francis J. Byrnes, who taught me to enjoy and appreciate the fields of science and engineering, and who encouraged me to set high standards in my professional career.*
- *My Portland State University professors, Dr. Michael L. Cummings, Dr. Paul E. Hammond, and Dr. Ansel G. Johnson, who set very high standards for my academic performance.*
- *My GoJu Ryu Nanadan karate instructor, Sensei Bob Davis, who for the past 18 years has refined my technical GoJu Ryu karate skills and has been a great support system for my professional career.*
- *My Level 5 Pickleball instructor, Dr. Ken L. Curry, MD, who always reminds me "you can't win unless you hit the ball over the net".*

Contents

Preface

The purpose of this book is to provide the reader with guidance on the development and implementation of cost-effective and defensible environmental sampling programs that are designed to collect data needed to:

- Define the nature and extent of contamination
- Calculate baseline human health and ecological risk
- Identify contaminant migration pathways
- Predict rates of contaminant migration
- Evaluate the viability of potential remedial alternatives
- Support treatability testing
- Assess the effectiveness of a remedial alternative during implementation
- Determine if remedial action objectives have been met
- Dispose of waste material properly

This book addresses both intrusive and nonintrusive sampling and surveying techniques that are effective in assessing a multitude of media types, such as soil, sediment, surface water, groundwater, sludge, soil gas, air, dust, and various types of building materials. It provides standard operating procedures (SOPs) and guidance to support the proper implement of these sampling methods. In addition, this book provides guidance on how to:

- Design a cost-effective and defensible sampling program
- Use aerial photography, surface geophysics, and airborne gamma surveys to define boundaries of study areas
- Use screening instruments to define the extent of radiological contamination inside and outside of buildings
- Use downhole logging to define maximum depth of contamination as well as soil properties
- Perform tracer tests and slug tests to define groundwater paths for contaminant movement and aquifer properties
- Properly prepare sample bottles for shipment
- Prepare and maintain defensible field documentation (e.g., logbooks, chain-of-custody forms)
- Use quality control sampling, data verification, data validation, and data quality assessment to assure the data collected is of adequate quality and quantity for its intended use
- Properly decontaminate drilling and field sampling equipment
- Determine appropriate health and safety requirements
- Manage investigation-derived waste

This book focuses on those methods that have both proven themselves to be effective and are acknowledged by the U.S. Environmental Protection Agency

(EPA), state regulatory agencies, U.S. Geological Survey, and/or U.S. Army Corps of Engineers as reputable techniques. Some of the primary references that were used to support the sampling approaches recommended by this book include:

- EPA (Region 4), 2020, *Soil Sampling Operating Procedure*, Laboratory Services and Applies Science Division, Athens, Georgia, LSASDPROC-300-R4, June, www.epa.gov/sites/production/files/2015-06/documents/Soil-Sampl ing.pdf
- EPA Region 4, 2020, *Field Equipment Cleaning and Decontamination*, LSASDPROC-205-R4
- EPA Region 4, 2020, *Groundwater Level and Well Depth Measurements*, LSASDPROC-105-R4
- EPA Region 4, 2020, *Waste Sampling*, LSASDPROC-302-R4
- U.S. Geological Survey, 2020, *Passive Sampling of Groundwater Wells for Determination of Water Chemistry, Chapter 8 of Section D Water Quality, Book 1 Collection of Water Data by Direct Measurement, Techniques and Methods 1-D8*, Reston, Virginia
- EPA, 2020, *Smart Scoping of an EPA-Lead Remedial Investigation/Feasibility Study*, EPA 542-F-19-006
- Ravansari, R., et al., 2020, *Portable X-Ray Fluorescence for Environmental Assessment of Soils: Not Just a Point and Shoot Method*, Environment International 134 (2020) 105250, www.sciencedirect.com/science/article/pii/S0160412019324614
- EPA, 2019, Method TO-15A, *Determination of Volatile Organic Compounds (VOCs) in Air Collected in Specially Prepared Canisters and Analyzed by Gas Chromatography-Mass Spectrometry (GC-MS), September*
- EPA Region 4, 2017, *Groundwater Sampling*, SESDPROC-301-R4, Science and Ecosystem Support Division, Athens, Georgia
- EPA Region 1, 2017, *Low Stress (Low Flow) Purging and Sampling Procedure for the Collection of Groundwater Samples from Monitoring Wells*, EQASOP-GW4
- EPA Region 4, 2017, *Field X-Ray Fluorescence Measurement*, SESDPROC-107-R4
- EPA Region 4, 2017, *Superfund X-Ray Fluorescence Field Operations Guide*, SFDGUID-001-R0
- EPA, 2017, *National Functional Guidance for Inorganic Superfund Methods Data Review*, EPA-540-R-2017-001
- EPA, 2017, *National Functional Guidance for Organic Superfund Methods Data Review*, EPA-540-R-2017-002
- EPA Region 1, 2017, *Standard Operating Procedure Calibration of Field Instruments (temperature, pH, dissolved oxygen, conductivity/specific conductance, oxidation/reduction potential [ORP], and turbidity*, EQASOP-FieldCalibrat3, March
- EPA Region 4, 2016, *Ambient Air Sampling*, SESDPROC-303-R5, Athens, Georgia

- EPA, 2016, *Quality Assurance Guidance Document 2.12, Monitoring $PM_{2.5}$ in Ambient Air Using Designated Reference or Class I Equivalent Methods*, EPA-454/B-16-001, January
- U.S. Geological Survey, 2015, *Monitoring-Well Installation, Slug Testing, and Groundwater Quality for Selected Sites in South Park, Park County, Colorado*, 2013, Open-File Report 2014-1231, Reston, Virginia
- EPA, 2014, *Sampler's Guide, Contract Laboratory Program Guidance for Field Samplers*, EPA-540-R-014-013
- EPA, 2014, *RCRA Orientation Manual 2014*, EPA530-F-11-003
- California EPA, 2014, *Well Design and Construction for Monitoring Groundwater at Contaminated* Sites, Department of Toxic Substance Control
- U.S. Army Corps of Engineers, 2014, *Demonstration of the AGI Universal Samplers (F.K.A the GORE® Modules) for Passive Sampling of Groundwater*, ERDC/CRREL TR-14-4
- EPA Region 4, 2013, *Design and Installation of Monitoring Wells*, SESDGUID-101-R1
- California EPA, 2012, *Guidelines for Planning and Implementing Groundwater Characterization of Contaminated Sites*, Department of Toxic Substance Control
- EPA (Region 1), 2011, *Canister Sampling Standard Operating Procedure, ECASOP-Canister Sampling SOP5*, September
- U.S. Geological Survey, 2011, *Groundwater Technical Procedures of the U.S. Geological Survey, Techniques and Methods 1-A1*, Reston, Virginia
- Ohio EPA, 2007, *Technical Guidance for Ground Water Investigations, Chapter 6 Drilling and Subsurface Sampling*, Columbus, Ohio
- EPA Region 4, 2007, *Soil Sampling*, SESDPROC-300-R1
- Sterrett, R.J., 2007, *Groundwater and Wells*, 3rd Edition, Johnson Screens, New Brighton, MN, ISBN 978-0978779306
- EPA, 2007, *Field Portable X-Ray Fluorescence Spectrometry for the Determination of Elemental Concentrations in Soil and Sediment*, Method 6200, Revision 0, February
- EPA, 2006, *Guidance on Systematic Planning Using the Data Quality Objectives Process*, EPA QA/G-4
- EPA, 2006, *Data Quality Assessment: A Reviewer's Guide*, EPA QA/G-9R
- EPA, 2006, *Data Quality Assessment: Statistical Methods for Practitioners*, EPA QA/G-9S
- U.S. Geological Survey, 2006, *National Field Manual for the Collection of Water-Quality Data, Chapter A4 Collection of Water Samples*, Reston, Virginia
- Ohio EPA, 2006, *Technical Guidance Manual for Ground Water Investigations, Chapter 4 Pumping and Slug Tests*, Columbus, Ohio
- EPA, 2005, *Groundwater Sampling and Monitoring with Direct Push Technologies, OSWER No. 9200.1-51, EPA540/R-04/005*, August
- Ohio EPA, 2005, *Technical Guidance for Ground Water Investigations, Chapter 15 Use of Direct Push Technologies for Soil and Ground Water* Sampling

- U.S. Geological Survey, 2004, *Use of Submersible Pressure Transducers in Water-Resources Investigations, Chapter A of Book 8, Instrumentation, Section A, Instruments for Measurement of Water Level*, Reston, Virginia

This book has been written for environmental scientist, field geologists, hydrogeologists, geochemists, environmental engineers, biologists, and ecologists who are involved in the design and implementation of environmental sampling programs supporting the cleanup of contaminated environmental sites.

The author would appreciate criticisms or suggestions should anyone note errors or have suggestions on how later editions may be improved. Although the author has attempted to provide detailed coverage of sampling principles and methods, some emerging technologies may have been overlooked.

Acknowledgments

The author would like to express his appreciation for the support provided during the preparation of this book by: his wife Karen Byrnes; daughters Christine Vallecillo and Kathleen Schuh; brother Gordon Byrnes, M.D.; sister-in-law Jane Crawford; son-in-laws Edras Vallecillo and Brock Schuh; nephew Ian Crawford; and mother Frieda Byrnes. The author is also grateful for the support provided by the following Central Plateau Cleanup Company staff: Alaa Aly, Craig Arola, Bill Barrett, Ryan Carter, Sam Cartmell, Chris Cearlock, Mark Cherry, Steve Churchill, Erika Cutsforth, Mary Day, Bonnie Dodge, Bob Evans, Bill Faught, Scot Fitzgerald, Randal Fox, Jim Geiger, Mary Hartman, Jason Hulstrom, Paul Humphreys, Marc Jewett, Chris Koerner, Art Lee, Kathy Leonard, Jeff Lerch, Paul Martin, Emily Macdonald, Byron Miller, Anthony Nagel, Darrell Newcomer, Laura O'Mara, Jennifer Richart, Virginia Rohay, Ana Sherwood, Sarah Springer, Dave St John, Meghann Stewart, Laine Sumner, Greg Thomas, Kelly Whitley, and Britt Wilkens, Rod Wurdeman.

Support and/or input provided by the following companies and individuals are also greatly appreciated:

Amplified Geochemical Imaging, LLC: Mark Arnold
Beacon Environmental Services: Harry O'Neill
C.C. Lynch & Associates, Inc.: Trey Campbell
EON Products, Inc.: Brad Varhol
FLUTe: Carl Keller
Freestone Environmental: Steve Airhart, Tracy Mallgren, Kim Schuyler, and Dan Tyler
GRAM Northwest, LLC: Nathan Bowles, Anna Radloff, and Michelle Riffe
In Situ, Inc.: Anne Oaks and Steve Sewell
Intera, Inc.: Alaa Aly, Marty Doornbos, and Greg Ruskauff
Pacific Northwest National Laboratory: Inci Demirkanli, Rob Mackley, and Vicky Freedman
QED Environmental Systems, Inc.: Sandy Britt, Sarah Conard, Ethan Cowling, and James Jones
Solinst Canada Ltd.: Chris Batt, Jason Redwood, and Mauricio Vasquez
S.S. Papadopulos & Associates, Inc.: Mashrur Chowdhury, Marinko Karanovic, and Matt Tonkin
Student: Connor Hrebeniuk
U.S. Department of Energy: Kate Amrhein, Mike Cline, and Jim Hanson
U.S. EPA: Emy Laija, Craig Cameron, and Geoff Schramm
Washington State Department of Ecology: Dib Goswami, Kim Welsch, and Beth Rochette

The author would also like to recognize his instructors and fellow GoJu Ryu practitioners: Sensei Teruo Chinen, Sensei Bruce McDavis, Sensei Bob Davis, Sensei

Mary Roe, Sensei Valeriy Kusiy, Sensei Clinton Streifling, Sensei John Lowrie, Sensei Brian Morrow, Sensei Art Lee, Sensei Wayne Currie, Sensei Don Lalonde, Sensei Daniel Thompson, Sensei Don Goodwin, Sensei Dennis Thomas, Sensei Bryan Thomas, Tod Dodge, Shane Young, and Kim Wilson.

About the Author

Mark Edward Byrnes, RG, RHG, HS-PMP, is a senior environmental scientist and project manager working for the Central Plateau Cleanup Company at the Hanford Nuclear Reservation, where he is responsible for the groundwater characterization of one of the largest chlorinated solvent plumes in the world. Mark has over 30 years of experience overseeing the design and implementation of statistical and nonstatistical environmental sampling programs to support site characterization, waste characterization, risk assessments, feasibility studies, and remedial designs. He received his Master of Science degree in geology/geochemistry from Portland State University (Oregon), and his Bachelor of Arts degree in geology from the University of Colorado (Boulder). Mark is a registered geologist and hydrogeologist in the State of Washington. He is the author of the earlier 2009 and 1994 editions of this book which have been used as textbooks at many universities across the country, including Georgia Institute of Technology. He is also the author of the 2003 Lewis Publishers book titled *Nuclear, Chemical, and Biological Terrorism: Emergency Response and Public Protection*, and the 2001 Lewis Publishers book titled *Sampling and Surveying Radiological Environments.*

About the Contributors

Matt Tonkin, PhD, is a senior hydrogeologist, principal and president of S.S. Papadopulos & Associates, Inc. Matt manages or advises on a wide variety of environmental projects, specializing in data synthesis and the development and application of models to support groundwater, surface water, and soil, studies on behalf of public, private, and legal clients. This includes planning strategic data acquisition efforts, collaborating with subject matter experts, and presenting to stakeholders. Matt received his PhD on the topic of environmental model calibration and uncertainty analysis and has instructed on these and other topics.

Mashrur A. Chowdhury, MASc., is a senior staff hydrogeologist at S.S. Papadopulos & Associates, Inc. Mashrur's primary experience is with Superfund sites where he provides expertise in the synthesis and interpretation of hydrogeologic data, and in the application of various groundwater modeling analyses in support of regulatory hydrogeological assessments and remediation studies. Mashrur completed his graduate degree in civil engineering with a specialization in Water Resources Engineering.

Darrell R. Newcomer, M.S., is a senior project scientist for Central Plateau Cleanup Company. He holds a Master of Science degree in hydrogeology from Montana Technological University, Butte, and a Bachelor of Science degree in geology from Western Washington University, Bellingham. He has approximately 30 years of applied and research experience in hydrogeologic characterization investigations, groundwater monitoring, and remediation activities for various U.S. Department of Energy and U.S. Department of Defense environmental cleanup projects.

Rob D. Mackley, M.S., is an environmental engineer at Pacific Northwest National Laboratory. His research has focused around the technical areas of groundwater and remediation monitoring, remedy development and implementation, and hydrogeologic characterization. He specializes in aquifer hydraulic testing, groundwater and vadose zone monitoring instrumentation, and remedy performance monitoring and evaluation. One of his favorite areas of research is the interaction of surface and groundwater. Rob has provided technical oversight of groundwater, surface water, and subsurface sediment sampling. He has overseen the drilling, design, and completion of groundwater wells. He is the proud father of four boys and enjoys coaching them in sports and exposing them to the adventures of the outdoors.

Kathleen M. Leonard, M.S., PMP, is a senior project scientist with the Central Plateau Cleanup Company supporting soil and groundwater remediation projects at the U.S. Department of Energy Hanford Site in Richland, Washington. She has over 20 years of experience serving as an environmental scientist and project manager, with experience in RCRA/CERCLA site characterization and remediation. In addition to having experience working at the Hanford Site, she also has experience working at the Advanced Mixed Waste Treatment Facility in Idaho Falls, Idaho, and the Waste Isolation Project Plant in Carlsbad New Mexico. She holds a B.A. degree from Central Washington University, an M.S. degree in Environmental Science from Washington State University, and is a Certified Project Manager through the Project Management Institute.

Harry O'Neill is the president of Beacon Environmental and has managed soil gas investigations for more than 20 years, working on U.S. Department of Defense, U.S. Department of Energy, EPA Superfund, and commercial projects. O'Neill has been on the forefront of the acceptance of passive soil gas sampling technologies at the national and international level, having supported projects in all 50 states and in over 40 countries spanning all 7 continents. He is a founding member of the Association of Vapor Intrusion Professionals and is the lead author of ASTM Standard D7758: Standard Practice for Passive Soil Gas Sampling in the Vadose Zone. He has published and presented findings throughout the United States, as well as internationally across five continents as an invited speaker. He received his B.A. degree with a focus on law and completed master's-level class work at Johns Hopkins University in geology.

Margo Aye, B.A., GISP, is a GIS cartographer who works for the Central Plateau Cleanup Company in Richland, Washington. She holds a Bachelor of Arts degree in geography from Central Washington University. She has 23 years of GIS experience along with technical graphic work focused on remediation work for various U.S. Department of Energy contractors.

Acronyms and Abbreviations

2D	Two-dimensional
3D	Three-dimensional
AEM	Analytical element method
ARAR	Applicable or relevant and appropriate requirement
ASTM	American Society for Testing Materials
BLM	Bureau of Land Management
BM	Bureau of Mines
BTEX	Benzene, toluene, ethylbenzene, and xylenes
C	Centigrade
CAA	Clean Air Act
CERCLA	Comprehensive Environmental Response, Compensation, and Liability Act
CGI	Combustible gas indicator
CI	Confidence interval
cm	Centimeter
COC	Contaminant of concern
COPC	Contaminant of potential concern
Cs-137	Cesium-137
CSM	Conceptual site model
CTS	Contaminant treatment system
CWA	Clean Water Act
CX	Categorical Exclusion
D&D	Decontamination and Decommissioning
DEFT	Decision Error Feasibility Trials
DNAPL	Dense nonaqueous phase liquid
DOD	Department of Defense
DOE	Department of Energy
DOT	Department of Transportation
DQA	Data Quality Assessment
DQO	Data Quality Objectives
DS	Decision statement
EA	Environmental Assessment
EDA	Environmental Data Collection
EIS	Environmental Impact Statement
ELR	Environmental Law Reporter
EPA	Environmental Protection Agency
F	Fahrenheit
FDR	Field Deployment Report
FEPs	Features, Events, and Processes
FFCA	Federal Facility Compliance Act
FID	Flame ionization detector

FLUTe	Flexible Liner Underground Technologies
FONSI	Finding of No Significant Impact
F&T	Fate and Transport
FS	Feasibility Study
ft	Feet
ft^2	Square feet
g	Gram
gal	Gallon
GC	Gas chromatograph
GC/MS	Gas chromatograph/mass spectrometer
gpm	Gallons-per-minute
GPERS-II	Global Positioning Environmental Radiological Surveyor
GPS	Global positioning system
H&S	Health and Safety
HCl	Hydrochloric acid
HEPA	High-efficiency particulate air
HNO_3	Nitric acid
HPGe	High-purity germanium
H_2SO_4	Sulfuric Acid
HSWA	Hazardous and Solid Waste Amendments
IAEA	International Atomic Energy Agency
IDLH	Immediately dangerous to life or health
IDW	Investigation-derived waste
in.	Inch
IS	Integrating Sensors
ISCO	In situ chemical oxidation
ISCR	In situ chemical reduction
ISOCS	In Situ Object Counting System
K-40	Potassium-40
Kd	Distribution coefficient
kg	Kilogram
L	Liter
LARADS	Laser-Assisted Ranging and Data System
lb	Pound
LDPFA	Land Disposal Program Flexibility Act
LNAPL	Light nonaqueous phase liquid
LRA	Local regulatory agency
m	Meter
m^2	Square meter
MARSSIM	Multi-Agency Radiation Survey and Site Investigation Manual
MCLs	Maximum contaminant levels
MDI	Model-data interaction
mg	Milligram
mi	Mile
mL	Milliliter

MNA	Monitor natural attenuation
mph	Miles-per-hour
MS	Mass Spectrometer
MTBE	Methyl tertiary butyl ether
NAAQS	National Ambient Air Quality Standards
NaOH	Sodium hydroxide
NAPL	Nonaqueous phase liquid
NCP	National Oil Hazardous Substance Pollution Contingency Plan
NEA	Nuclear Energy Agency
NEPA	National Environmental Policy Act
NESHAPs	National Emission Standards for Hazardous Air Pollutants
NIOSH	National Institute for Occupational Safety and Health
NOI	Notice of Intent
NPDES	National Pollutant Discharge Elimination System
NRC	Nuclear Regulatory Commission
OCS	Object Counting System
OSHA	Occupational Safety and Health Administration
PAHs	Polycyclic aromatic hydrocarbons
PCBs	Polychlorinated biphenyls
PCE	Tetrachloroethylene
PID	Photoionization detector
ppb	Parts-per-billion
PPE	Personal protective equipment
ppm	Part-per-million
PRPs	Potentially responsible parties
psi	Pounds-per-square-inch
PSQ	Principal study question
P&T	Pump and treat
PVC	Polyvinylchloride
QA	Quality assurance
RAO	Remedial action objective
RCRA	Resource Conservation and Recovery Act
RFI	Resource Conservation and Recovery Act Facility Investigation
RI	Remedial Investigation
RI/FS	Remedial Investigation/Feasibility Study
SARA	Superfund Amendments and Reauthorization Act
SCBA	Self-contained breathing apparatus
SDWA	Safe Drinking Water Act
SOPs	Standard operating procedures
SRA	State regulatory agency
SVOC	Semi-volatile organic compound
SWDA	Solid Waste Disposal Act
TCE	Trichloroethylene
TCLP	Toxic Characteristic Leaching Procedure
Th-230	Thorium-230

TSCA	Toxic Substance Control Act
U.S.	United States
USC	United States Code
μg	Microgram
U-238	Uranium-238
USGS	U.S. Geological Survey
UXO	Unexploded ordnance
VOA	Volatile organic analysis
VOC	Volatile organic compound
VSP	Visual Sample Plan
α	False rejection decision error
β	False acceptance decision error
Δ	Width of the gray region
σ	Estimated standard deviation

1 Introduction

The purpose of this book is to provide the reader with guidance on the development and implementation of cost-effective and defensible environmental sampling programs that are designed to collect data needed to

- Define the nature and extent of contamination
- Calculate baseline human health and ecological risk
- Identify contaminant migration pathways
- Predict rates of contaminant migration
- Evaluate the viability of potential remedial alternatives
- Support treatability testing
- Assess the effectiveness of a remedial alternative during implementation
- Determine if remedial action objectives have been met
- Dispose of waste material properly

This book addresses both intrusive and nonintrusive sampling and surveying techniques that are effective in assessing a multitude of media types, such as soil, sediment, surface water, groundwater, sludge, soil gas, air, dust, and various types of building materials. It also provides standard operating procedures (SOPs) and guidance to support the proper implementation of these methods. This book focuses on those methods that have both proven themselves to be effective and are acknowledged by the U.S. Environmental Protection Agency (EPA) as reputable techniques.

The guidance provided in this book is slanted toward the Comprehensive Environmental Response, Compensation, and Liability Act (CERCLA) Remedial Investigation/Feasibility Study (RI/FS) program but is also useful for the Resource Conservation and Recovery Act (RCRA) Facility Investigation (RFI) program and a multitude of other environmental laws that are discussed in detail in Chapter 2.

Guidance is provided in Chapter 3 on how to use the EPA's seven-step Data Quality Objectives (DQO) process to identify all of the decisions that need to be made to complete an environmental study along with the types and quantity of data needed to resolve those decisions. This chapter provides guidance for deciding whether a statistical or nonstatistical sampling design is most appropriate for a project. It helps the reader understand the strategies for implementing nonstatistical designs, and presents a variety of statistical sampling design options for consideration along with guidance on the advantages and disadvantages of each. This chapter explains statistical concepts (e.g., false acceptance, false rejection, gray region) in easy-to-understand language and uses graphical illustrations for easy visualization. Details are provided on the capability of a number of current statistical sampling design software packages (e.g., Visual Sample Plan) that can be downloaded at no cost over the

DOI: 10.1201/9781003284000-1

Internet to support statistical designs. Suggested document outlines are provided for a DQO Summary Report and Sampling and Analysis Plan.

Chapter 4 presents field screening and sampling methods that should be considered when developing environmental sampling designs. To assist in selecting the most appropriate site-specific sampling technique, summary tables have been provided for each media type that rate techniques against their effectiveness in collecting samples for specific laboratory analyses, sample types, and sampling depths. Standard operating procedures have been provided in this chapter for most of these methods to assist the appropriate implementation of these techniques. Chapter 4 highlights the advantages and limitations of each sampling method.

Proper techniques for sample preparation, documentation, and shipping are presented in Chapter 5. These include details on how sample bottles should be preserved and packaged in preparation for shipment, and how to properly complete and maintain a sample logbook. This chapter includes examples of various types of forms that must be completed (e.g., sample labels, chain-of-custody forms and seals, shipping airbills) as well as forms that should be considered to simplify the data collection process. Guidance is provided on developing a site-specific sample identification numbering system.

Chapters 6, 7, and 8 provide guidance on the types of field and laboratory quality control samples that should be collected and analyzed, and data verification/validation and data quality assessment methods that should be performed, to assure that the data collected is usable and meets the data quality objectives for the project. Equipment decontamination procedures for both chemical and radiological contaminants are provided in Chapter 9. Chapters 10 and 11 provide guidance on health and safety and the proper management of investigation-derived waste.

To add to the usefulness of this book as a field or office reference manual, soil classification and general reference tables have been provided in Appendix.

2 Summary of Major Environmental Laws and Regulations

The purpose of this chapter is to provide the readers with general guidance on the types of issues addressed under the major environmental laws and federal regulations. This information is not intended to be cited as law but rather is provided to give the reader general information about the issues addressed by each of the cited laws and regulations.

An environmental law is first introduced as a bill in the U.S. House of Representatives or the U.S. Senate. The bill is then passed on to a committee where it undergoes a detailed evaluation. As part of this evaluation, it is not uncommon for committee hearings to be held where expert witnesses are called to testify on the key technical aspects of the bill. If the bill is passed, it becomes an act that is then sent to the President of the United States to sign into law. Once an environmental law is passed, administrative agencies (e.g., U.S. Environmental Protection Agency) develop and promulgate regulations that are then enforced at the federal level. Individual states often choose to promulgate their own regulations, which are required either to meet or to exceed the federal standards.

2.1 ENVIRONMENTAL LAWS

This section discusses the major environmental laws that pertain to performing environmental studies. These laws include the following:

- Comprehensive Environmental Response, Compensation, and Liability Act (CERCLA)
- National Oil Hazardous Substance Pollution Contingency Plan (NCP)/ Superfund Amendments and Reauthorization Act (SARA)
- Resource Conservation and Recovery Act (RCRA)
- Toxic Substance Control Act (TSCA)
- National Environmental Policy Act (NEPA)
- Clean Water Act (CWA)
- Safe Drinking Water Act (SDWA)
- Clean Air Act (CAA)

Although each of these laws is different in terms of the types of materials or activities that it regulates, together they are the United States' means of controlling the quality of its environment (Figure 2.1).

DOI: 10.1201/9781003284000-2

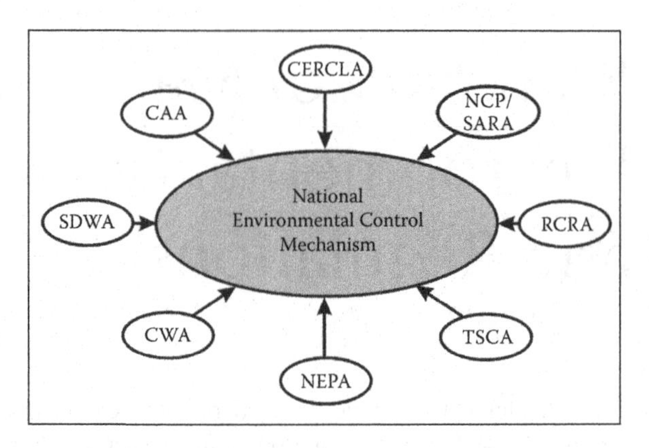

FIGURE 2.1 Major environmental laws protecting environmental quality.

2.1.1 CERCLA, NCP, and SARA Compliance

In 1980, Congress passed the Comprehensive Environmental Response, Compensation, and Liability Act (CERCLA, 40 CFR 300–312), which regulates the cleanup of abandoned or closed waste sites across the country. CERCLA

- Establishes prohibitions and requirements concerning closed and abandoned hazardous waste sites
- Provides for liability of persons responsible for releases of hazardous waste at these sites
- Establishes a trust fund to provide for cleanup when no responsible party can be identified

The law authorizes two kinds of response actions:

- Short-term removals, where actions may be taken to address releases or threatened releases requiring prompt response
- Long-term remedial response actions, which permanently and significantly reduce the dangers associated with releases (or threats of releases) of hazardous substances that are serious but not immediately life threatening

CERCLA also enabled the revision of the National Oil Hazardous Substance Pollution Contingency Plan (NCP, 40 CFR 300). The NCP provides the guidelines and procedures needed to respond to releases and threatened releases of hazardous substances, pollutants, or contaminants. It established the National Priorities List.

CERCLA is commonly referred to as "Superfund" because CERCLA was amended in 1986 by the SARA (EPA 2021b). SARA reflected the EPA's experience in administering CERCLA during its first 6 years and made several important changes and additions. For example, SARA

- Stresses the importance of permanent remedies and innovative treatment technologies in cleaning up hazardous waste sites

- Requires remedial actions to consider the standards and requirements found in other state and federal environmental laws and regulations
- Increases state involvement in every phase of the cleanup program
- Increases the focus on human health problems posed by hazardous waste sites
- Encourages greater citizen participation

SARA also required the EPA to revise the "Hazard Ranking System" to ensure that it accurately assessed the relative degree of risk to human health and the environment posed by uncontrolled hazardous waste sites that may be placed on the National Priorities List.

2.1.1.1 Remedial Investigation/Feasibility Study Process

CERCLA utilizes the Remedial Investigation & Feasibility Study (RI/FS) process (40 CFR 300.430) to evaluate the environmental conditions at a site and to select the final remedial alternative that will be implemented to remediate the site (Figures 2.2 and 2.3). From Figure 2.3, the implementation sequence for the RI/FS process begins (Step 1) by going through the EPA's Data Quality Objectives (DQO) process (see Chapter 3, Section 3.2). This process begins by performing a thorough scoping (see Chapter 3, Section 3.2.2) followed by a site visit. Scoping involves the evaluation of existing analytical data and historical records that may include but is not limited to

- Media sampling results (e.g., analytical, geochemical, and geotechnical)
- Maps (e.g., groundwater elevation, contaminant contour, and soil type)
- Borehole drilling logs (e.g., lithology and downhole geophysical surveying)
- Health physics or industrial hygiene records (e.g., air sampling results)
- Historical documents related to process operations, chemicals used, spills, and disposal practices
- Photographs and video footage

A site visit is often performed to familiarize the project team with the current site conditions. During this visit, interviews should be scheduled and performed to verify that the information gathered during the scoping process is accurate and to assist in filling in any data gaps. Section 3.2.2 (Chapter 3) discusses who should be interviewed and the type of information that should be gathered during the interviews. Once the scoping, site visit, and interviews have been completed, the EPA's seven-step DQO process should be implemented in accordance with the guidance provided in Sections 3.2.5.

Step 2 of the RI/FS process is to prepare an RI/FS Work Plan (Figure 2.3). This plan summarizes site background information gathered during the scoping process (Step 1), describes the environmental setting, summarizes the results from previous investigations, and then provides details on the additional investigations still needing to be performed. These additional investigations are taken from Step 7 of the DQO Process (Chapter 3, Section 3.2.5.7). The RI/FS Work Plan includes a schedule for implementing the RI/FS process (including field investigations). The author's suggested outline for the RI/FS Work Plan is presented in Figure 2.4.

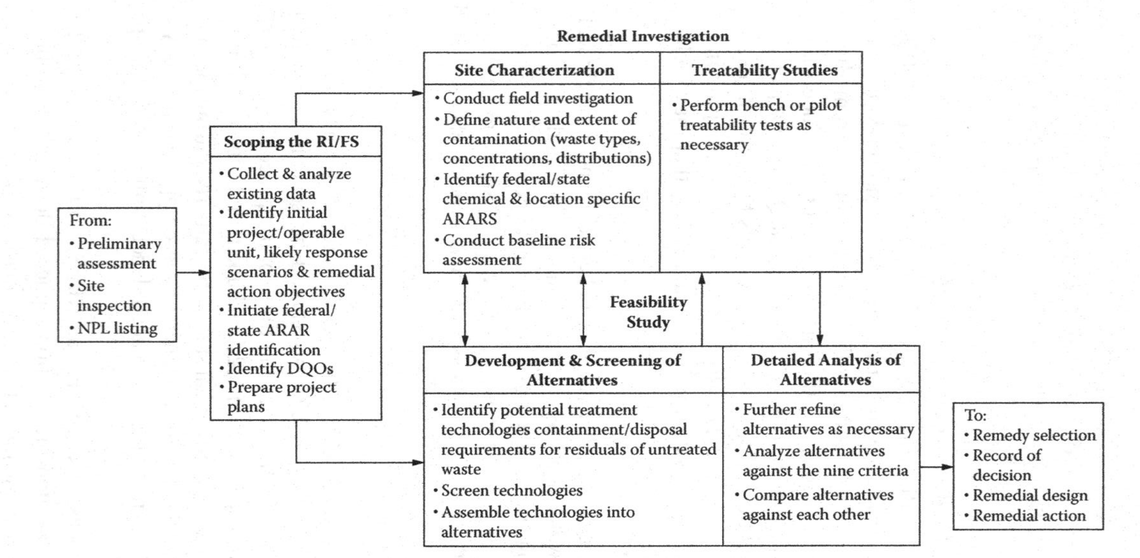

FIGURE 2.2 CERCLA RI/FS process from EPA Guidance Manual. (From EPA, 1988. *Guidance for Conducting Remedial Investigations and Feasibility Studies under CERCLA,* EPA/540/G-89/004, OSWER Directive 9355.3-01, Interim Final, October.)

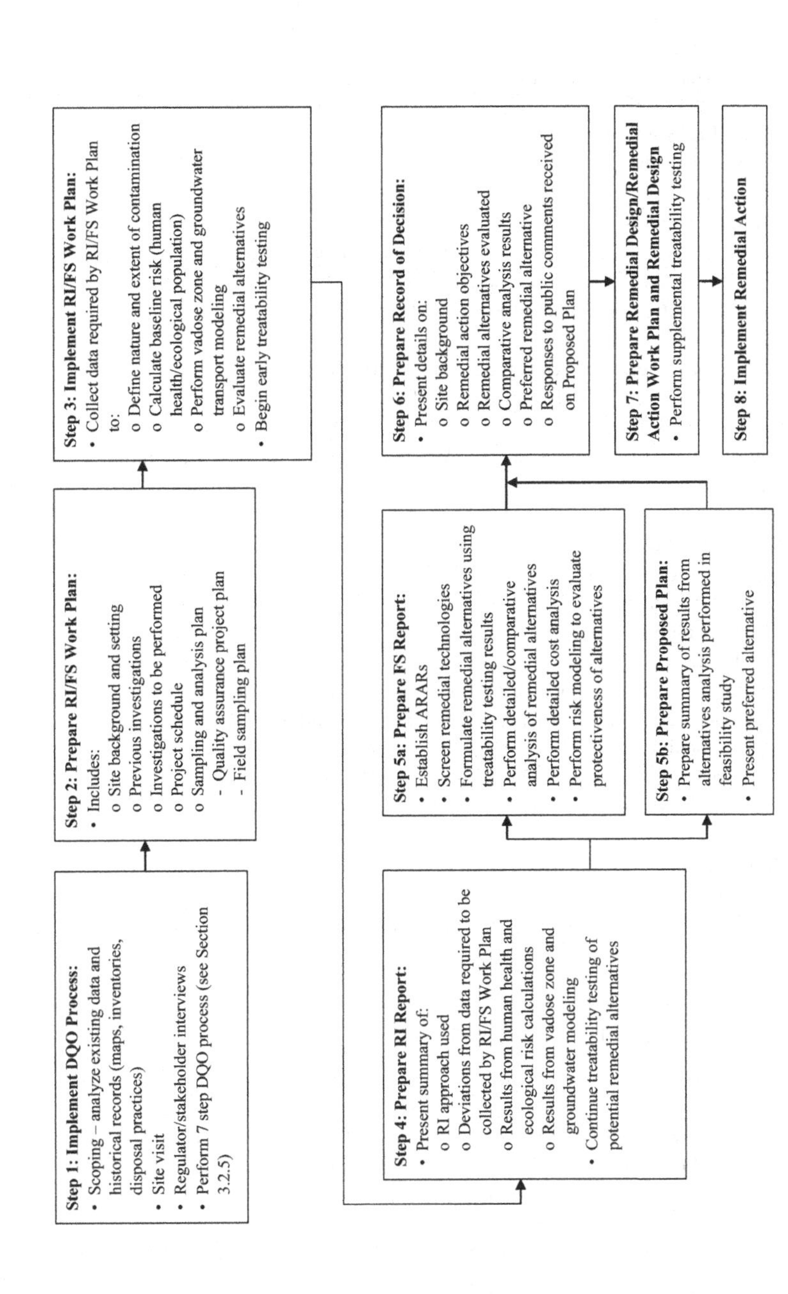

FIGURE 2.3 Implementation sequence for CERCLA RI/FS process.

RI/FS Work Plan

1. Introduction
   a. Scope and Objectives
   b. CERCLA Process
2. Operable Unit Background and Environmental Setting
   a. History of Operations
   b. Waste Site/Facility Descriptions
   c. Environmental Setting
3. Initial Evaluation
   a. Contaminant Sources
   b. Previous Investigations, Monitoring, and Remediation Activities
   c. Nature and Extent of Contamination
   d. Development of Contaminants of Potential Concern
   e. Land and Groundwater Use
   f. Potential Applicable or Relevant and Appropriate Requirements
   g. Conceptual Exposure Models for Fate and Transport Evaluation
   h. Conceptual Site Model
   i. Preliminary Risk Assessment/Remedial Action Objectives/Remedial Technology Options
4. Approach
   a. Data Strategy
   b. Treatability Studies
5. Remedial Investigation/Feasibility Study Tasks
   a. Task 1 – Project Planning
   b. Task 2 – Community Relations
   c. Task 3 – Field Investigations
   d. Tasks 4 and 5 – Sample Analysis/Validation and Data Evaluation
   e. Task 6 and 7 – Assessment of Risk and Treatability Studies
   f. Task 8 – Remedial Alternatives Development and Screening
   g. Task 9 – Detailed Analysis of Alternatives
   h. Task 10 – RI/FS Report and Post-Final RI/FS Support
6. Project Schedule
7. Project Management Considerations
   a. Project Organization
   b. Project Coordination, Decision-Making and Documentation
   c. Change Control and Dispute Resolution
8. References

FIGURE 2.4 Suggested outline for an RI/FS work plan.

A Sampling and Analysis Plan, composed of a Quality Assurance Project Plan and Field Sampling Plan, is commonly attached as an appendix to the RI/FS Work Plan to assist field implementation. The Sampling and Analysis Plan provides information such as

- Maps showing specific sampling locations
- Tables identifying detection limit and precision and accuracy requirements for analytical methods

- Sample bottle and preservation requirements
- Quality control sampling to be performed
- Quality assurance requirements
- Data quality assessment requirements

The author's suggested outline for a Sampling and Analysis Plan is presented in Figure 2.5.

The next step in the RI/FS process (Step 3, Figure 2.3) is the implementation of the field work identified in the RI/FS Work Plan. This data is most often collected to support defining the nature and extent of contamination, risk assessment calculations, transport modeling, and evaluation of remedial alternatives. Treatability testing of potential remedial alternatives should be considered at this time.

Following the implementation of fieldwork, the Remedial Investigation Report is prepared (Step 4, Figure 2.3). This report provides a summary of the background information about the site presented in the RI/FS Work Plan, details on the remedial investigation approach implemented, results from remedial investigations performed, deviations from data required to be collected by RI/FS Work Plan, results from human and ecological risk calculations, and results from groundwater and vadose zone modeling. The results from the remedial investigations and risk evaluation should be summarized in easy-to-read figures, graphs, and tables. Treatability testing for potential remedial alternatives should continue during the preparation of the Remedial Investigation Report.

Step 5a of the RI/FS process is the preparation of a Feasibility Study report (Figure 2.3). This report establishes applicable or relevant and appropriate requirements (ARARs), performs a screening of remedial technologies, joins remedial technologies into a broad range of remedial alternatives (e.g., no action through complete remediation), and performs a detailed and comparative analysis of alternatives against nine evaluation criteria including:

- Overall protectiveness to human health and the environment
- Compliance with ARARs
- Long-term effectiveness
- Reduction of toxicity, mobility, or volume
- Short-term effectiveness
- Implementability
- Cost
- State acceptance
- Community acceptance

Risk modeling is performed to evaluate whether or not each remedial alternative is adequately protective.

Step 5b of the RI/FS process is the preparation of a Proposed Plan report (Figure 2.3). This report is a very short document designed to summarize results from alternatives analysis in feasibility study, and to communicate to the public what the preferred remedial alternative is from the feasibility study.

Sampling and Analysis Plan

1. Introduction
    a. Project Scope and Objective
    b. Background
        i. Site Geology/Hydrogeology
    c. Data Quality Objectives Summary
    d. Contaminants of Concern
    e. Project Schedule
2. Quality Assurance Project Plan
    a. Project Management
        i. Project Organization
        ii. Quality Objectives and Criteria for Measurement Data
        iii. Methods-Based Analysis
        iv. Analytical Priority
        v. Special Training Requirements/Certifications
        vi. Documents and Records
    b. Data Generation and Acquisition
        i. Analytical Methods Requirements
        ii. Field Analytical Methods
        iii. Quality Control
        iv. Measurement Equipment
        v. Instrument and Equipment Testing, Inspection, and Maintenance
        vi. Instrument/Equipment Calibration and Frequency
        vii. Inspection/Acceptance of Supplies and Consumables
        viii. Non-direct Measurements
        ix. Data Management
    c. Assessment and Oversight
    d. Data Review and Usability
        i. Data Review/Verification/Validation
        ii. Reconciliation with User Requirements
3. Field Sampling Plan
    a. Sampling Objectives and Sampling Design
    b. Sample Location, Frequency, and Constituents to be Monitored
    c. Sampling Methods
        i. Chemical and Radiological Sampling Methods
        ii. Decontamination of Sampling Equipment
    d. Documentation of Field Activities
        i. Corrective Actions and Deviations for Sampling Activities
    e. Calibration of Field Equipment
    f. Sample Handling
        i. Sample Containers, Labeling, Custody, Transportation
4. Management of Waste
5. Health and Safety
6. References

FIGURE 2.5 Suggested outline for a sampling and analysis plan.

The next step in the RI/FS process is the preparation of a Record of Decision report (Step 6, Figure 2.3). This report is designed to present remedial action objectives, summarize remedial alternatives evaluated, and identify what the final selected remedy is for a site. This document is often prepared by the lead regulatory agency.

Once the Record of Decision has been issued, the Remedial Design/Remedial Action Work Plan and Remedial Design are prepared (Step 7, Figure 2.3). This report and design provide all the engineering details needed to support the installation and operation of the selected remedy. The remedy is then implemented (Step 8, Figure 2.3).

2.1.1.2 CERCLA Five-Year Review

Following the issuance of the Record of Decision, CERCLA requires remedial actions to be reevaluated every 5 years to ensure they are effective. If for some reason changes need to be made to the Record of Decision, the type of change must be categorized as either

- *Nonsignificant or minor change* – May affect things such as the type or cost of material, equipment, facilities, services, and/or supplies used to implement the remedy. This type of change will not have a significant impact on the scope, performance, or cost of the remedy.
- *Significant change* – Generally involves a change to a component of a remedy that does not fundamentally alter the overall cleanup approach.
- *Fundamental change* – Involves an appreciable change or changes in the scope, performance, and/or cost, or may be a number of significant changes that together have the effect of a fundamental change (EPA 1999).

2.1.2 RCRA Compliance

The Resource Conservation and Recovery Act of 1976 (RCRA, 40 CFR 260–282) requires the "cradle-to-grave" management of hazardous waste (EPA 2021c). This act

- Regulates solid and hazardous waste
- Prevents new uncontrolled hazardous sites from developing
- Protects human health and the environment
- Reduces the amount of waste generated
- Ensures that wastes are managed in an environmentally sound manner

RCRA is actually a combination of the first federal solid waste statutes and all subsequent amendments. In 1965, Congress enacted the Solid Waste Disposal Act, the first statute that specifically focused on improving solid waste disposal methods. The Solid Waste Disposal Act established economic incentives for states to develop planning, training, research, and demonstration projects for the management of solid waste. It was amended in 1976 by RCRA, which substantially remodeled the nation's solid waste management system and laid out the basic framework of the current hazardous waste management program.

RCRA, which has been amended several times since 1976, continues to evolve as Congress alters it to reflect changing waste management needs. It was amended most recently in 2016 to:

- Reorganize the hazardous waste generator regulations to make them more user-friendly
- Provide better understanding about how RCRA hazardous waste generator regulatory program works
- Address gaps in existing regulations to strengthen environmental protection
- Provide greater flexibility for hazardous waste generators to manage their hazardous waste in a cost-effective and protective manner

The Land Disposal Program Flexibility Act of 1996 amended RCRA to provide regulatory flexibility for the land disposal of certain wastes. Congress also revised RCRA in 1992 by passing the Federal Facility Compliance Act, which strengthened the authority to enforce RCRA at federal facilities. On November 8, 1984, the Hazardous and Solid Waste Amendments (HSWA) were expanded, which modified the scope and requirements of RCRA. HSWA was created largely in response to citizen concerns that existing methods of hazardous waste disposal, particularly land disposal, were not safe (EPA 2006).

2.1.2.1 Solid Waste

Under RCRA, the term *waste* refers to any discarded material that is abandoned, disposed of, burned, or incinerated, or stored in lieu of being abandoned. A *solid waste* material under RCRA includes:

- Garbage (e.g., milk cartons and coffee grounds)
- Refuse (e.g., metal scrap, wall board, and empty containers)
- Sludges from waste treatment plants, water supply treatment plants, or pollution control facilities (e.g., scrubber slags)
- Industrial wastes (e.g., manufacturing process wastewaters and nonwastewater sludges and solids)
- Other discarded materials, including solid, semisolid, liquid, or contained gaseous materials resulting from industrial, commercial, mining, agricultural, and community activities (e.g., boiler slags)

The definition of solid waste is *not* limited to wastes that are physically solid. Many solid wastes are liquid, whereas others are semisolid or gaseous. The regulations further define solid waste as any material that is discarded by being either abandoned, inherently waste-like, military munitions (unused or defective), or recycled.

The materials in Table 2.1 have been *excluded* from the definition of a solid waste. 40 CFR Part 257, Subpart A presents provisions designed to ensure that wastes disposed of in solid waste disposal units will not threaten endangered species, surface water, groundwater, or flood plains (EPA 2006).

TABLE 2.1
Materials Excluded from Definition of a Solid Waste

Domestic sewage and mixtures of domestic sewage	Hazardous secondary material generated/transferred for purposes of reclamation
Point source discharge	Solvent-contaminated wipes that are sent for cleaning and reuse
Irrigation return flows	Radioactive waste
In situ mining waste	Pulping liquors
Spent sulfuric acid	Closed-loop recycling[a]
Spent wood preservatives	Coke by-product wastes[b]
Splash condenser dross residue[c] Reclamation in enclosed tanks	Hazardous oil-bearing secondary materials and recovered oil from petroleum-refining operations
Condensates from kraft mill steam strippers	Used cathode ray tubes
	Hazardous secondary material generated and transferred to another person for remanufacturing
Processed scrap metal	Shredded circuit boards
Spent materials from mineral processing	Petrochemical recovered oil
	Hazardous secondary materials used to make zinc fertilizers
Spent caustic solutions from petroleum refining	Zinc fertilizers made from recycled hazardous secondary materials

Source: From EPA, 2021j. www.epa.gov/hw/criteria-definition-solid-waste-and-solid-and-hazardous-waste-exclusions

[a] Spent materials that are reclaimed and returned to the original process in an enclosed system of pipes and tanks.
[b] Coke is used in the production of iron and is made by heating coal in high-temperature ovens.
[c] The treatment of steel production pollution control sludge generates a zinc-laden residue called a dross. This material, generated from a splash condenser in a high-temperature metal recovery process, is known as a splash condenser dross residue.

2.1.2.2 Hazardous Waste

Once a material has been determined to be a "solid waste," it must next be determined whether or not it is a *hazardous waste.* A solid waste may also be a hazardous waste if it has properties that make it dangerous or capable of having a harmful effect on human health or the environment.

A material cannot be a hazardous waste if it does not meet the definition of a solid waste. Thus, wastes that are excluded from the definition of solid waste (see-Section 2.1.2.1) are not subject to RCRA Subtitle C hazardous waste regulation (EPA 2006).

2.1.2.2.1 Materials Excluded From Hazardous Waste Definition

The materials listed in Table 2.2 have been *excluded* from the definition of a hazardous waste.

TABLE 2.2
Materials Excluded from the Definition of a Hazardous Waste

Household hazardous waste[a]	Agricultural waste
Mining overburden	Bevill and Bentsen wastes[b]
Fossil fuel combustion waste	Oil, gas, and geothermal wastes
Mining and mineral-processing wastes	Cement kiln dust
Trivalent chromium wastes	Arsenically treated wood
Petroleum-contaminated media, and debris from underground storage tanks	Spent chlorofluorocarbon refrigerants
Used oil filters	Used oil distillation bottoms
Landfill leachate or gas condensate derived from certain listed wastes	Project XL pilot-project exclusions[c]

Source: From EPA, 2021j. www.epa.gov/hw/criteria-definition-solid-waste-and-solid-and-hazardous-waste-exclusions.

[a] Examples of household hazardous waste include old solvents, paints, pesticides, fertilizers, or poisons.
[b] *Bentsen wastes* include oil, gas, and geothermal exploration, development, and production wastes. *Bevill wastes* include fossil fuel combustion wastes, mining and mineral-processing wastes, and cement kiln dust wastes.
[c] EPA has provided three facilities with site-specific hazardous waste exclusions pursuant to the Project XL pilot program.

2.1.2.2.2 Listed Hazardous Wastes

Listed hazardous wastes are wastes that are dangerous enough to warrant full Subtitle C regulation based on their origin. Any waste fitting a narrative listing description is considered a listed hazardous waste. Before developing each hazardous waste listing, EPA thoroughly studies a particular waste-stream and the threats that it can pose to human health and the environment. If the waste poses sufficient threat, EPA includes a precise description of that waste on one of four hazardous waste lists within the regulations. These four hazardous waste listings include:

- *F list* – The F list includes wastes from certain common industrial and manufacturing processes. Because the processes generating these wastes can occur in different sectors of industry, the F list wastes are known as wastes from non-specific sources. The F list is codified in the regulations in 40 CFR §261.31.
- *P list* – This list includes pure or commercial-grade formulations of specific unused chemicals. Chemicals are included in the *P* list if they are acutely toxic. A chemical is acutely toxic if it is fatal to humans in low doses, if scientific studies have shown that it has lethal effects on experimental organisms, or if it causes serious irreversible or incapacitating illness. P list chemicals are codified in 40 CFR §261.33.
- *K list* – The *K* list includes wastes from specific industries. As a result, *K* list wastes are known as wastes from specific sources. The *K* list is found in 40 CFR §261.32.
- *U list* – This list includes pure or commercial-grade formulations of specific unused chemicals. The *U* list is generally comprised of chemicals that are toxic,

but it also includes chemicals that display other characteristics, such as ignitability or reactivity. *U* list chemicals are codified in 40 CFR §261.33.

Each list includes anywhere from 30 to a few hundred listed hazardous wastestreams. All of the wastes on these lists are assigned an identification number (i.e., a waste code) consisting of the letter associated with the list (i.e., F, K, P, or U) followed by three numbers. For example, wastes on the F list may be assigned a waste code ranging from F001 to F039, whereas wastes on the K list may be assigned a waste code ranging from K001 to K148. These waste codes are an important part of the RCRA regulatory system because waste code assignment has important implications for the future management standards that will apply to the waste (EPA 2006).

2.1.2.2.3 Characteristic Hazardous Wastes

Even if a waste is a listed hazardous waste, the facility must also determine if the waste exhibits a hazardous characteristic by testing or applying knowledge of the waste.

Characteristic hazardous wastes are wastes that exhibit measurable properties that indicate that a waste poses enough of a threat to warrant regulation as hazardous waste. EPA tried to identify characteristics that, when present in a waste, can cause death or injury to humans or lead to ecological damage (EPA 2006).

EPA decided that the characteristics of hazardous waste should be detectable by using a standardized test method or by applying general knowledge of the waste's properties. Given these criteria, EPA established four hazardous waste characteristics:

- Ignitability
- Corrosivity
- Reactivity
- Toxicity

2.1.2.2.3.1 Ignitability The *ignitability characteristic* identifies wastes that can readily catch fire and sustain combustion. Many paints, cleaners, and other industrial wastes pose such a hazard. Liquid and nonliquid wastes are treated differently by the ignitability characteristic.

Most ignitable wastes are liquid in physical form. EPA selected a flash point test as the method for determining whether a liquid waste is combustible enough to deserve regulation as hazardous. The flash point test determines the lowest temperature at which the fumes above a waste will ignite when exposed to flame. Liquid wastes with a flash point of less than 60°C (140°F) in closed-cup test are ignitable.

Many wastes in solid or nonliquid physical form (e.g., wood or paper) can also readily catch fire and sustain combustion, but EPA did not intend to regulate most of these nonliquid materials as ignitable wastes. A nonliquid waste is considered ignitable only if it can spontaneously catch fire or catch fire through friction or absorption of moisture under normal handling conditions and can burn so vigorously that it creates a hazard. Certain compressed gases are also classified as ignitable. Finally, substances meeting the DOT's definition of oxidizer are classified as ignitable wastes. Ignitable wastes carry the waste code D001 and are among some of the most common hazardous wastes. The regulations describing the characteristic of ignitability are codified in 40 CFR §261.21.

2.1.2.2.3.2 Corrosivity The *corrosivity characteristic* identifies wastes that are acidic or alkaline (basic). Such wastes can readily corrode or dissolve flesh, metal, or other materials. They are also among some of the most common hazardous wastes. An example is waste sulfuric acid from automotive batteries. EPA uses two criteria to identify liquid and aqueous corrosive hazardous wastes. The first is a pH test. Aqueous wastes with a pH $\geq$ 12.5 or $\leq$ 2 are corrosive. A liquid waste may also be corrosive if it has the ability to corrode steel under specific conditions. Physically solid, nonaqueous wastes are not evaluated for corrosivity. Corrosive wastes carry the waste code D002. The regulations describing the corrosivity characteristic are found in 40 CFR §261.22.

2.1.2.2.3.3 Reactivity The *reactivity characteristic* identifies wastes that readily explode or undergo violent reactions or react to release toxic gases or fumes. Common examples are discarded munitions or explosives. In many cases, there is no reliable test method to evaluate a waste's potential to explode, react violently, or release toxic gas under common waste-handling conditions. Therefore, EPA uses narrative criteria to define most reactive wastes. The narrative criteria, along with knowledge or information about the waste properties, are used to classify a waste as reactive. A waste is reactive if it meets any of the following criteria:

- It can explode or violently react when exposed to water or under normal handling conditions.
- It can create toxic fumes or gases at hazardous levels when exposed to water or under normal waste-handling conditions.
- It can explode if heated under confinement or exposed to a strong igniting source, or it meets the criteria for classification as an explosive under DOT rules.
- It generates toxic levels of sulfide or cyanide gas when exposed to a pH range of 2 through 12.5.

Wastes exhibiting the characteristic of reactivity are assigned the waste code D003. The reactivity characteristic is described in the regulations in 40 CFR §261.23.

2.1.2.2.3.4 Toxicity EPA developed the *toxicity characteristic* to identify wastes likely to leach dangerous concentrations of toxic chemicals into groundwater. In order to predict whether any particular waste is likely to leach chemicals into groundwater at dangerous levels, EPA designed a laboratory procedure to estimate the leaching potential of waste when disposed in a municipal solid waste landfill. This laboratory procedure is known as the Toxicity Characteristic Leaching Procedure (TCLP). The TCLP requires a generator to create a liquid leachate from its hazardous waste samples. This leachate would be similar to the leachate generated by a landfill containing a mixture of household and industrial wastes. Once this leachate is created via the TCLP, the waste generator must determine whether it contains any of 40 different toxic chemicals in amounts above the specified regulatory levels (see Table 2.3). If the leachate sample contains a concentration above the regulatory limit for one of the specified chemicals, the waste exhibits the toxicity characteristic and

TABLE 2.3
Maximum Concentration of Contaminants for the Toxicity Characteristic

EPA HW No.[a]	Contaminant	CAS No.[b]	Regulatory Level (mg/L)
D004	Arsenic	7440-38-2	5.0
D005	Barium	7440-39-3	100.0
D018	Benzene	71-43-2	0.5
D006	Cadmium	7440-43-9	1.0
D019	Carbon tetrachloride	56-23-5	0.5
D020	Chlordane	57-74-9	0.03
D021	Chlorobenzene	108-90-7	100.0
D022	Chloroform	67-66-3	6.0
D007	Chromium	7440-47-3	5.0
D023	*o*-Cresol	95-48-7	200.0[c]
D024	*m*-Cresol	108-39-4	200.0[c]
D025	*p*-Cresol	106-44-5	200.0[c]
D026	Cresol	—	200.0[c]
D016	2,4-D	94-75-7	10.0
D027	1,4-Dichlorobenzene	106-46-7	7.5
D028	1,2-Dichloroethane	107-06-2	0.5
D029	1,1-Dichloroethylene	75-35-4	0.7
D030	2,4-Dinitrotoluene	121-14-2	0.13
D012	Endrin	72-20-8	0.02
D031	Heptachlor (and its epoxide)	76-44-8	0.008
D032	Hexachlorobenzene	118-74-1	0.13
D033	Hexachlorobutadiene	87-68-3	0.5
D034	Hexachloroethane	67-72-1	3.0
D008	Lead	7439-92-1	5.0
D013	Lindane	58-89-9	0.4
D009	Mercury	7439-97-6	0.2
D014	Methoxychlor	72-43-5	10.0
D035	Methyl ethyl ketone	78-93-3	200.0
D036	Nitrobenzene	98-95-3	2.0
D037	Pentachlorophenol	87-86-5	100.0
D038	Pyridine	110-86-1	5.0
D010	Selenium	7782-49-2	1.0
D011	Silver	7440-22-4	5.0
D039	Tetrachloroethylene	127-18-4	0.7
D015	Toxaphene	8001-35-2	0.5
D040	Trichloroethylene	79-01-6	0.5
D041	2,4,5-Trichlorophenol	95-95-4	400.0
D042	2,4,6-Trichlorophenol	88-06-2	2.0
D017	2,4,5-TP (Silvex)	93-72-1	1.0
D043	Vinyl chloride	75-01-4	0.2

Source: 40 CFR Part 261.24.

[a] Hazardous waste number.

[b] Chemical abstracts service number.

[c] If *o*-, *m*-, and p-cresol concentrations cannot be differentiated, the total cresol (D026) concentration is used. The regulatory level of total cresol is 200 mg/L.

carries the waste code associated with that compound or element. The regulations describing the toxicity characteristic are codified in 40 CFR §261.24.

2.1.2.2.4 Mixture Rule

The *mixture rule* is intended to ensure that mixtures of listed wastes with nonhazardous solid wastes are regulated in a manner that minimizes threats to human health and the environment. The mixture rule regulates a combination of any amount of a nonhazardous solid waste and any amount of a listed hazardous waste as a listed hazardous waste. Even if a small vial of listed waste is mixed with a large quantity of nonhazardous waste, the resulting mixture bears the same waste code and regulatory status as the original listed component of the mixture, unless the generator obtains a delisting. A mixture involving characteristic wastes is hazardous only if the mixture itself exhibits a characteristic (EPA 2006).

2.1.2.2.5 The Derived-From Rule

Hazardous waste treatment, storage, and disposal processes often generate residues that may contain high concentrations of hazardous constituents. In order to adequately protect human health and the environment from the threats posed by these potentially harmful wastes, the *derived-from rule* governs the regulatory status of such listed waste residues (EPA 2006).

2.1.2.2.6 Contained-In/Contained-Out Policy

Sometimes, listed and characteristic wastes are spilled onto soil, or contaminate equipment, buildings, or other structures. The mixture and derived-from rules do not apply to such contaminated soil and materials because these materials are not actually wastes. Soil is considered *environmental media* (e.g., soil, groundwater, and sediment), whereas the equipment, buildings, and structures are considered debris (e.g., a broad category of larger manufactured and naturally occurring objects that are commonly discarded). In order to address such contaminated media and debris, EPA created the *contained-in policy* to determine when contaminated media and debris must be managed as RCRA hazardous wastes.

EPA considers contaminated media or debris to no longer contain hazardous waste when it no longer exhibits a characteristic of hazardous waste. This applies when the hazardous waste contained within the media or debris is a characteristic waste. Otherwise, when dealing with listed waste contamination, EPA or states can determine that media and debris no longer contain hazardous waste by determining that the media or debris no longer poses a sufficient health threat to deserve RCRA regulation. Once this *contained-out* determination is made, the media and debris are generally no longer regulated under RCRA Subtitle C (EPA 2006).

2.1.2.2.7 Mixed Waste

RCRA specifically exempts certain radioactive mixed materials from the definition of solid waste. However, some radioactive material may be mixed with hazardous wastes that are regulated under RCRA. In addition, a facility may generate a hazardous waste that is also radioactive. Because the material in both of these situations contains both radioactive material and RCRA hazardous waste, it is referred to as mixed waste under RCRA. RCRA and the Atomic Energy Act regulate these mixed wastes jointly (EPA 2006).

2.1.2.2.8 Land Disposal Restrictions

The *land disposal restriction* program does not mandate physical barriers to protect groundwater from disposed-of hazardous waste, but instead requires that hazardous wastes undergo fundamental physical or chemical changes so that they pose less of a threat to groundwater, surface water, and air when disposed. The obvious advantage of such hazardous waste treatment is that it provides a longer-lasting form of protection than simply hazardous waste containment. Although synthetic barriers designed to prevent the migration of leachate can break down and fail over time, physical and chemical changes to the waste itself provide a more permanent type of protection. All hazardous wastes, except under certain circumstances, must meet a specific treatment standard before they can be disposed (EPA 2006).

2.1.3 TSCA Compliance

The Toxic Substances Control Act (TSCA, 40 CFR 195, 700–766) became law on October 11, 1976, and was enacted by Congress in an effort to prevent unreasonable risks of injury to health or the environment associated with the manufacture, processing, distribution in commerce, use, or disposal of chemical substances (EPA 2021f). Congress was particularly concerned with preventing unreasonable risks to health and the environment before both the manufacture of chemicals and their entrance to the market. Therefore, much of the legislation includes requirements that apply early on in the life of a chemical, such as premarket testing and notification. TSCA covers all organic and inorganic chemical substances and mixtures, both synthetic and naturally occurring, with the exception of food, food additives, drugs, cosmetics, nuclear materials, tobacco, and pesticides, which are all covered by other legislations (EPA 2007a).

TSCA requires all manufacturers, processors, and distributors to maintain records of the hazards that each of their products poses to human health and the environment. It also requires the EPA to compile and publish a list of each chemical substance manufactured or processed in the United States. The statute authorizes the EPA to conduct limited inspections of areas where substances are processed or stored, and of conveyances used to transport the substances.

TSCA requires all manufacturers and processors of new substances, or substances that will be applied to a significant new use, to notify the administrator of the EPA that they intend to manufacture or process the substance. If analytical testing is required, the manufacturer or processor must provide the results along with the notification.

If the EPA finds the analytical testing to be insufficient, a proposed order is written to restrict the manufacturing of the substance until adequate testing is completed. If the testing data indicate that the substance may present a significant risk of cancer, gene mutations, or birth defects, the EPA will promulgate regulations concerning the distribution, handling, and labeling of the substance. In the case of an imminently hazardous substance, the EPA may commence a civil action for seizure of the substance, and possibly a recall and repurchase of the substance previously sold.

The requirements of this statute generally do not apply to toxic substances distributed for export unless they would cause an unreasonable risk of harm within the United States. On the other hand, imported substances are subject to the requirements of the statute, and any substances that do not comply will be refused entry into the United States. Violations of this statute can result in both civil and criminal penalties, and the violating substance may be seized.

Some important regulations under TSCA (40 CFR 761) govern the manufacture, use, and disposal of polychlorinated biphenyls (PCBs). PCBs are found in many substances, such as oils, paints, and contaminated solvents. The regulations establish concentration limits and define acceptable methods of disposal.

TSCA was revised in 1986 to require that asbestos inspections (40 CFR 763) be performed in school buildings to define the appropriate level of response actions. The statute requires the implementation of maintenance and repair programs, and periodic surveillance of school buildings where asbestos is located, as well as prescribing standards for the transportation and disposal of this material. For those school buildings containing asbestos, local educational agencies are required to develop an asbestos management program, which must include plans for response actions, long-term surveillance, and use of warning labels for asbestos remaining in the buildings.

TSCA was revised in 1988 to require the EPA to determine the extent of radon contamination in the nation's schools, develop model construction standards and techniques for controlling radon levels within new buildings, and make grants available to states to assist them in the development and implementation of their radon programs. TSCA was also revised in 1992 to reduce environmental exposure to lead contamination and prevent the adverse health effects caused by it (40 CFR 745). Exposure of children to lead was the primary concern. Provisions of the act included exposure studies, determination of lead levels in products, establishing state programs for monitoring and abatement, and training and certification requirements for lead abatement workers (EPA 2007a).

2.1.4 NEPA Compliance

The National Environmental Policy Act (NEPA, 40 CFR 1500–1508) was passed in 1969 and was one of the first statutes directed specifically at protecting the environment (EPA 2021d). NEPA documentation is necessary when any "major Federal action" that may have a significant impact on the environment may be undertaken. The NEPA process places heavy emphasis on public involvement. Public notice must be provided for NEPA-related hearings, public meetings, and to announce the availability of environmental documents. In the case of a NEPA action of national concern, notice is included in the *Federal Register*, and notice is made by mail to national organizations reasonably expected to be interested in the matter. NEPA's purpose is not to generate paperwork but rather to optimize the environmental decision-making process through public involvement. The NEPA process is intended to help public officials make decisions that are based on understanding of environmental consequences, and take actions that protect, restore, and enhance the environment (40 CFR 1500).

The primary documents prepared under the NEPA process are the Notice of Intent (NOI), Environmental Assessment (EA), Environmental Impact Statement (EIS), Finding of No Significant Impact (FONSI), and Categorical Exclusion (CX). Any NEPA environmental document may be combined or integrated with any other agency document to reduce duplication and paperwork. All NEPA documents should be written so that the public can readily understand them.

An EA is a concise public document that determines whether or not to prepare an EIS (40 CFR 1508.9). If there are no significant impacts on the environment, a FONSI is published. An EA can facilitate preparation of an EIS when one is needed. However, an EA need not be prepared if it is already known that there will be significant impacts to the environment, and an EIS must be prepared (40 CFR 1502).

Before preparing an EIS, an NOI must be issued for public review. The NOI describes the proposed action and possible alternatives, describes the federal agency's proposed scoping process including whether, when, and where any public scoping meetings will be held, and finally states the name and address of a person within the agency who can answer questions about the proposed action.

The EIS serves as an action-forcing device to ensure that the policies and goals defined in NEPA are infused into the ongoing programs and actions of the federal government. The objective of the EIS is to provide a full and fair discussion of significant environmental impacts, and is used to inform decision-makers and the public of the reasonable alternatives that would avoid or minimize adverse impacts or enhance the quality of the human environment (40 CFR 1502.1). The EIS is meant to serve as the means of assessing the environmental impact of proposed federal agency actions, but is not used to justify decisions that have already been made.

A FONSI is a document prepared by a federal agency to describe briefly the reasons why an action will not have a significant effect on the human environment, and for which an EIS is not needed (40 CFR 1508.13). This document includes the EA or a summary of this study, and notes any other environmental documents related to it. If the EA is included, the finding need not repeat any of the discussion in the assessment but may incorporate it by reference.

A CX refers to a category of actions that do not individually or cumulatively have a significant effect on human health or the environment and that have been found to have no such effect in procedures adopted by a federal agency in implementation of these regulations (40 CFR 1508.4). Consequently, there is no need for the preparation of an EA or an EIS.

2.1.5 CWA Compliance

The Clean Water Act (CWA) (33 U.S.C. §1251 et seq.) establishes the basic structure for regulating discharges of pollutants into the waters of the United States and regulating quality standards for surface waters (EPA 2021e). The basis of the CWA was enacted in 1948 and was called the Federal Water Pollution Control Act, but the Act was significantly reorganized and expanded in 1972. "Clean Water Act" became the Act's common name with amendments in 1972.

The CWA prohibits discharging pollutants through a "point source" into a water of the United States unless they have a National Pollutant Discharge Elimination

System (NPDES) permit (EPA 2020a, EPA 2020b, EPA 2020c, EPA 2014). The NPDES permit identifies limits on what you can discharge, monitoring and reporting requirements, and other provisions to ensure that the discharge does not negatively impact water quality or human health.

A "point source" is defined as any discernible, confined, and discrete conveyance such as a pipe, ditch, channel, tunnel, conduit, discrete fissure, or container. It also includes floating vessels from which pollutants are (or may be) discharged, as well as concentrated animal feeding operations. Agricultural stormwater discharges and return flows from irrigated agriculture are *not* "point sources".

Under the NPDES program, EPA regulates discharges of pollutants from municipal and industrial wastewater treatment plants, sewer collection systems, and stormwater discharges from industrial facilities and municipalities. Industries must pretreat pollutants in their wastes in order to protect local sanitary sewers and wastewater treatment plants. Industrial discharges of metals, oil and grease, and other pollutants can interfere with the operation of local sanitary sewers and wastewater treatment plants, leading to the discharge of untreated or inadequately treated pollutants into local waterways.

Compliance monitoring of the NPDES Program takes place to protect human health and the environment by ensuring that the regulated community follows environmental laws/regulations. EPA has authorized all but four states to implement their own NPDES programs to control water pollution. EPA oversees authorized state programs and has direct implementation responsibilities for the unauthorized states (Idaho, New Mexico, Massachusetts, and New Hampshire) as well as federal facilities. Compliance monitoring activities include reviewing Discharge Monitoring Reports, performing on-site compliance evaluations, and providing assistance to enhance compliance with NPDES permits.

The primary goals of compliance monitoring are to:

- Assess and document compliance with permits and regulations
- Support the enforcement process through evidence collection and case development
- Monitor compliance with enforcement orders and decrees
- Deter noncompliance

A Compliance Monitoring Strategy (CMS) document is used to provide compliance monitoring goals for the CWA NPDES program. EPA regional NPDES programs work closely with each of their states and internally to plan compliance monitoring activities for all NPDES sources covered by this policy and to ensure an effective inspection presence in each direct implementation program area. To provide for consistent implementation of the national CMS goals, each EPA region and state should develop a written CMS plan on an annual basis. Inspections should be performed every few years, and no less than every 5 years. On-site inspections should be conducted by an authorized inspector.

2.1.6 SDWA Compliance

The Safe Drinking Water Act (SDWA, 42 USC §§300f–300j-26, 40 CFR 141–148) was originally passed by Congress in 1974 to protect public health by regulating the nation's public drinking water supply. The law was amended in 1986 and 1996 and requires many actions to protect drinking water and its sources, which include rivers, lakes, reservoirs, springs, and groundwater wells. The SDWA applies to every public water system in the United States. However, this Act does not regulate private wells that serve fewer than 25 individuals. The SDWA authorizes the EPA to set national health-based standards for drinking water to protect against both naturally occurring and man-made contaminants that may be found in drinking water. The EPA, state regulatory agencies, and local water systems then work together to make sure that these standards are met (EPA 2007b).

Originally, the SDWA focused primarily on treatment as the means of providing safe drinking water at the tap. The 1996 amendments greatly enhanced the existing law by recognizing source water protection, operator training, funding for water system improvements, and public information as important components of safe drinking water. This approach is more comprehensive and ensures the quality of drinking water by protecting it from the source to the tap.

The EPA sets national standards for drinking water based on sound science to protect against health risks, considering available technology and costs. These National Primary Drinking Water Regulations set enforceable maximum contaminant levels (MCLs) for particular contaminants in drinking water or require ways to treat water to remove contaminants. The SDWA also includes requirements for water systems to test for contaminants in the water to make sure standards are achieved.

The most direct oversight of water systems is conducted by state drinking water programs. States can apply to the EPA for "primacy," the authority to implement SDWA within their jurisdictions, if they can show that they will adopt standards at least as stringent as the EPA's and make sure water systems meet these standards. All states and territories, except Wyoming and the District of Columbia, have received primacy. States, or the EPA acting as a primacy agent, make sure that water systems test for contaminants, review plans for water system improvements, conduct on-site inspections and sanitary surveys, provide training and technical assistance, and take action against water systems not meeting standards.

To ensure that drinking water is safe, the SDWA sets up multiple barriers against pollution. These barriers include source water protection, treatment, distribution system integrity, and public information. Public water systems are responsible for ensuring that contaminants in tap water do not exceed the standards. Water systems treat the water, and must test their water frequently for specified contaminants and report the results to states. If a water system is not meeting these standards, it is the water supplier's responsibility to notify its customers. Many water suppliers now are also required to prepare annual reports for their customers. The public is responsible for helping local water suppliers to set priorities, make decisions on funding and system improvements, and establish programs to protect drinking water sources.

The SDWA mandates that states have programs to certify water system operators and make sure that new water systems have the technical, financial, and managerial capacity to provide safe drinking water. The SDWA also sets a framework for the Underground Injection Control program to control the injection of wastes into groundwater. All of these programs help prevent the contamination of drinking water.

2.1.7 CAA Compliance

The Clean Air Act (CAA) (42 U.S.C. §7401 et seq.) of 1963 established a federal program to research techniques for monitoring and controlling air pollution (EPA 2021g, EPA 2021h, EPA 2021i, and EPA 2007c). However, there was no comprehensive federal response to address air pollution until Congress passed a much stronger CAA in 1970. This stronger CAA authorized the development of comprehensive federal and state regulations to limit emissions from both stationary (industrial) and mobile air pollution sources. Four major regulatory programs affecting stationary sources were initiated: the National Ambient Air Quality Standards (NAAQS), State Implementation Plans, New Source Performance Standards, and National Emission Standards for Hazardous Air Pollutants (NESHAPs). Furthermore, the enforcement authority for the EPA was substantially expanded.

Major amendments were added to the CAA in 1977. These amendments included provisions for the prevention of significant deterioration of air quality in areas attaining the NAAQS as well as requirements pertaining to sources in non-attainment areas for NAAQS. A non-attainment area is a geographic area that does not meet one or more of the federal air quality standards. Both of these 1977 amendments established major permit review requirements to ensure attainment and maintenance of the NAAQS.

Another set of major amendments to the CAA occurred in 1990. These amendments provided EPA a broader authority to implement and enforce regulations reducing air pollutant emissions.

New regulatory programs were authorized for control of acid rain and for the issuance of stationary source operating permits. The NESHAPs were incorporated into a greatly expanded program for controlling toxic air pollutants. The provisions for attainment and maintenance of NAAQS were substantially modified and expanded. Other revisions included provisions regarding stratospheric ozone protection, increased enforcement authority, and expanded research programs. In its 1990 revision of the CAA, Congress recognized that Indian Tribes have the authority to implement air pollution control programs. Tribes were given the ability to develop air quality management programs, write rules to reduce air pollution and implement and enforce their rules on their land.

The CAA requires EPA to set NAAQS (40 CFR part 50) for six principal pollutants which can be harmful to public health and the environment. The CAA identifies two types of national ambient air quality standards. ***Primary standards*** provide public health protection, including protecting the health of "sensitive" populations such as asthmatics, children, and the elderly. ***Secondary standards*** are designed to protect the public welfare from adverse effects, including those related to effects on soils, water, crops, vegetation, man-made (anthropogenic) materials, animals, wildlife, weather, visibility, and climate; damage to property; transportation hazards; economic values, and personal comfort and well-being.

The most recently established NAAQS standards are listed in Table 2.4 (EPA 2021a). Units of measure for the standards are parts per million (ppm) by volume, parts per billion (ppb) by volume, and micrograms per cubic meter of air ($\mu g/m^3$).

TABLE 2.4
National Ambient Air Quality Standards for Six Principal Pollutants

Pollutant		Primary/ Secondary	Averaging Time	Level	Form
Carbon Monoxide (CO)		primary	8 hours	9 ppm	Not to be exceeded more than once per year
			1 hour	35 ppm	
Lead (Pb)		primary and secondary	Rolling 3 month average	0.15 μg/m³ (1)	Not to be exceeded
Nitrogen Dioxide (NO_2)		primary	1 hour	100 ppb	98th percentile of 1-hour daily maximum concentrations, averaged over 3 years
		primary and secondary	1 year	53 ppb (2)	Annual Mean
Ozone (O_3)		primary and secondary	8 hours	0.070 ppm (3)	Annual fourth-highest daily maximum 8-hour concentration, averaged over 3 years
Particle Pollution (PM)	$PM_{2.5}$	primary	1 year	12.0 μg/m³	annual mean, averaged over 3 years
		secondary	1 year	15.0 μg/m³	annual mean, averaged over 3 years
		primary and secondary	24 hours	35 μg/m³	98th percentile, averaged over 3 years
	PM_{10}	primary and secondary	24 hours	150 μg/m³	Not to be exceeded more than once per year on average over 3 years
Sulfur Dioxide (SO_2)		primary	1 hour	75 ppb (4)	99th percentile of 1-hour daily maximum concentrations, averaged over 3 years
		secondary	3 hours	0.5 ppm	Not to be exceeded more than once per year

(1) In areas designated nonattainment for the Pb standards prior to the promulgation of the current (2008) standards, and for which implementation plans to attain or maintain the current (2008) standards have not been submitted and approved, the previous standards (1.5 μg/m3 as a calendar quarter average) also remain in effect.

(2) The level of the annual NO_2 standard is 0.053 ppm. It is shown here in terms of ppb for the purposes of clearer comparison to the 1-hour standard level.

(*continued*)

TABLE 2.4 (Continued)
National Ambient Air Quality Standards for Six Principal Pollutants

(3) Final rule signed October 1, 2015, and effective December 28, 2015. The previous (2008) O_3 standards are not revoked and remain in effect for designated areas. Additionally, some areas may have certain continuing implementation obligations under the prior revoked 1-hour (1979) and 8-hour (1997) O_3 standards.

(4) The previous SO_2 standards (0.14 ppm 24-hour and 0.03 ppm annual) will additionally remain in effect in certain areas: (1) any area for which it is not yet 1 year since the effective date of designation under the current (2010) standards, and (2) any area for which an implementation plan providing for attainment of the current (2010) standard has not been submitted and approved and which is designated nonattainment under the previous SO_2 standards or is not meeting the requirements of a SIP call under the previous SO_2 standards (40 CFR 50.4(3)). A SIP call is an EPA action requiring a state to resubmit all or part of its State Implementation Plan to demonstrate attainment of the required NAAQS.

TABLE 2.5
Summary of Code of Federal Regulations Addressing Radionuclides

Title	Part	Agency	Summary
10	20	NRC	*Standards for protection against radiation:* Establishes standards for protection against ionizing radiation resulting from activities conducted under licenses issued by the NRC. Addresses occupational dose limits for various exposure pathways, dose limits for members of the public, surveying and monitoring requirements, etc.
	71	NRC	*Packaging and transportation of radioactive material:* Establishes requirements for packaging, preparation for shipment, and transportation of licensed material. Establishes procedures and standards for NRC approval of packaging and shipping procedures for fissile material.
10	835	DOE	*Occupational radiation protection:* Establishes radiation protection standards, limits, and program requirements for protecting individuals from ionizing radiation resulting from the conduct of DOE activities.
10	835	DOE	10 CFR 835.209 *Concentrations of radioactive material in air:* The derived air concentration (DAC) values given in appendix A and C (of the CFR reference) shall be used in the control of occupational exposures to airborne radioactive material.
40	141	EPA	141.15 *Maximum contaminant levels for Ra-226, Ra-228, and gross alpha particle radioactivity in community water systems*: The following are the maximum contaminant levels for Ra-226, Ra-228, and gross alpha particle radioactivity: (1) combined Ra-226 and Ra-228 = 5 pCi/L; (2) gross alpha particle activity (including Ra-226 but excluding radon and uranium) = 15 pCi/L.

TABLE 2.5 (Continued)
Summary of Code of Federal Regulations Addressing Radionuclides

Title	Part	Agency	Summary
			141.16 *Maximum contaminant levels for beta particle and photon radioactivity from man-made radionuclides in community water systems:* (1) The average annual concentration of beta particle and photon radioactivity from man-made radionuclides in drinking water shall not produce an annual dose equivalent to the total body or any internal organ greater than 4 millirem/yr.
			141.16 *Table A, Average annual concentration assumed to produce a total body or organ dose of 4 mrem/yr:* Tritium (total body) 20,000 pCi/L; Sr-90 (bone marrow) 8 pCi/L.

2.2 REGULATIONS PERTAINING TO RADIONUCLIDES

Although several of the environmental laws discussed in Section 2.1 address some radionuclides, the majority of the regulations pertaining to radionuclides are published in Title 10 of the Code of Federal Regulations (Nuclear Regulatory Commission, Department of Energy) as well as Department of Energy directives. Table 2.5 presents a summary of the Code of Federal Regulations that are the most pertinent to radionuclides and environmental studies.

BIBLIOGRAPHY

Environmental Law Reporter (ELR), Statute Administrative Proceeding 001, Statute Binder, Environmental Law Institute, 1990.

EPA, 2021a, EPA Web Page: National Ambient Air Quality Standards, www.epa.gov/criteria-air-pollutants/naaqs-table.

EPA, 2021b, EPA Web Page: CERCLA and Superfund overview, www.epa.gov/superfund/superfund-cercla-overview.

EPA, 2021c, EPA Web Page: RCRA, www.epa.gov/rcra.

EPA, 2021d, EPA Web Page: NEPA, www.epa.gov/nepa.

EPA, 2021e, EPA Web Page: CWA, www.epa.gov/cwa-401.

EPA, 2021f, EPA Web Page: TSCA, www.epa.gov/enviro/tsca-overview.

EPA, 2021g, EPA Web Page: Air Pollutants, www.epa.gov/criteria-air-pollutants/naaqs-table.

EPA, 2021h, EPA Web Page: CAA, www.epa.gov/clean-air-act-overview/evolution-clean-air-act.

EPA, 2021i, EPA Web Page: National Ambient Air Quality Standards, www.epa.gov/naaqs.

EPA, 2021j, EPA Web Page: RCRA Solid and Hazardous Waste Exclusions, www.epa.gov/hw/criteria-definition-solid-waste-and-solid-and-hazardous-waste-exclusions

EPA, 2020a, EPA Web Page: CWA NPDES Permit, www.epa.gov/npdes/npdes-permit-basics.

EPA, 2020b, EPA Web Page: CWA, www.epa.gov/enforcement/water-enforcement#cwacompliance.

EPA, 2020c, EPA Web Page: CWA, www.epa.gov/laws-regulations/summary-clean-water-act.

EPA, 2014, *Clean Water Act National Pollutant Discharge Elimination System Compliance Monitoring Strategy*, Office of Enforcement and Compliance Assurance, Office of Compliance, U.S. Environmental Protection Agency, Washington, D.C. Available at: www.epa.gov/sites/production/files/2013-09/documents/npdescms.pdf.

EPA, 2007a, EPA Web Page: *TSCA Statute, Regulations & Enforcement,* www.epa.gov/compliance/civil/tsca/tscaenfstatreq.html#regs.

EPA, 2007b, EPA Web Page: *Safe Drinking Water Act (SDWA),* www.epa.gov/safewater/sdwa/30th/factsheets/understand.html.

EPA, 2007c, *The Plain English Guide to the Clean Air Act,* EPA-456/K-07-001, Office of Air Quality Planning and Standards, Research Triangle Park, North Carolina, April. Available at: www.epa.gov/sites/production/files/2015-08/documents/peg.pdf.

EPA, 2006, *RCRA Orientation Manual 2006—Resource Conservation and Recovery Act,* EPA-530-R-06-003, March.EPA, 1988, *Guidance for Conducting Remedial Investigations and Feasibility Studies under CERCLA,* EPA/540/G-89/004, OSWER Directive 9355.3-01, Interim Final, October.

EPA, 1999, Memorandum transmittal of final guidance entitled *"A Guide to Preparing Superfund Proposed Plans, Records of Decision, and Other Remedy Selection Decision Document,"* From S.D. Luftig to Superfund National Policy Managers Regions 1–10, OSWER 9200.1-23P, EPA 540-R-98-031, PB98-963241, July.

3 Designing a Defensible Sampling Program

Some of the more common sites that require extensive environmental sampling programs include abandoned or operating chemical plants, oil refineries, fuel storage and transfer stations (including gas stations), mine sites, landfills, military installations, nuclear or chemical weapon production and storage facilities, manufacturing facilities, and facilities associated with nuclear reactors. The primary objectives of performing remedial investigations at these types of sites often include collecting adequate data to

- Define the nature and extent of contamination
- Identify contaminant migration pathways
- Predict rates of contaminant migration
- Assess the risk to human health and ecological population receptors
- Evaluate potential remedial alternatives
- Support treatability testing
- Determine if an implemented remedy is working
- Define whether cleanup objectives have been met
- Properly dispose of waste material

The objective of this chapter is to provide the reader with guidance on how to design cost-effective sampling programs for these types of sites that are both comprehensive and defensible. This guidance emphasizes the use of the Environmental Protection Agency's (EPA's) Data Quality Objectives (DQO) process (EPA 2006a) to assist in the development of defensible sampling programs that meet all the sampling objectives.

The DQO process is used to develop performance and acceptance criteria (or DQO) that clarify study objectives, define the appropriate type of data needed, and specify tolerable levels of potential decision errors that will be used as a basis for establishing the quality and quantity of data needed to support environmental decisions. While the EPA (2006a) guidance focuses on statistical sampling designs, this book expands remedial investigation approaches to also include judgmental sampling designs and the use of field screening methods (e.g., soil gas sampling, gamma flyover surveys) to very cost-effectively identify those portions of the site showing the highest levels of contamination. Judgmental sampling designs can provide a defensible estimate of the maximum concentration of contamination at the site. Field screening methods help reduce cost by saving higher-cost analytical methods for those portions of the site that are known to be contaminated.

This chapter presents details on several useful statistical sample design software packages and guidance for preparing a Sampling and Analysis Plan. See Chapter 4

DOI: 10.1201/9781003284000-3

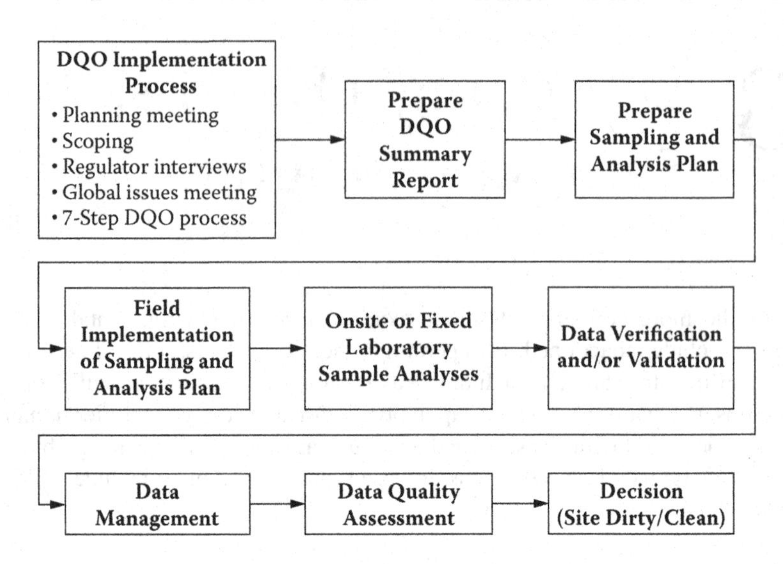

FIGURE 3.1 Data life cycle. (From Byrnes, 2001. *Sampling and Surveying Radiological Environments,* Lewis Publishers, Boca Raton, FL.)

for details on many of the most effective sampling and field screening methods for supporting environmental investigations.

3.1 GENERAL CONSIDERATIONS WHEN DESIGNING A SAMPLING PROGRAM

When designing a sampling program for an environmental study, the goal should be to collect data of sufficient quality and quantity to resolve all of the decisions that need to be made to complete the entire study. Because substantial cost is incurred with the mobilization and demobilization of a field sampling team, every effort should be made to perform all of the required environmental investigations under one mobilization.

Figure 3.1 identifies all of the key steps that are required to develop and implement a defensible sampling program that supports the environmental decision-making process.

This chapter provides guidance on implementing the first three of the nine components of the data life cycle identified in Figure 3.1, and Chapters 4 through 8 address the remaining components.

3.2 DQO IMPLEMENTATION PROCESS

Prior to implementing the seven-step EPA DQO process, a number of preparatory steps must first be implemented. These steps include holding a project planning meeting, performing a thorough scoping effort, holding interviews with the regulatory agency (or agencies) that will be involved in the decision-making process, and holding a Global Issues Meeting to resolve any disagreements with the requirements specified by the regulators, or disagreements between two or more regulatory agencies. If these preparatory steps are *not* implemented prior to beginning the seven-step process, the seven-step process could drag on for months because all of the required information needed to support the process will not be available.

3.2.1 Planning Meeting

The project manager should schedule and conduct a planning meeting with one or more technical advisors who have experience in performing projects with a similar scope. The purpose of this meeting is to identify the project schedule, budget, staffing needs (planning team), regulatory agencies that will be involved in the project, and anticipated procurement requirements. The size of the planning team will vary between projects and is dependent on the complexity of the problem to be resolved. Examples of technical backgrounds that may be needed on the planning team include:

- Environmental science
- Environmental engineering
- Remedial design
- Chemistry
- Radiochemistry
- Hydrogeology
- Groundwater/surface water/air modeling
- Risk assessment
- Statistics
- Quality assurance
- Waste management
- Regulatory compliance

Once the objectives of the planning meeting have been met, the project manager may begin the scoping process.

3.2.2 Scoping

An essential component to designing a defensible sampling program is scoping. Scoping involves the review and evaluation of all applicable historical documents, records, data sets, maps, diagrams, and photographs related to process operations, spills and releases, waste-handling and disposal practices, and previous environmental investigations. The results from the scoping process are used in Step 1 (Section 3.2.5.1) of the DQO process to:

- Identify the contaminants of concern (COCs)
- Support the development of a conceptual site model
- Develop a clear description of the problem

Because the results from the scoping process are used as the foundation upon which the sampling program will be designed, a project team should never attempt to rush through the scoping process in an effort to save money. Doing so could lead to the misidentification of the COCs, and the development of a severely flawed conceptual site model and description of the problem.

A site visit should be scheduled following the completion of the scoping effort to familiarize the project team with the current site conditions. In addition, interviews should be scheduled and performed to verify that the information gathered during the scoping process is accurate, and to assist in filling in any informational data gaps. These interviews should include:

- Historical site workers/managers/owners
- Owners of other property in the near vicinity of the site
- Federal, state, and/or local regulatory agencies (see Section 3.2.3)
- Potentially responsible parties (PRPs)

3.2.3 Regulatory Agency Interviews

The regulatory agencies for a project may include federal, state, and/or local agency representatives who are responsible for making decisions on issues such as cleanup guidelines, deliverable milestone dates, waste classification and disposal requirements, preferred remedial alternatives, and/or favored sampling methods.

Prior to the commencement of the DQO process, the project manager should contact federal, state, and local regulatory agencies to identify the names of the representatives who will be involved with the project. Once these agency representatives have been identified, a one- to two-hour interview should be scheduled with each of them individually. The purpose of the interview is to identify all of the key issues and concerns that need to be addressed by the DQO process. The project manager should consider bringing a few key technical experts to these interviews to answer any technical questions that may arise.

Interviews performed with federal, state, and/or local regulatory agencies should identify specific regulatory requirements that must be taken into consideration, and general concerns that they have related to the project. Examples of requirements and/or concerns expressed by regulatory agencies include:

- Cleanup guidelines
- Enforceable deadlines
- Waste classification and disposal requirements
- Preferred remedial action alternatives for cleaning up the site
- Favored sampling and/or survey methods

The purpose of meeting with agency representatives one-on-one is to assure that those being interviewed feel free to express their specific concerns without worrying about who else may be present in the room. It is not uncommon for two regulatory agencies (e.g., EPA and State) to have very different views about what needs to be done at a particular site. As a result, it is your job to help understand the concerns of each regulatory agency and to eventually help find common ground that both sides can live with. The purpose of these meetings is also to build relationships. If you want to ensure your project is going to be successful, the regulatory agencies must learn to trust that you have the project's best interest in mind.

3.2.4 Global Issues Meeting

A Global Issues Meeting is held whenever you have a disagreement with any of the requirements specified by the regulatory agencies, or when the requirements from one regulatory agency contradict the requirements of another. For example, the EPA may require the site under investigation to be remediated under the Comprehensive Environmental Response, Compensation, and Liability Act (CERCLA) process, whereas the state regulatory agency may require the site to be remediated under the Resource Conservation and Recovery Act (RCRA) process. This meeting should be attended by the project manager, key technical project staff, and regulatory agencies associated with the disagreement. All agreements made at the conclusion of the Global Issues Meeting should be carefully documented in a memorandum that is then entered into the document record.

3.2.5 Seven-Step DQO Process

The seven-step DQO process is a strategic planning approach developed by the EPA (EPA, 2006a) to prepare for data collection activities. This process provides a systematic procedure for defining the criteria that a data collection design should satisfy, including when/where/how to collect samples/measurements, tolerable limits on decision errors, and how many samples/measurements to collect. One of the advantages of the DQO process is that it enables data users and relevant technical experts to participate in the data collection planning process, where they can specify their data needs and data quality requirements prior to data collection. The DQO process provides a method for defining quality and performance requirements appropriate to the intended use of the data by considering the consequences of drawing incorrect conclusions and then placing tolerable limits on them.

The DQO process should be implemented during the planning stage of an investigation prior to data collection. Using this process will ensure that the type, quantity, and quality of the environmental data used in the decision-making process will be appropriate for the intended application. The DQO process is intended to minimize expenditures related to data collection by eliminating unnecessary, duplicative, or overly precise data. In addition, the DQO process ensures that resources will not be committed to a data collection effort that does not support a defensible decision.

The DQO process consists of the seven steps identified in Figure 3.2. The output from each step influences the choices that will be made later in the process. Even though the DQO process is depicted as a linear sequence of steps, in practice, the process is iterative. In other words, the outputs from one step may lead to reconsideration of prior steps. This iterative approach should be encouraged because it will ultimately lead to a more efficient data collection design. During the first six steps of the process, the Planning team will develop the decision performance criteria, otherwise

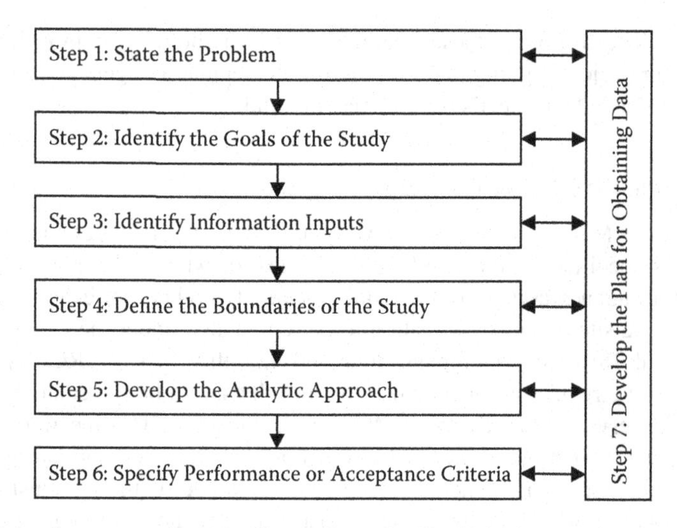

FIGURE 3.2 Seven steps that comprise the Data Quality Objectives (DQO) process.

referred to as DQOs. The final step of the process (Step 7) involves developing the data collection design based on the DQOs.

The first six steps must be completed prior to developing the data collection design. In Figure 3.2, the iterative link between DQO Steps 1 through 6 and DQO Step 7, "Develop the Plan for Obtaining Data," is illustrated by double arrows, which signify that it may be necessary to revisit any one or more of the first six steps to develop a feasible and appropriate data collection design. The results from the DQO process should then be used to support the preparation of a Sampling and Analysis Plan (see Section 3.2.7).

The guidance provided in this book focuses on the collection of data needed to support decision-making, such as determining if a site poses a human/ecological health risk or if cleanup levels have been achieved. For guidance on performing less common data collection efforts that support scientific research estimation studies, see the guidance provided in EPA (2006a).

3.2.5.1 Step 1: State the Problem

Objective: The objective of DQO Step 1 is to define the problem that needs to be resolved. The problem should be stated in a way that is easy to understand.

Activities:

- Identify members of the planning team (staff resources) needed to support the project.
- Identify the key decision-makers (e.g., regulatory agency representatives).
- Identify project budget constraints and deadlines associated with the project.
- Identify other constraints that could affect the project (e.g., staff availability).

Identify deadlines for planning, data collection, and data evaluation.
Identifying the contaminants of concern.
Develop a conceptual site model that identifies the sources of contamination, contamination release mechanisms, pathways for contaminant migration, exposure routes, and potential receptors.
Develop a concise description of the problem.

3.2.5.1.1 Background

In the process of defining the problem to be resolved, a combination of administrative and technical activities needs to be performed. The administrative activities include identifying the members of the planning team; the key decision-makers (e.g., regulatory agency representatives); the list of staff resources needed to support the various aspects of the project; budget and other constraints that could affect the project; and relevant deadlines for planning the project, collecting the data, and evaluating the data. The technical activities include identifying the contaminants of concern, developing a conceptual site model, and developing a concise statement of the problem.

3.2.5.1.2 Identify Members of the Planning Team

The project manager should identify the members of the planning team in the planning meeting (Section 3.2.1). The planning team should be comprised of technical staff with a broad range of technical backgrounds that covers the entire project from site characterization, risk assessment, evaluation of remedial alternatives, treatability testing, and remedial design. By doing so, data needs of all future aspects of the project will be taken into consideration at one time. One should attempt to collect a full data set under one field mobilization because this will save a considerable amount of money in the project budget.

The number of members on the planning team should be directly related to the size and complexity of the problem. Complex tasks may require a team of 10–15 members, whereas simpler tasks only require a few members. The required technical backgrounds for the planning team members will vary depending on the scope of the project, but often include:

- Environmental science
- Environmental engineering
- Remedial design
- Chemistry
- Radiochemistry
- Hydrogeology
- Groundwater/surface water/air modeling
- Risk assessment
- Statistics
- Quality assurance
- Waste management
- Regulatory compliance

3.2.5.1.3 *Identify the Key Decision-Makers*

The project manager should identify in the planning meeting the regulatory agencies that will be responsible for project oversight (Section 3.2.1). These agencies are those that have authority over the study and are responsible for making final decisions based on the recommendations of the planning team. For this reason they are referred to as Key Decision-Makers. When working on a Department of Energy, Department of Defense, or other sites under the jurisdiction of the federal government, the project manager from that federal agency is also considered a Key Decision-Maker. Examples of Key Decision-Makers might include representatives of the:

- Environmental Protection Agency
- State regulatory agency
- Local regulatory agency
- Department of Energy
- Department of Defense
- Nuclear Regulatory Commission
- Bureau of Mines
- Bureau of Land Management

The project manager should encourage early Key Decision-Maker involvement because this will help ensure that they are fully integrated into the team and develop ownership of the project.

3.2.5.1.4 *Identify Project Budget Constraints and Deadlines*

The project manager should identify any budget constraints and relevant deadlines in the planning meeting (Section 3.2.1). The project budget should include the cost of

- Implementing the DQO process
- Preparing a DQO Summary Report
- Preparing a Sampling and Analysis Plan
- Implementing drilling and sampling activities
- Performing laboratory analyses
- Performing data verification/validation
- Performing data quality assessment
- Evaluating the resulting data
- Preparing a report summarizing the results

Identifying all deadlines for completion of the study will assist the project manager in prioritizing work activities.

3.2.5.1.5 *Identify Other Constraints that Could Affect the Project*

It is important that the project manager consult with the planning team to identify all of the constraints that could potentially impact the project budget and/or schedule. Examples of a few of these constraints include:

- Staff availability
- Drill rig availability
- Potential for drill rig breakdown
- Possibility for encountering high levels of contamination
- Potential for severe weather conditions
- Potential for problems with sample recovery
- Potential for encountering archeological materials

Identifying these potential constraints early in the planning phase will allow the project manager and project team time to identify potential ways to avoid or minimize these constraints, or at least be prepared to address them if they are encountered.

3.2.5.1.6 *Identify the Contaminants of Concern*

Based on the results from the thorough review of historical site documents/records and site interviews performed as part of the scoping process (see Section 3.2.2), a comprehensive list of contaminants of potential concern (COPCs) is identified. At complex sites, this list of COPCs is often comprised of a hundred or more compounds. This list is then reduced in size by removing those compounds that are:

- Noncarcinogenic and nontoxic
- Simply reported as part of a suite of compounds screened by an analytical method but were never detected above detection limits
- Only reported in very few instances at concentrations just above detection limits (e.g., suspected false-positive readings)
- Known to be laboratory contaminants (e.g., compounds also detected in laboratory blank samples)
- Known to have been used in such small quantities that they pose no threat
- Detected in concentrations below background
- Radioactive and have a half-life of less than 2 years

The items in this reduced list of compounds are referred to as the contaminants of concern (COCs) and are carried through the remainder of the DQO process.

3.2.5.1.7 *Develop a Conceptual Site Model*

A conceptual site model is prepared for the purpose of helping the planning team better understand what is currently understood about the site conditions and where there may still be significant data gaps. A conceptual site model may be prepared in a tabular and/or graphical format that identifies the contamination sources, contaminant release mechanisms, pathways for contaminant migration, exposure routes, and potential receptors. The process of developing a conceptual site model will help the planning team identify areas within the site where significant uncertainty still exists. Areas showing significant uncertainty should be identified as data gaps that need to be filled.

The following paragraph describes an example conceptual site model that is illustrated in tabular and graphical format in Table 3.1 and Figure 3.3. This example

TABLE 3.1
Example of a Tabular Conceptual Site Model

Primary Source of Contamination	Secondary Source of Contamination	Release Mechanisms	Migration Pathways	Exposure Routes	Receptors
Chipped metal pile located in the northwest corner of the site	Contaminated soil beneath and surrounding the chipped metal pile	Wind, rain, human/ecological (e.g., burrowing animals) receptors disturbing contaminated soil, percolation of rain water to underlying groundwater, groundwater movement	Air, surface water, groundwater	Direct exposure, inhalation, ingestion, dermal contact	Humans, rodents, birds of prey, farm animals, natural vegetation, and agricultural crops
	Contaminated sediment within drainage channels leading from chipped metal pile to pond, and contaminated sediment surrounding edges of pond	Wind, rain, human/ecological (e.g., burrowing animals) receptors disturbing contaminated sediment, percolation of drainage channel water to underlying groundwater, groundwater movement	Air, surface water, groundwater	Direct exposure, inhalation, ingestion, dermal contact	Humans, rodents, birds of prey, farm animals, and natural vegetation
	Contaminated sediment at the base of pond	Ecological receptors (e.g., burrowing animals and aquatic organisms) disturbing contaminated sediment, percolation of pond water to underlying groundwater, groundwater movement	Surface water, groundwater	Ingestion, dermal contact	Humans, fish, burrowing aquatic organisms, water fowl, birds of prey, farm animals, and natural vegetation

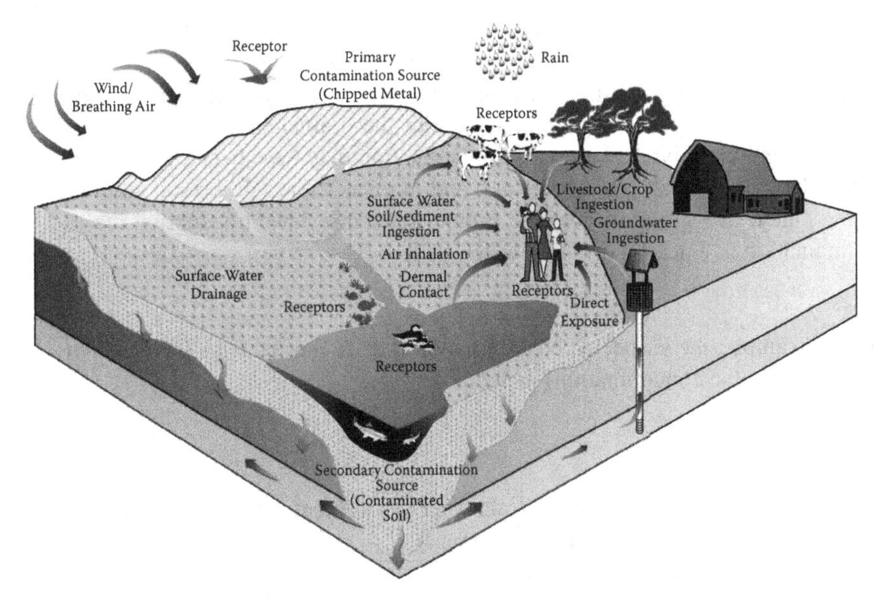

FIGURE 3.3 Example of a graphical conceptual site model.

conceptual site model will be referred to throughout the remainder of this chapter and is intended to help the reader understand what information is needed for the various steps of the DQO process.

Example Conceptual Site Model

Known Information

At a site under investigation, the primary source of contamination is a chipped metal pile located in the northwest corner of the site. From historical data, this chipped metal is known to contain high concentrations of lead, arsenic, and chromium and is also known to be radiologically contaminated primarily with the isotope uranium-238. Historical data has also shown the chipped metal pile to contain low concentrations of trichloroethylene. A residential home and drinking water well are currently on the property. This site could be used in the future for agricultural purposes or to raise farm animals.

Suspected Information

The secondary sources of contamination at the site are suspected to be contaminated soil and sediment beneath and surrounding the chipped metal piles and found within the surface water drainage channel and pond. The contamination from the metal chips is suspected of being transported to the surrounding soil by rain water, wind, and human/ecological receptors disturbing the soil. Rain water has likely carried the contamination down the surface water drainage channels to a nearby pond, contaminating the drainage channel water, pond water, and underlying sediment. Contamination may have been transported by pond water, drainage channel water, and rainwater through the soil column to the underlying groundwater. The groundwater may have transported the contamination to a nearby drinking water well.

In this example, the suspected contamination release mechanisms (mechanisms that will mobilize the contamination) include:

- Wind
- Rain
- Surface water movement (rainwater movement, drainage channel water movement, and pond water movement)
- Groundwater movement (from natural groundwater flow and pumping from the drinking water well)
- Human/ecological (e.g., burrowing animals and aquatic organisms) receptors disturbing contaminated soil and sediment

The suspected pathways for contaminant migration include:

- Air
- Surface water
- Groundwater

The suspected routes for exposure include receptors:

- Receiving direct radiation exposure from the chipped metal
- Breathing air containing contaminated dust particles and volatilized trichloroethylene
- Receiving dermal contact with contaminated chipped metal, soil, sediment, surface water, and groundwater
- Ingesting contaminated chipped metal, soil, sediment, surface water, and/or groundwater
- Ingesting products from contaminated farm animals
- Ingesting contaminated agricultural crops grown on the site

Potential receptors in this example include:

- Humans
- Rodents
- Birds of prey
- Farm animals
- Fish

- Burrowing aquatic organisms
- Water fowl
- Natural vegetation
- Agricultural crops

The conceptual site model is continuously refined throughout the implementation of the sampling program resulting from the implementation of the DQO process.

3.2.5.1.8 Develop a Concise Description of the Problem

Once the conceptual site model has been established, the last activity in DQO Step 1 is developing a concise description of the problem as it is currently understood. One should consider preparing the description of the problem using the following format:

> **In order to** [determine/achieve/support/understand/establish/confirm/prevent] [some issues or objectives of this study], **data regarding** [contaminant distributions/media characteristics/physical-chemical-biological parameters] **are needed** [for some intended purpose].

For the example conceptual site model provided in Table 3.1 and Figure 3.3, the problem statement should read something like the following:

> **In order to** determine whether the chipped metal pile present at the study site has impacted the surrounding environmental media (air, soil, sediment, surface water, and groundwater) and human/ecological population to the point of requiring some form of remedial action, **data regarding** contaminant distributions, contaminant concentration ranges, aquifer characteristics, receptor uptake, and/or various modeling input parameters **are needed** to support the preparation of a remedial investigation report, baseline risk assessment, screening of remedial alternatives, and remedial design.

3.2.5.1.9 Outputs from DQO Step 1

The primary outputs resulting from the implementation of this step include:

- List of planning team members
- List of key decision-makers (e.g., regulatory agency representatives)
- List of staff resources needed throughout the study
- Summary of project budget and other constraints (e.g., staff availability)
- Relevant deadlines for planning, data collection, and data evaluation
- Conceptual site model
- Concise description of the problem

3.2.5.2 Step 2: Identify the Goals of the Study

Objective: The objective of DQO Step 2 is to develop decision statements that need to be addressed to resolve the problem identified in DQO Step 1.

Activities:

Identify the principal study questions (PSQs) (key questions) that need to be answered.
Define the alternative actions that may be taken based upon answering each PSQ.
Join the PSQs and alternative actions into decisions statements.
Organize the decision statements into an order of sequence or priority.

3.2.5.2.1 Background

Step 2 of the DQO process involves identifying the PSQs (key questions) that the study will attempt to address, along with the alternative actions or outcomes that may result based on the answers to these questions. The PSQs and the alternative actions are then combined into decision statements.

3.2.5.2.2 Identify the Principal Study Questions

The first activity to be performed under DQO Step 2 is identifying all of the PSQs needed to resolve the problem identified in DQO Step 1. The PSQs identify key unknown conditions or unresolved issues that need to be resolved to address the problem. Note that only questions that require environmental data to answer should be included as PSQs. For example, questions such as "Should a split-spoon or solid-tube sampler be used to collect the soil samples?" should not be included in the DQO process because this decision should be made based on experience and requires no analytical data to answer. The answers to the PSQs will provide the basis for determining what course of action should be taken to resolve the problem.

For the example conceptual site model provided in Table 3.1 and Figure 3.3, the following are examples of some PSQs that might need to be answered to address the problem:

PSQ 1: Have COCs from the primary contamination source (e.g., chipped metal pile) migrated laterally to contaminate surrounding surface soils, sediment within the surface water drainage channels, and/or sediment within the pond?
PSQ 2: Have COCs from the primary (e.g., chipped metal pile) and secondary contamination sources (e.g., surrounding contaminated soil/sediment) migrated vertically to contaminate deep soil and groundwater?
PSQ 3: Do concentrations of COCs in the on-site drinking water well exceed drinking water standards?
PSQ 4: What are the concentrations and vertical and lateral distribution of COCs within the contaminated soil, sediment, and groundwater?
PSQ 5: Has the health of the ecological population within the study area been impacted by the COCs through direct exposure, ingestion, inhalation, or dermal contact (e.g., rodents, birds of prey, water fowl, fish, burrowing aquatic organisms, farm animals, natural/agricultural vegetation)?

PSQ 6: Has the health of the human population within the study area been impacted by the COCs through direct exposure, ingestion, inhalation, or dermal contact?
PSQ 7: Will the future migration of COCs likely impact the human or ecological population?
PSQ 8: Are unsaturated zone modeling input parameters known well enough to support defensible modeling of the future movement of COCs from ground surface to groundwater?
PSQ 9: Are groundwater modeling input parameters known well enough to support defensible modeling of future COC plume movement through the groundwater (e.g., groundwater elevations, gradients, distribution coefficients, hydraulic properties)?
PSQ 10: Is sufficient data available to support an evaluation of remedial action alternatives (e.g., range of COC concentrations, depth and distribution of contamination, chemical form, aquifer properties)?
PSQ 11: Is sufficient data available to evaluate risk reduction offered by various remedial alternatives?

3.2.5.2.3 Define the Alternative Actions

For each PSQ identified, the planning team should identify at least two possible actions (e.g., no action, remediate and dispose of soil at an appropriately licensed landfill) that may be taken once the PSQ has been answered. Keep in mind that alternative actions are not taken to resolve the PSQ, but rather are taken only after the PSQ has been resolved. One of the alternative actions may be to take no action.

Once the alternative actions have been identified, perform a qualitative assessment of the consequences of unintentionally taking the wrong action if a decision error is made. In other words, if one implements the no-action alternative when one should have implemented a soil remediation action, what are the potential consequences of this mistake? When making this assessment, one should consider how the site may be used in the future (e.g., industrial, residential, farming).

Express the consequences using the following terms:

- Low
- Moderate
- Severe

When assessing consequences, take the following aspects into consideration:

- Human health
- Ecological health (flora/fauna)
- Political consequences
- Economic consequences
- Legal consequences

TABLE 3.2
Examples of a Principal Study Questions and Alternative Actions

Principal Study Question	Alternative Actions
PSQ 1: Have COCs from the primary contamination source (e.g., chipped metal pile) migrated laterally to contaminate surrounding surface soils, sediment within the surface water drainage channels, and/or sediment within the pond?	**If the answer to PSQ 1 is YES:** Perform additional soil and sediment characterization to define the vertical extent of the contamination beneath the chipped metal pile as well as other areas showing surface soil/sediment contamination. **If the answer to PSQ 1 is NO:** Perform additional soil characterization only beneath the chipped metal pile to define the vertical extent of the soil contamination.
PSQ 2: Have COCs from the primary (e.g., chipped metal pile) and secondary contamination sources (e.g., surrounding contaminated soil/sediment) migrated vertically to contaminate the underlying groundwater?	**If the answer to PSQ 2 is YES:** Perform additional field work to define the vertical and lateral extent of deep soil and groundwater contamination. **If the answer to PSQ 2 is NO:** Perform only routine quarterly groundwater monitoring from the on-site well.
PSQ 3: Do concentrations of COCs in the on-site drinking water well exceed drinking water standards?	**If the answer to PSQ 3 is YES:** Remove the pump from the drinking water well, begin routine quarterly groundwater monitoring program, and implement institutional controls to prevent the future use of the well while COCs concentrations remain above drinking water standards. **If the answer to PSQ 3 is NO:** Perform routine quarterly groundwater monitoring.

The results from this consequence assessment will be used later in the DQO process to assist in the selection between a statistical or nonstatistical sampling design.

Table 3.2 presents examples of alternative actions that may be taken after the PSQ has been resolved.

3.2.5.2.4 Develop Decision Statements

Decision statements (DS) are developed by linking the PSQs with their corresponding alternative actions. The following format should be used when formulating the decision statements:

> Determine whether [PSQ 1] and therefore requires [Alternative Action A] otherwise perform [Alternative Action B].

TABLE 3.3
Examples of Decision Statements

DS 1: Determine whether contaminants of concern (COCs) from the primary contamination source (e.g., chipped metal pile) have migrated laterally to contaminate surrounding surface soils, sediment within the surface water drainage channels, and/or sediment within the pond and therefore require additional soil and sediment characterization to define both the vertical extent of the contamination beneath the chipped metal pile as well as other areas showing surface soil/sediment contamination; otherwise, only perform vertical soil characterization beneath the chipped metal pile to define the maximum depth of soil contamination.

DS 2: Determine whether COCs from the primary (e.g., chipped metal pile) and secondary contamination sources (e.g., surrounding contaminated soil/sediment) have migrated vertically to contaminate deep soil and groundwater and therefore require additional fieldwork to define the vertical and lateral extent of deep soil and groundwater contamination; otherwise, perform only routine quarterly groundwater monitoring from the on-site well.

DS 3: Determine whether concentrations of COCs in the on-site drinking water well exceed drinking water standards and therefore require removal of the pump from the drinking water well, beginning a routine quarterly groundwater monitoring program, and implementing institutional controls to prevent the future use of the well while COCs concentrations remain above drinking water standards; otherwise, only routine quarterly groundwater monitoring is required.

Table 3.3 takes the PSQs from Table 3.2 and combines them with their corresponding alternative actions to form several example decision statements.

3.2.5.2.5 Organizing Decision Statements

Once all of the decision statements have been identified, they should be ordered based on proper sequence or priority. For example, in Section 3.2.5.2.2, the decision statement related to PSQ 2 should be sequenced before the decision statement related to PSQ 3. This is because PSQ 3 may not apply if the contaminants of concern have not yet migrated to the groundwater. Ordering the decision statements based on priority could be useful if funding impacts the ability to perform all of the proposed field work at one time. In this case, sampling to address the higher priority decisions could be funded as part of the first field effort, whereas sampling to support lower priority decisions could be funded at a later date.

3.2.5.2.6 Outputs from DQO Step 2

The primary outputs resulting from the implementation of this step include:

- A list of PSQs (key questions) that need to be answered
- A list of alternative actions that may be taken based upon answering each PSQ
- A list of decision statements that are developed by joining the PSQs and their corresponding alternative actions

3.2.5.3 Step 3: Identify Information Inputs

Objective: The objectives of DQO Step 3 are to identify the types of data needed to resolve each of the decision statements identified in DQO Step 2 and to determine whether or not existing data are of adequate quality and quantity for use in resolving these decisions.

Activities:

Identify data required to address the decision statements identified in DQO Step 2.
Identify the potential data sources.
Define the action level for each of the COCs.
Determine the level of quality required for the data.
Evaluate the appropriateness of existing data.
Confirm that appropriate sampling and analytical methods exist to provide the necessary data.

3.2.5.3.1 Identify Data Required to Address the Decision Statements

Generally, all of the decision statements identified in DQO Step 2 will be resolved by data (existing or new) from either environmental measurements or scientific literature. The most common data needed to support environmental investigations include physical and chemical properties of various types of media. Although modeling may be used to resolve some decisions, all models require some input data to run the model.

Table 3.4 provides examples of data that would likely be required to resolve several example decision statements.

3.2.5.3.2 Identify the Potential Data Sources

It is important to identify and list all of the potential sources of data that may be able to address each of the data needs identified in Section 3.2.5.3.1. Examples of potential data sources include:

- New data collection activities
- Previous data collection activities
- Historical records
- Scientific literature
- Regulatory guidance
- Professional judgment
- Modeling

3.2.5.3.3 Define the Action Level for Each of the Contaminants of Concern

Prior to defining the level of quality of data required to resolve each decision statement, it is necessary to first define what the action level is for each of the COCs. Knowing the action levels is essential for establishing the detection limits for any new data that needs to be collected, and is needed to determine if detection limits for existing data are low enough to meet data quality requirements. For groundwater and surface water, these action levels are often drinking water standards (e.g., maximum contaminant levels [MCLs]). For soil and sediment, these action levels are typically calculated based on risk or are compared to background.

TABLE 3.4
Example of Data Needed to Resolve Decision Statements

Decision Statements	Data Needed
DS 1: Determine whether COCs from the primary contamination source (e.g., chipped metal pile) have migrated laterally to contaminate surrounding surface soils, sediment within the surface water drainage channels, and/or sediment within the pond and therefore require additional soil and sediment characterization to define both the vertical extent of the contamination beneath the chipped metal pile and other areas showing surface soil/sediment contamination; otherwise, only perform vertical soil characterization beneath the chipped metal pile to define the maximum depth of soil contamination.	Concentrations of lead, arsenic, chromium, trichloroethylene, and uranium-238 contamination in shallow soil and sediment
DS 2: Determine whether COCs from the primary (e.g., chipped metal pile) and secondary contamination sources (e.g., surrounding contaminated soil/sediment) have migrated vertically to contaminate deep soil and groundwater and therefore require additional fieldwork to define the vertical and lateral extent of deep soil and groundwater contamination; otherwise, perform only routine quarterly groundwater monitoring from the on-site well.	Concentrations of lead, arsenic, chromium, trichloroethylene, and uranium-238 contamination in deep soil and groundwater
DS 3: Determine whether concentrations of COCs in the on-site drinking water well exceed drinking water standards and therefore require removal of the pump from the drinking water well, beginning a routine quarterly groundwater monitoring program, and implementing institutional controls to prevent the future use of the well while COCs concentrations remain above drinking water standards; otherwise, only routine quarterly groundwater monitoring is required.	Concentrations of lead, arsenic, chromium, trichloroethylene, and uranium-238 contamination in water collected from the drinking water well

3.2.5.3.4 Determine the Level of Quality Required for the Data

When determining the level of data quality required to resolve each decision statement, one should take into consideration the potential consequences of making an incorrect decision. One should consider the human health, ecological, political, cost, and legal impacts if the incorrect alternative action (see DQO Step 2, Section 3.2.5.2.3) is selected and implemented.

If human health or ecological impacts could occur by accidentally selecting and implementing the wrong alternative action, then standard laboratory analytical methods (e.g., EPA SW-846 methods) should be run in combination with robust quality control analyses (e.g., duplicates, spikes, and blanks) to assure that the resulting data is representative. Another lower-cost option more recently accepted by many regulatory agencies is to collect a large number of lower-cost field-screening measurements or on-site laboratory measurements in combination with 5 to 10% of the samples being split and sent for standard laboratory testing for confirmation analyses. Only the confirmation analyses would have robust quality control analyses run to verify the quality of the data. This approach is only affective if the detection limit requirements can be met by field-screening measurements or on-site laboratory methods, and it assumes that the standard laboratory confirmation analyses will confirm that the field screening measurements are representative.

When there are very little or no potential human health or ecological impacts if the wrong alternative action is implemented by accident, then less expensive analytical methods (e.g., field screening, on-site laboratory measurements) may be all that is required.

The specific analytic methods to be run and analytical performance requirements (e.g., detection limits, precision/accuracy requirements) are specified in this step of the DQO process along with quality control requirements. These analytical performance requirement details are also later documented in a Sampling and Analysis Plan that is often approved by the lead regulatory agency. See Section 3.2.7 for guidance on preparing a Sampling and Analysis Plan.

For the example decision statements presented earlier in Table 3.3, Table 3.5 provides an example of data quality requirements for the analytical methods to be performed.

3.2.5.3.5 Evaluate the Appropriateness of Existing Data

To determine whether an existing data set is of adequate quality to resolve one or more decision statements, one must compare the detection limits and the results from the accompanying laboratory quality control data with the data quality requirements specified in Section 3.2.5.3.4. Laboratory quality control data should include the results from the analysis of

- Spike samples
- Duplicate samples
- Blank samples

The results from spike and duplicate samples are used to estimate the accuracy and precision of the analytical methods, respectively. Blank samples are analyzed to check that the instrument was not contaminated from the analysis of previous samples.

When evaluating the appropriateness of existing data, one should also take into consideration:

- Instrument detection limits
- Types of samples collected (e.g., grab, composite, integrated)
- Sample collection design (e.g., random, systematic, judgmental)

Remove data that are of poor quality, that does not have low-enough detection limits, or that are not representative of the population. Data should not be used that

TABLE 3.5
Example of Data Quality Requirements Table

COC	Survey or Analytical Method	Detection Limit	Action Level	Precision Required	Accuracy Required
Soil/Sediment analyses					
Lead	SW-846 Method 6010-B (trace)	500 μg/kg	1000 μg/kg	± 20%	80–120%
Arsenic	SW-846 Method 6010-B (trace)	1000 μg/kg	2000 μg/kg	± 20%	80–120%
Chromium (total)	SW-846 Method 6010-B (trace)	200 μg/kg	400 μg/kg	± 20%	80–120%
Trichloroethylene	SW-846 Method 8260	5 μg/kg	20 μg/kg	± 20%	80–120%
Uranium-238	Alpha Energy Analysis (aEa)	1 pCi/kg	5 pCi/g	± 30%	70–130%
Water analyses					
Lead	SW-846 Method 6010-B (trace)	5 μg/L	15 μg/L	± 20%	80–120%
Arsenic	SW-846 Method 6010-B (trace)	5 μg/L	10 μg/L	± 20%	80–120%
Chromium (total)	SW-846 Method 6010-B (trace)	2 μg/L	100 μg/L	± 20%	80–120%
Trichloroethylene	SW-846 Method 8260	2 μg/L	5 μg/L	± 20%	80–120%
Uranium-238	Alpha Energy Analysis (aEa)	1 pCi/L	5 pCi/L	± 30%	70–130%

have no corresponding quality control data accompanying it, as there is no way to determine if the data are representative. Existing data may only be used to support statistical analyses if they were collected using a statistical design (see Section 3.2.5.7.2).

3.2.5.3.6 Confirm that Appropriate Sampling and Analytical Methods Exist

For any new analytical data that is needed, develop a comprehensive list of potentially appropriate sampling and analytical methods.

3.2.5.3.7 Outputs from DQO Step 3

The primary outputs resulting from the implementation of this step include:

- Summary of the types of environmental data needed to resolve the decision statements identified in DQO Step 2 (Section 3.2.5.2.4)
- List of all of the potential sources for the needed data
- List of action levels for each of the COCs
- Data quality and analytical performance requirements for the needed data
- Determination of whether or not data of adequate quality and quantity already exist
- List of new data required to be collected along with analytical performance requirements

3.2.5.4 Step 4: Define the Boundaries of the Study

Objective: The objectives of DQO Step 4 are to define the temporal and spatial boundaries of the study.

Activities:

Define the target population of interest.
Define the spatial boundaries of the study.
Define the temporal boundaries of the study.
Define when conditions are most favorable for collecting data.
Define time frame over which each decision applies.
Define practical constraints associated with sample/data collection.
Specify the smallest unit on which decisions or estimates will be made.

3.2.5.4.1 Define the Target Population of Interest

It is difficult to make a decision with data that have not been drawn from a well-defined population. The term *population* refers to the total universe of potential individual sampling units within the study area. For example, if one is collecting surface sediment samples from the pond shown in Figure 3.3 to determine the concentrations of the contaminants of concern in the sediment, the population is the total number of potential 1-kg sediment samples that could be collected to a depth of 6 inches within the perimeter of the pond.

Because it would be cost-prohibitive to sample and analyze every member of the population (e.g., thousands of samples) for COC concentrations, a statistical sample of the population is collected to provide an estimate of the COC concentrations in the population. Keep in mind that this estimate will have error associated with it.

Other examples of the population of interest include:

- All subsurface soil samples (1 kg) within the study area to a depth of the first confining layer

- All surface water samples (1 L) from all depths within the perimeter boundaries of the pond
- All direct surface radioactivity measurement areas (100 cm^2) on the inside and outside surfaces of a residential home

3.2.5.4.2 Define the Spatial Boundaries of the Study

One needs to define the entire geographical area where data are to be collected using unambiguous coordinates (e.g., latitude, longitude, elevation) or distinctive physical features described in terms of length, area, volume, or legal boundaries.

Some examples of geographic areas include:

- Area: Sediment within the perimeter of a pond
- Volume: Soil to a depth of 30 feet beneath the chipped metal pile
- Length: Length of a discharge pipeline
- Other identifiable boundary – The natural habitat range of a particular animal/ plant species

It is often an advantage to divide the study area into subareas (often referred to as "strata") that have relatively homogeneous environmental characteristics based on process history. Many times, dividing the site up in this manner can significantly reduce the number of samples that need to be collected for statistical sampling designs. This is because there is less variability in the population when a site is divided into subareas that have the same process history. Section 3.2.5.7 discusses other factors that influence the number of samples required to be collected for statistical sampling designs. Dividing the site into subareas (or strata) also has the advantage of reducing the complexity of the problem by breaking it up into more manageable pieces.

The Planning Team should use its knowledge of the conceptual site model (see Section 3.2.5.1.7) to consider whether or not the study area should be divided into subareas (or strata). Figure 3.4 provides a sample of how the example conceptual site model (described in Section 3.2.5.1.7) could be divided up into relatively homogeneous subareas (strata) based on process history.

3.2.5.4.3 Define the Temporal Boundaries of the Study

When defining the temporal boundaries of the study, one needs to define when conditions are most favorable for collecting data that are representative of the target population as well as the time frame over which each decision applies.

3.2.5.4.3.1 Define When Conditions Are Most Favorable for Collecting Data

For each decision statement identified in DQO Step 2 (see Section 3.2.5.2.4) it is important that one determines when conditions are most favorable for collecting data that are representative of the target population.

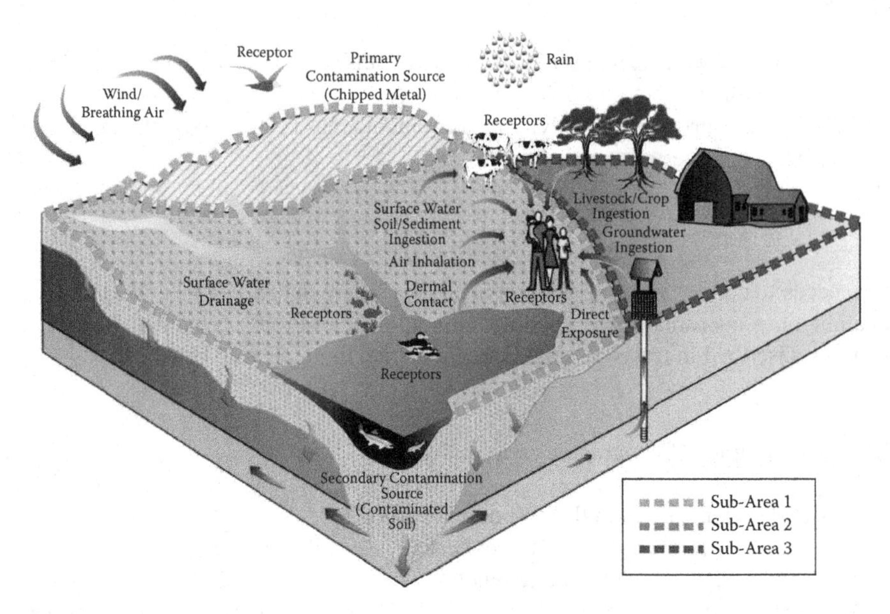

FIGURE 3.4 Example of dividing study area into subareas using process history.

For the example conceptual site model presented in Figure 3.3 and Table 3.1, if one were attempting to measure the human receptor's maximum exposure to the contaminants of concern (lead, arsenic, chromium, trichloroethylene, and uranium-238) through the air inhalation pathway, one would want to collect air samples for lead, arsenic, chromium, and uranium-238 analysis during a windy day that was relatively dry. Collecting air samples on a rainy day with little wind would result in a significant underestimation of maximum exposure to these COCs because they are most mobile on dry windy days. On the other hand, one would want to collect additional air samples in the heat of a calm midsummer day to estimate the maximum exposure to trichloroethylene because this chemical is most volatile under these conditions. Also, if some type of ground-disturbing work activity was performed in the vicinity of the chipped metal pile that could potentially mobilize the COCs, one would want to collect the air samples while work was being performed to assess the maximum exposure.

3.2.5.4.3.2 Define the Time Frame over Which Each Decision Applies Each decision statement identified in DQO Step 2 (see Section 3.2.5.2.4) will have a specific time frame over which it applies. For many decisions related to large environmental cleanup projects, this time frame often ranges between 7 and 10 years, which is the average length of time required to define the nature and extent of contamination, assess the risk to human health and ecological population, select a preferred remedial alternative, perform remedial design, and implement the final remedy. However, some decision statements may apply for the shorter or much longer period of time.

3.2.5.4.4 Define Practical Constraints Associated with Sample/Data Collection
Identify any constraints or obstacles that could potentially interfere with the full implementation of the data collection design. These should be taken into consideration in the sampling design and when developing implementation schedules.

Examples of these constraints or obstacles include:

- Seasonal or meteorological conditions when sampling is not possible
- Inability to gain site access or informed consent
- Unavailability of personnel, time, or equipment
- Presence of building or other structure that could prevent access to sampling locations
- Security clearance requirements to access site

3.2.5.4.5 *Specify the Smallest Unit on Which Decisions Will Be Made (Scale of Decision-Making)*

The "scale of decision-making" needs to be defined by the planning team. This term refers to the smallest unit of area or volume over which data will be collected, analyzed, and interpreted to make a decision. One defines the "scale of decision-making" in order to help control decision error. It is defined by merging the spatial and temporal boundaries discussed earlier (see Sections 3.2.5.4.2 and 3.2.5.4.3). Table 3.6 presents examples of scales of decision-making that apply to the example conceptual site model presented in Table 3.1 and Figure 3.3.

3.2.5.4.6 *Outputs from DQO Step 4*

The primary outputs resulting from the implementation of this step include:

- A clear definition of the target population
- A definition of the spatial and temporal boundaries of the study
- The conditions that are most favorable for collecting data
- The time frame over which each decision applies
- A list of constraints (or obstacles) that could potentially interfere with the full implementation of the data collection design
- The scale for decision-making

3.2.5.5 Step 5: Develop the Analytic Approach

Objective: The objective of DQO Step 5 is to define the parameter of interest (e.g., mean, median, percentile, maximum) to be used to represent the contaminants of concern, then combine these with outputs from the previous DQO steps (e.g., action levels, scale of decision-making, and alternative actions) to produce decision rules that provide a logical basis for choosing between alternative actions.

Activities:

- Define the parameter of interest (e.g., mean, median, percentile, maximum) considered to be most appropriate for making inferences about the contaminants of concern.
- Construct decision rules ("If … then … else …" statements) for each of the decision statements identified in DQO Step 2 (see Section 3.2.5.2) by combining the parameter of interest (e.g., mean, median, percentile, maximum), the action level, the scale of decision-making, and alternative actions.

TABLE 3.6
Examples of Scale of Decision-Making

Decision Statements	Scale of Decision-Making
DS 1: Determine whether COCs from the primary contamination source (e.g., chipped metal pile) have migrated laterally to contaminate surrounding surface soils, sediment within the surface water drainage channels, and/or sediment within the pond and therefore require additional soil and sediment characterization to define both the vertical extent of the contamination beneath the chipped metal pile and other areas showing surface soil/sediment contamination; otherwise perform only vertical soil characterization beneath the chipped metal pile to define the maximum depth of soil contamination.	Soil and sediment (to depth of 1 ft) between the chipped metal pile and the on-site pond (see Figure 3.3) over the next 8 years (sampling to be performed under clear weather conditions with wind speeds <10 mph and temperature < 80°F)
DS 2: Determine whether COCs from the primary (e.g., chipped metal pile) and secondary contamination sources (e.g., surrounding contaminated soil/sediment) have migrated vertically to contaminate deep soil and groundwater and therefore require additional field work to define the vertical and lateral extent of deep soil and groundwater contamination; otherwise perform only routine quarterly groundwater monitoring from the on-site well.	Soil (to a depth of 30 ft) and groundwater between the chipped metal pile and the on-site pond over the next 8 years (sampling to be performed under clear weather conditions with wind speeds <10 mph and temperature < 80°F)
DS 3: Determine whether concentrations of COCs in the on-site drinking water well exceed drinking water standards and therefore require removal of the pump from the drinking water well, beginning a routine quarterly groundwater monitoring program, and implementing institutional controls to prevent the future use of the well while COCs concentrations remain above drinking water standards; otherwise, only routine quarterly groundwater monitoring is required.	Groundwater (throughout the thickness of the aquifer) in the vicinity of the on-site drinking water well over the next 20 years (sampling to be performed when wind speeds are <10 mph and temperature < 80°F).

3.2.5.5.1 Define Parameters of Interest to be Used for Making Decisions

In this step, the planning team must choose the parameter of interest (e.g., mean, median, percentile, maximum) that will be used to resolve the decision statements identified in DQO Step 2 (see Section 3.2.5.2.4). EPA guidance manuals, particularly those associated with performing baseline risk assessments and feasibility studies, should be consulted in addition to a project statistician when choosing the most appropriate parameter of interest. The client and lead regulatory agency should be consulted prior to the final selection.

The parameter most commonly selected to characterize a site is the "mean" because it is frequently used to model random exposure to environmental contamination. The maximum concentration is also often selected to identify the worst-case situation at the site. The parameter of interest that is selected will influence the type of sampling designed and developed in DQO Step 7 (see Section 3.2.5.7). For example, a random or systematic sampling approach would be selected if the parameter of interest was the "mean," whereas a judgmental design would be selected if the parameter of interest was the "maximum."

The mean parameter is the average of the population. Selecting the mean parameter would compare the average of the population to the action level. This parameter is appropriate for chemicals that could cause cancer after long-term chronic exposure. For skewed distributions with long right tails, the geometric mean may be more relevant than the arithmetic mean. Either mean may not be useful if a large proportion of values are below the detection limit (EPA 2006b).

The "median" parameter refers to the middle of the population where half of the data is above and half is below. The median provides a better estimate of central tendency for a population that is highly skewed (nonsymmetrical). It may also be preferred if the population contains many values that are less than the measurement detection limit. The median is not a good choice if more than 50% of the population is less than the detection limit because a true median does not exist in this case (EPA 2006b).

The "percentile" parameter specifies the percent of the sample that is below the given value. This parameter is often selected for cases where only a small portion of the population can be allowed to exceed the action level. This parameter is often selected for a decision rule if a chemical can cause acute health effects. It is also useful when a large part of the population contains values less than the detection limit (EPA 2006b).

3.2.5.5.2 *Constructing Decision Rules*

After the parameter of interest (e.g., mean, median, percentile, maximum) has been selected, this information is combined with the scale of decision-making (Section 3.2.5.4.5), action level (Section 3.2.5.3.3), and alternative actions (see Section 3.2.5.2.3). The decision rule should follow the following general format:

> If the [parameter of interest] within the [scale of decision] is greater than the [action level], then take [alternative action A]; otherwise take [alternative action B].

Table 3.7 presents example decision rules that correspond to the decision statements presented in Table 3.3, and the conceptual site model presented in Table 3.1 and Figure 3.3.

3.2.5.5.3 *Outputs from DQO Step 5*

The primary outputs resulting from the implementation of this step include:

- The parameter of interest (e.g., mean, median, percentile, maximum) that is most appropriate for making decisions related to the target population
- A separate decision rule for each decision statement identified in DQO Step 2 (see Section 3.2.5.2)

TABLE 3.7
Examples of Decision Rules

DR 1: If the mean concentration of the COCs in soil and sediment (to a depth of 1 ft) within the area between the chipped metal pile and the on-site pond (over the next 8 years) is greater than their corresponding action levels specified in Table 3.5, then perform additional soil and sediment characterization to define both the vertical extent of the contamination beneath the chipped metal pile and other areas showing surface soil/sediment contamination; otherwise perform only vertical soil characterization beneath the chipped metal pile to define the maximum depth of soil contamination.

DR 2: If the mean concentration of the COCs in deep soil (to a depth of 30 ft) and groundwater between the chipped metal pile and the on-site pond (over the next 8 years) is greater than their corresponding action levels specified in Table 3.5, then perform additional fieldwork to define the vertical and lateral extent of deep soil and groundwater contamination; otherwise perform only routine quarterly groundwater monitoring from the on-site well.

DR 3: If the mean concentration of the COCs in groundwater (throughout the thickness of the aquifer) in the vicinity of on-site drinking water well (over the next 20 years) is greater than their corresponding action levels specified in Table 3.5, then remove the pump from the drinking water well, beginning a routine quarterly groundwater monitoring program, and implementing institutional controls to prevent the future use of the well while COCs concentrations remain above drinking water standards; otherwise, only routine quarterly groundwater monitoring is required.

3.2.5.6 Step 6: Specify Performance or Acceptance Criteria

Objective: The objective of DQO Step 6 is to differentiate between those decision rules that will be addressed using a statistical sample design and those that are to be addressed using a nonstatistical design (e.g., judgmental design that may utilize field screening/surveying). For the former, DQO Step 6 must specify each decision rule as a statistical hypothesis test, examine the consequences of making incorrect decisions from the test, and place acceptable limits on the likelihood of making decision errors. For the latter, proceed directly to DQO Step 7 (see Section 3.2.5.7).

Activities:

- Differentiate between those decision rules to be addressed using a statistical and those to be addressed using a nonstatistical (judgmental) sampling design.
- For those decision rules to be addressed using a judgmental design, proceed to DQO Step 7 (see Section 3.2.5.7).
- For those decision rules to be addressed using a statistical approach:
  - Specify each decision rule as a statistical hypothesis test.
  - Examine the consequences of making an incorrect decision from each statistical hypothesis test.
  - Place acceptable limits on the likelihood of making a decision error for each statistical hypothesis test.

3.2.5.6.1 Differentiating between a Statistical and Nonstatistical (Judgmental) Sampling Design

The first activity to be performed in DQO Step 6 is differentiating between those decision statements that will be addressed using a statistical sampling design and those that will be addressed using a judgmental design (that may utilize field screening/surveying). Although the EPA's QA/G-4 DQO guidance manual (EPA 2006a) focuses on statistical sampling designs, there are many situations where judgmental designs are more appropriate. Important factors that must be taken into consideration when selecting between a statistical and judgmental design include:

- The qualitative consequences (low/moderate/severe) of making a decision error
- The time frame over which a decision statement applies
- The accessibility of the site or facility if resampling is required

One should seek guidance from a statistician when performing this step in the DQO process.

3.2.5.6.1.1 Qualitative Consequences of Making a Decision Error The consequences of making a decision error must be taken into consideration when selecting between a statistical and judgmental sampling design. For example, if a child day-care center is planned to be built at an environmental site following the remediation of chemical or radioactive contamination, the consequences of decision error (e.g., concluding the site is clean when in reality the site is still contaminated) are "severe" because children and workers would be exposed to the remaining contamination. In this example, a statistical sampling design is warranted for site close-out sampling. As a general rule, when the consequences of decision error are "moderate" to "severe," a statistical sampling design is often warranted.

However, this is not always the case. For example, it is not uncommon for a project to have the need to determine the maximum risk that a site poses to human health and the environment. Because maximum risk can only be determined using a judgmental design, for this example, a judgmental design is needed. This judgmental design would likely use some form of field screening to help identify hotspots from which samples would be collected for analytical testing.

As a general rule, when the consequences of decision error are "low" to "moderate," a judgmental sampling design is often selected over a statistical design. The following are some examples of situations where a judgmental sampling designs may be appropriate. These examples refer to the sample study area shown in Figure 3.3.

Example 1: Dividing a Study Site into Subareas

As discussed in Section 3.2.5.4.2, when performing an environmental investigation, it is often beneficial to divide a study area up into subareas that have similar environmental characteristics (e.g., study area broken into subareas that have a similar process history, similar contaminants of concern, and similar contaminant concentrations). Although historical process knowledge can help support making these types of divisions, it is preferred that some preliminary field screening/survey data be collected from the study area and used in combination with historical knowledge when making these divisions. In the example study area shown in Figure 3.3, soil gas surveying (Section 4.2.2) could be used in combination with a Global Positioning Environmental Radiological Surveyor-II (GPERS-II) gamma walkover survey (Section 4.1.4) and historical process knowledge to divide the site into subareas. Once these subareas have been established, separate detailed sampling designs would be developed for each subarea.

A decision error in this example might incorrectly define one subarea as "less contaminated" when in reality it is "moderately contaminated." The qualitative consequences of this type of error are relatively low because this division is simply used to assist the selection of the optimum sampling design for characterizing the contamination present within the subarea. When the second round of sampling is performed, this error will be identified. For this reason, a judgmental sampling design is appropriate.

Example 2: Disposal of Waste Drums Containing Soil Cuttings

Assume that, during remedial investigations at the example study area (shown in Figure 3.3), a total of ten 55-gallon drums are filled with soil cuttings that resulted from the drilling of one deep borehole. In order to ensure the proper disposal of these drums, they must first be characterized to determine their content and characteristics. For a problem like this, it is not uncommon in a project to select the waste drum showing the highest contaminant levels based on field screening measurements (e.g., downhole logging, head-space analysis) to be sampled and laboratory tested to determine the proper disposal of all ten drums. This judgmental approach basically treats all the drums as equivalent to the worst-case drum and only requires the analytical testing of one sample from one drum. This approach is most often implemented in an effort to minimize analytical costs.

In this example, a decision error could incorrectly conclude that the sampled waste drum exceeds one or more RCRA characteristics (e.g., ignitability, corrosivity, reactivity, toxicity) and therefore requires the drums to be disposed of in an RCRA-permitted landfill as opposed to a less expensive disposal option (e.g., CERCLA landfill). This type of error results in spending more dollars on waste disposal than necessary (generally considered a low-to-moderate risk). If this situation really occurred in the field, a project would likely spend additional dollars to collect a separate sample from each of the remaining nine drums for analytical testing because sending all the drums to an RCRA-permitted landfill would be much more expensive than running the extra analyses.

As the worst-case drum was selected for testing in the foregoing example, it is very unlikely that the opposite type of decision error would occur (e.g., incorrectly concluding that the drums are below RCRA characteristics when in reality they are above). For this example, a judgmental sampling design is reasonable.

Example 3: Storing Waste Prior to Disposal

If samples are being collected from a waste material simply to determine the most appropriate storage area for the waste while awaiting final characterization and disposal, a judgmental sampling design is appropriate. The consequence of decision error in this case is just the inconvenience of moving the misidentified waste material after final characterization to the proper area to await disposal.

3.2.5.6.1.2 Time Frame over Which a Decision Statement Applies The time frame over which a decision statement applies should also be considered when selecting between a statistical and a judgmental design. If it is many years, the consequences of a decision error are more severe than if the decision only applies for a shorter period of time. Decisions that apply over a longer period of time favor the selection of a statistical sampling design. For example, deciding whether or not a site has been remediated to below an action level (so that it can be released to the public) applies forever, and therefore a statistical sampling design is most appropriate.

3.2.5.6.1.3 Accessibility of the Site if Resampling Is Required If a site is only accessible for sampling for a short period of time, a robust statistical sampling design may be warranted. For example, if a Walmart Superstore was proposed to be built over the top of the example study site (shown in Figure 3.3) immediately after remediation activities were completed, a very robust statistical postremediation sampling design should be performed to assure remedial action objectives have been fully achieved across the site. This is because, if it were determined at a later date that additional remediation is still required, in addition to the human health risk this scenario would pose, it would be much more expensive to remove the contaminated soil from beneath the foundation of the Walmart store.

3.2.5.6.2 Decision Rules to be Addressed Using a Nonstatistical (Judgmental) Approach

For those decision rules that are to be addressed using a judgmental sampling approach, proceed directly to DQO Step 7 (Section 3.2.5.7) as the remainder of DQO Step 6 applies only to statistical sampling approaches.

3.2.5.6.3 Decision Rules to be Addressed Using a Statistical Approach

The purpose in using a statistical sample design is to reduce the chances of making a decision error to a tolerable level. DQO Step 6 provides a mechanism for the regulators to define tolerable limits on the probability of making a decision error. Examples of the most common decision errors that occur in the environmental industry include:

- Walking away from a dirty site
- Cleaning up a clean site

The decision error that causes one to walk away from a dirty site is the more serious of the two consequences because it could negatively impact human health and the

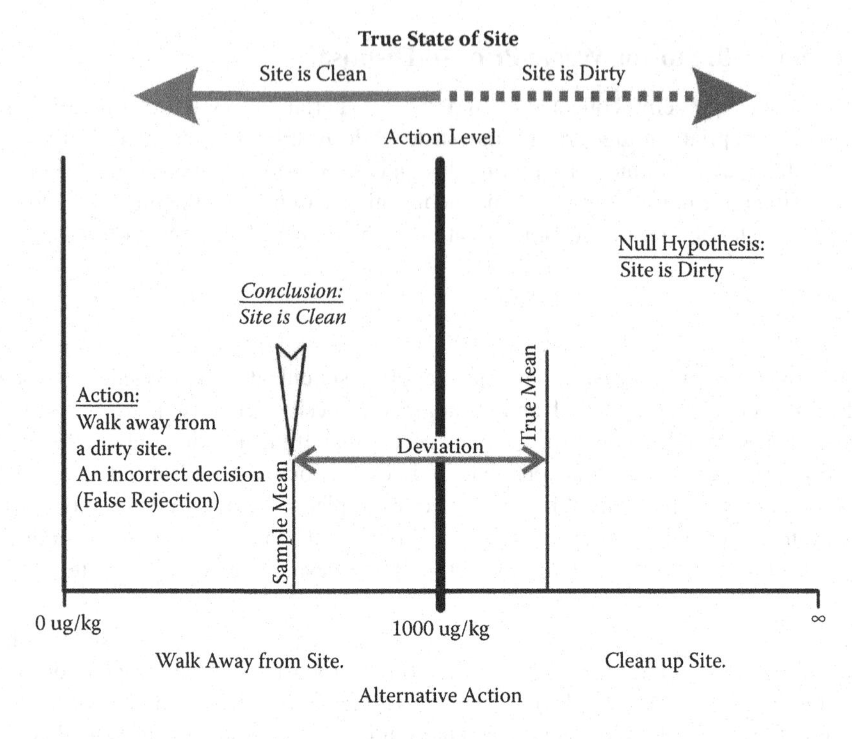

FIGURE 3.5 Decision error causing one to walk away from a dirty site.

environment. The decision error that causes one to clean up a clean site results in higher remediation costs.

Figure 3.5 illustrates the decision error of walking away from a dirty site. In this example, the true mean lead concentration (which only God knows) in soil is above the action level of 1000 μg/kg, whereas the sample mean (calculated based on a statistical sampling of the population) is below the action level. Because decisions are made based on the sample mean, the site is incorrectly determined to be clean. The action is to walk away from a dirty site. When the null hypothesis assumes the site to be contaminated until shown to be clean, the error is referred to as a "false rejection." Section 3.2.5.6.3.3 provides more detail on this type of decision error.

Figure 3.6 illustrates the decision error of cleaning up a clean site. In this example, the true mean concentration (which only God knows) of lead is below the action level of 1000 μg/kg, whereas the sample mean (calculated based on a statistical sampling of the population) is above the action level. Because decisions are made based on the sample mean, the site is incorrectly determined to be contaminated. The action is to clean up a clean site. When the null hypothesis assumes the site to be contaminated until shown to be clean, the error is referred to as a "false acceptance." Section 3.2.5.6.3.3 provides more detail on this type of decision error.

To control the amount of decision error for a sample design it is necessary to define the possible range of concentrations for the parameter of interest (e.g., mean, median),

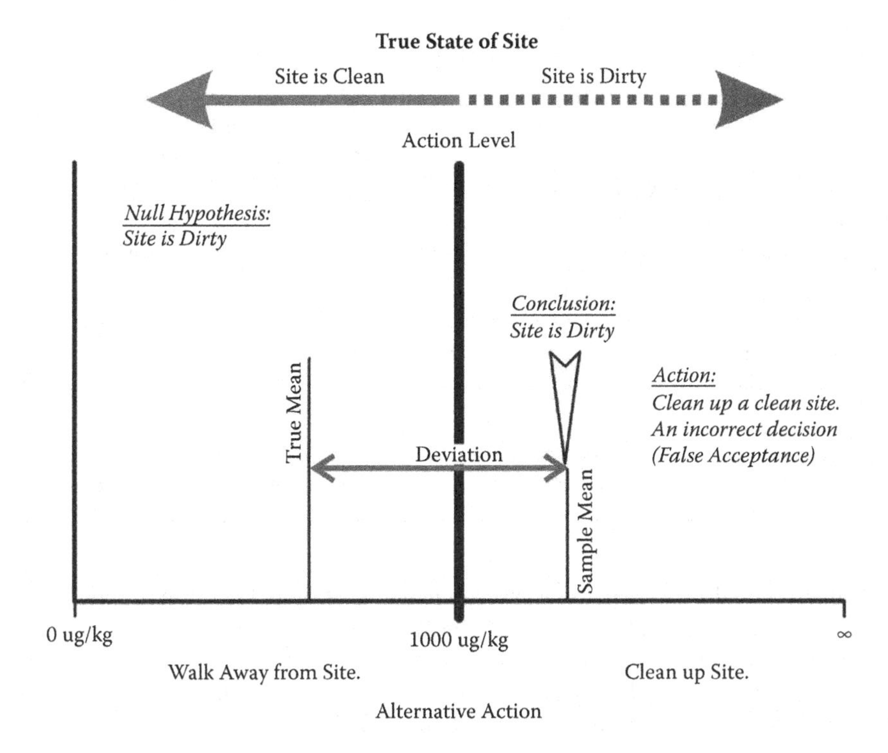

FIGURE 3.6 Decision error causing one to clean up a clean site.

types of decision errors, boundaries of the gray region, and tolerable limits on decision error. These criteria are needed to support statistical calculations performed in DQO Step 7 when determining the required number of samples/measurements to collect. They are discussed in further detail in the following sections.

3.2.5.6.3.1 Determine the Possible Range of the Parameter of Interest An initial step in the process of establishing a statistically based sample design is to define the expected range of concentrations for the statistical parameter of interest (e.g., mean, median) for each contaminant of concern. Recall that the statistical parameter of interest was already selected in DQO Step 5 (see Section 3.2.5.5.1). The expected range (lowest to highest values) of the statistical parameter of interest should be defined using the results from historical analytical data. If no historical data are available, process knowledge should be used to estimate the expected range.

3.2.5.6.3.2 Choosing the Null Hypothesis In the process of establishing a statistically based sample design, it is necessary to define the null hypothesis or baseline condition of the site. The two possible null hypotheses for environmental sites are

- The site is assumed to be *contaminated* until shown to be *clean.*
- The site is assumed to be *clean* until shown to be *contaminated.*

TABLE 3.8
Four Possible Outcomes When Using a Statistical Sampling Approach

Decision You Make by Applying the Statistical Hypothesis Test to the Collected Data	True Condition of the Site (Only God Knows This)	
	Site is Truly Contaminated	**Site is Truly Clean**
You decide the site is contaminated	Correct decision	Decision error (False acceptance[a,c])
You decide the site is clean	Decision error (False rejection[b,c])	Correct decision

[a] False acceptance: You decided the site is contaminated when in reality it is clean.
[b] False rejection: You decided the site is clean when in reality it is contaminated.
[c] Null hypothesis (baseline condition) assumes the site is contaminated until it is shown to be clean.

When selecting the null hypothesis, keep in mind that it should state the "opposite" of what the project eventually hopes to demonstrate. Because, for an environmental site, the objective is almost always to show that a contaminated site is clean after remediation is complete, all of the examples provided in this book assume a null hypothesis that the "site is contaminated until shown to be clean."

3.2.5.6.3.3 Identify the Decision Errors As discussed earlier in Section 3.2.5.6.3, the two types of decision error that can occur are walking away from a dirty site (false rejection) and cleaning up a clean site (false acceptance). Figures 3.5 and 3.6 provide a graphical illustration of what these decision errors look like. When the null hypothesis assumes the site to be contaminated until shown to be clean, the decision error that causes one to walk away from a dirty site (false rejection) is the more serious of the two consequences because it could negatively impact human health and the environment. The decision error that causes one to clean up a clean site (false acceptance) results in higher remediation costs.

Table 3.8 identifies the four possible outcomes from a statistical sampling approach. As you can see from this table, two of the four outcomes lead to a correct decision. The other two outcomes represent the two types of decision errors.

3.2.5.6.3.4 Specify the Boundaries of the Gray Region The gray region is a range of possible parameters of interest (e.g., mean, median) values (contaminant concentrations) where the consequences of a decision error are relatively minor. It is bounded on one side by the action level, and on the other side by the parameter of interest (contaminant concentration) value where the consequences of the decision error begin to be significant (Figure 3.7). It is necessary to specify the gray region because variability in the population and unavoidable imprecision in the measurement system combine to produce variability in the data such that a decision may be "too close to call" when the true parameter value (contaminant concentration) is very near the action level.

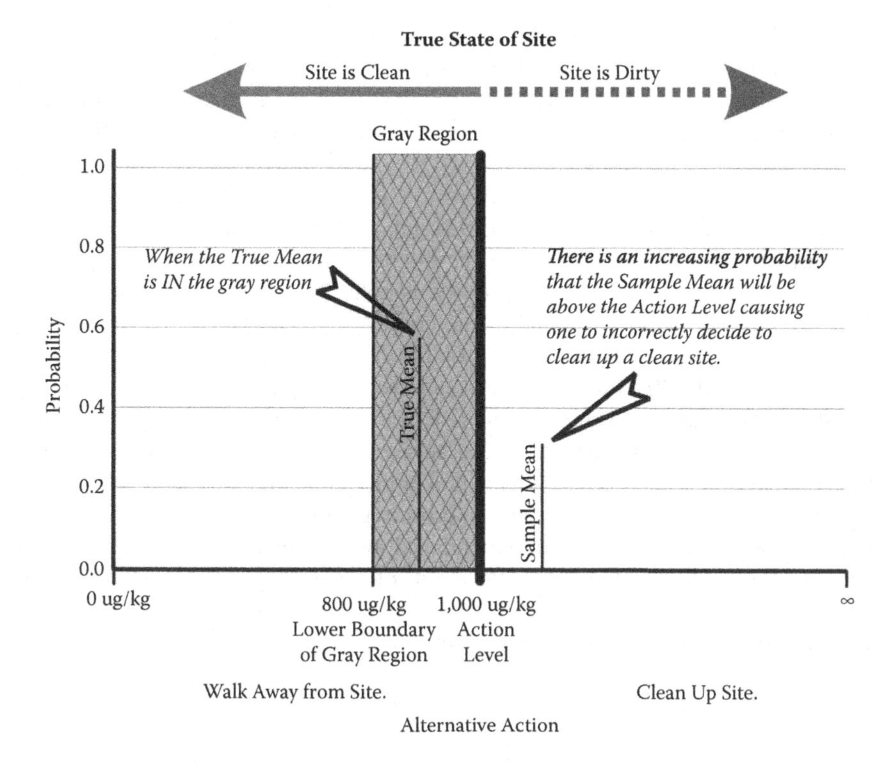

FIGURE 3.7 Boundaries of the gray region.

In the example provided in Figure 3.7, the lower bound of the gray region (LBGR) for lead concentrations in soil has been set at 800 μg/kg initially, and the upper bound of the gray region is set at the action level (1000 μg/kg). The LBGR is one of many inputs used in statistical calculations to determine the total number of samples needed to resolve a decision. Setting a narrower gray region for lead concentrations in soil in this example (900–1000 μg/kg) would result in lower uncertainty but higher sampling costs (larger number of samples required). On the other hand, a wider gray region (e.g., 700–1000 μg/kg) would result in lower sampling costs (fewer samples required) but higher uncertainty. It is important to select a multitude of widths of the gray region and see what effect it has on the sample size. Initially, the author typically sets the LBGR at 80% of the action level. When setting the LBGR, keep in mind the consequences of decision error. For example, a less stringent gray region (wider gray region) may be acceptable for some decisions (e.g., closing out a site to be used strictly for industrial purposes), whereas a more stringent gray region (narrower gray region) may be required for other decisions (e.g., closing out a site to be used for residential or farming purposes).

The *Multi-Agency Radiation Survey and Site Investigation Manual (MARSSIM)* (EPA 2000) recommends initially setting the LBGR arbitrarily at 50% of the action level, then adjusting it to provide a "relative shift" value of between 1 and 3. The term *relative shift* (Δ/σ) is the width of the gray region (Δ) divided by the estimated standard deviation (σ) of the population. The estimated standard deviation (σ) can often be calculated using existing historical data from the site (e.g., scoping or characterization data). If no historical data exist, it may be necessary to either perform some limited preliminary measurements (e.g., 10–20) to estimate the distributions, or to make a reasonable estimate based on available site knowledge (EPA 2000).

3.2.5.6.3.5 Assign Tolerable Limits on Decision Error The next step in preparing a statistical sampling design is to assign probability values to points above and below the gray region that reflect the tolerable limits for making an incorrect decision. At a minimum, one should specify the tolerable decision error limits at the action level and at the lower bound of the gray region (Figure 3.8). Although EPA guidance recommends a starting value of 0.05 (5%) for the false rejection decision error limit and 0.20 (20%) for the false acceptance decision error limit, the planning team should fully explore balancing the risk of making incorrect decisions with the potential consequences associated with these risks. It is recommended that, initially, very stringent limits should be set on the probability of making a false rejection (e.g., 0.02 or 2%) and false acceptance decision error (e.g., 0.1 or 10%). In an iterative process, these requirements should then be relaxed until a balance is achieved between decision error and cost. Table 3.9 is an example of a table that can be used to assist in evaluating what effects the varying of the lower bound of the gray region, and false rejection and false acceptance decision error limits, have on sample size.

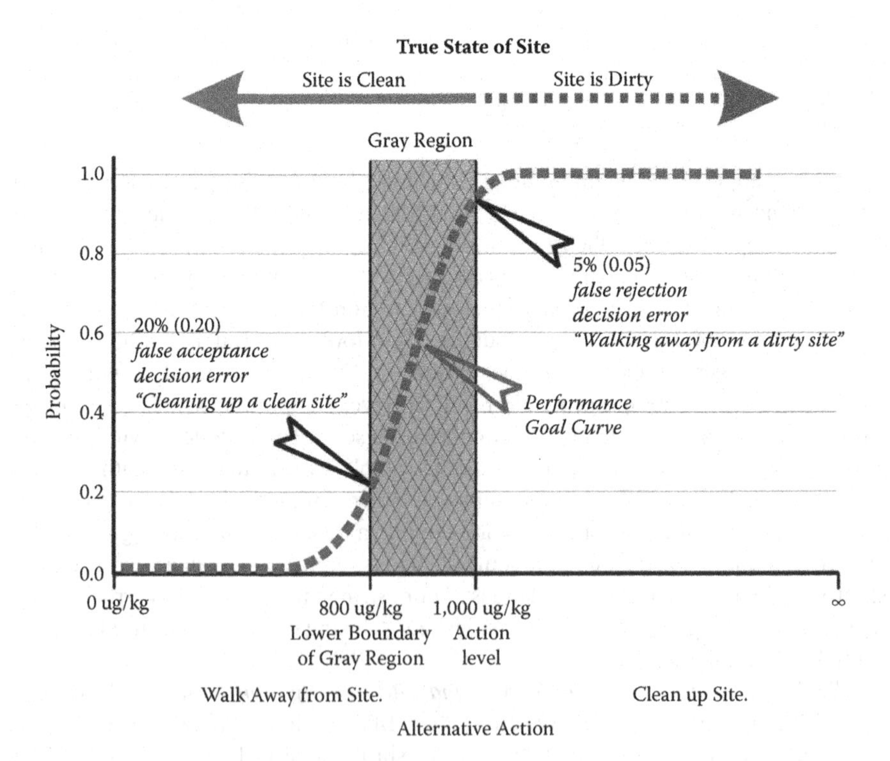

FIGURE 3.8 Example of a decision performance goal diagram.

TABLE 3.9
Sample Size Based on Varying Error Tolerances and Lower Bound of Gray Region

		Mistakenly Concluding < Action Level		
		False Rejection Error (α) =	False Rejection Error (α) =	False Rejection Error (α) =
DS #1				
Lower bound of gray region =				
Mistakenly concluding > action level	False acceptance error (β) =			
	False acceptance error (β) =			
	False acceptance error (β) =			
Lower bound of gray region =				
Mistakenly concluding > action level	False acceptance error (β) =			
	False acceptance error (β) =			
	False acceptance error (β) =			
Lower bound of gray region =				
Mistakenly concluding > action level	False acceptance error (β) =			
	False acceptance error (β) =			
	False acceptance error (β) =			

One should consider evaluating the severity of the potential consequences of decision errors at different points within the domains of each type of decision error because the severity of the consequences may change as the parameter moves farther away from the action level. This results in the creation of the performance goal curve shown in Figure 3.8.

3.2.5.6.4 Outputs from DQO Step 6

The primary outputs resulting from the implementation of this step include:

- A list of the decision statements that will be addressed using a statistical versus a judgmental sampling approach
- For those decision statements being addressed using a statistical approach, this step identifies
  - The possible range for the parameter of interest (e.g., mean, median)
  - The null hypothesis

- The false rejection and false acceptance decision errors
- The boundaries of the gray region
- Tolerable limits on decision error

3.2.5.7 Step 7: Develop the Plan for Obtaining Data

Objective: The goal of Data Quality Objective Step 7 is to first use the information gathered earlier in the DQO process to identify a number of alternative sampling and analysis designs that could be performed to meet the minimum performance and acceptance criteria specified in DQO Steps 1 through 6. After this, one shall select and document the sample design that will best achieve the project performance and acceptance criteria while keeping cost in mind.

Activities:

- Outputs from DQO Steps 1 through 6 shall be used to identify alternative sampling and analysis designs for resolving each of the decision statements.
- From these design options, one design shall be selected that achieves all performance and acceptance criteria that also takes cost into consideration.

3.2.5.7.1 Nonstatistical (Judgmental) Designs

Nonstatistical sampling designs utilize information gathered during the scoping process or utilize field screening/surveying (e.g., soil gas sampling, airborne gamma spectrometry) to collect data that help focus the investigation on those areas that have the highest likelihood of being contaminated. This type of sampling is often referred to as *judgmental sampling.* Judgmental sampling designs often provide data that represent the worst-case conditions for a site. For this reason, this type of sampling is most commonly used to support site characterization activities where the objective is to define the nature and extent of chemical or radiological contamination. This sampling is also sometimes used to support risk assessments by estimating the worst-case exposure.

One of the advantages of a judgmental sampling design is that it can be less expensive to implement than a statistical design if there is a good historical knowledge of the site. A judgmental design is also generally easier to implement than a statistical design. On the other hand, data collected using a judgmental design *cannot* be used to support statistical decision-making.

Judgmental sampling should *not* be used when collecting data to support site/facility closeout decisions because these data cannot be evaluated statistically. Even when performing site characterization activities, it is not uncommon to perform judgmental sampling in combination with one or more statistical sampling approaches (e.g., simple random sampling, systematic sampling) because these data are often needed to support risk calculations, modeling studies, etc.

When a judgmental design is determined to be most appropriate to resolve one or more decision statements, the next step is to identify all potential surveying technologies or judgmental sampling methods that could potentially be used to provide the required data for each type of media (e.g., soil, concrete, paint). Chapter 4, Section 4.1, provides a number of scanning, direct measurement, and sampling methods that

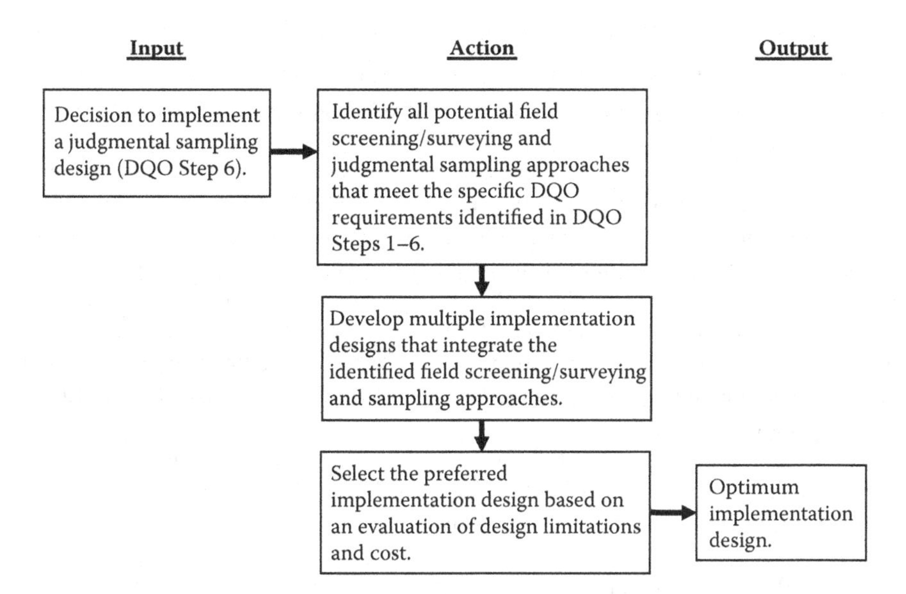

FIGURE 3.9 Implementation process for nonstatistical (judgmental) sampling design.

should be taken into consideration. The identified surveying and sampling methods should then be joined to form multiple alternative implementation designs.

Finally, the limitations and cost associated with each implementation design should be used to support the selection of the preferred implementation design. A Sampling and Analysis Plan is prepared following the completion of the DQO process in accordance with guidance provided in Section 3.2.7. Figure 3.9 provides a flowchart showing the implementation process for a judgmental sampling design.

3.2.5.7.2 Statistical Designs

The purpose of this section is to provide general information on statistical sampling concepts. Reference is made to commonly used statistical hypothesis tests for calculating the number of samples required and confidence limits. However, this section is not intended as a technical discussion of these topics; nor is it intended to provide formulas needed to perform various statistical hypothesis tests. Rather, the purpose is to make the readers aware of what is commonly used, what is available, and where to find more detailed information on topics of interest. In Section 3.2.5.7.3, the reader is introduced to statistical sampling design software that is available to support the development of statistical sampling designs.

As discussed earlier in Section 3.2.5.6.1.1, a statistical sampling design should be considered whenever the consequences of decision error are moderate or severe. Several commonly used statistical sampling designs are described in the following text. However, before discussing particular sampling schemes, it is important to understand what happens in general when sampling occurs.

Suppose the concentration of lead needs to be determined for the surface soil present at a site that measures 1,000 ft × 1,000 ft. Further assume that the site is divided into 1-ft^2 sections for sampling purposes. That is, each 1 ft^2 is considered to be one sampling unit. In this example there are 1,000,000 possible surface soil samples that could be taken from this site. These 1,000,000 samples comprise the population. Because it is cost- and time-prohibitive to collect and analyze all 1,000,000 samples from the population, an alternative strategy is to select some smaller number of samples and to use a single number such as the mean concentration of lead from this to represent the population. In a nutshell, this is statistical sampling.

One important thing to note at this point is that statistical sampling provides an incomplete picture of the population. As only a few of all possible samples are taken from the population, the data obtained from these samples are incomplete. Because the data are incomplete, the population will not be represented *exactly*, which could therefore lead one unknowingly to making an incorrect decision about the status of the population.

Regardless of the history of the site under investigation, it is extremely unlikely that every sample (every square-foot section) would have exactly the same concentration of lead. Two questions now come to mind:

- If different sample units have different concentrations of lead, what is the true concentration of lead for the site as a whole?
- How well can the mean of a subset of all possible samples represent the true mean of the site as a whole?

To answer these questions, a brief discussion of some basic statistical concepts is needed. This discussion is not intended to be a course in statistics but rather a general and intuitive discussion of some basic concepts that underlie statistical thinking.

3.2.5.7.2.1 What Is the True Concentration of Lead in the Surface Soil at the Site? Suppose every possible sample could be taken from the surface soil at the site and measured for lead. As stated earlier, not all measured concentrations would be the same – even if there were no analytical error. Some samples would truly contain a higher concentration of lead than others. To get one number that represents the site as a whole, the mean of all samples could be calculated. This number would be the *true* mean concentration of lead for the site. Although it is convenient to have one number to represent the entire site, the mean does not give a complete picture of the concentration of lead for this site.

If measured concentrations from all possible samples were ordered from lowest to highest, a pattern would likely be observed. For example, there may be a few very low and a few very high concentrations, but the great majority of the concentrations could cluster around a single point in the middle. Figure 3.10 is a graphical representation of such a pattern. (Concentration level is on the x-axis, and frequency of occurrence is on the y-axis.) This kind of graphical representation is called a distribution. It reflects the fact that the measured concentrations obtained from all possible samples vary from one another. A distribution is a more complete way of describing the concentration of lead for the site. It provides information that cannot be deduced from the mean alone. First, by looking at a distribution, the full range of measured concentrations can be seen. Second, it is often easy to identify a central point around which most of

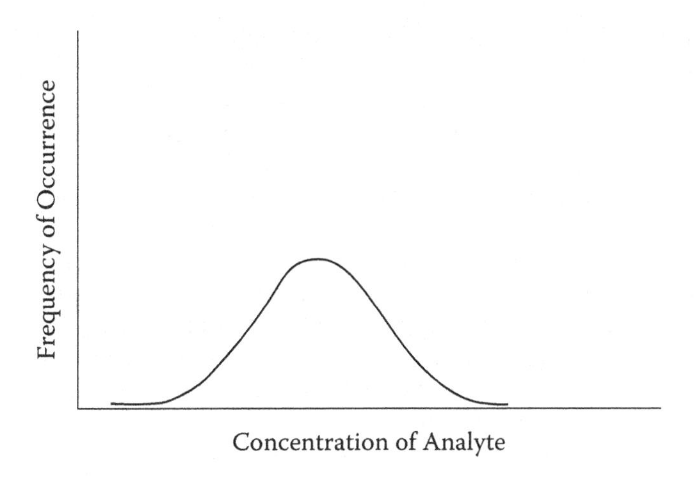

FIGURE 3.10 Graphical illustration of a distribution. (From Byrnes, 2001. *Sampling and Surveying Radiological Environments,* Lewis Publishers, Boca Raton, FL.)

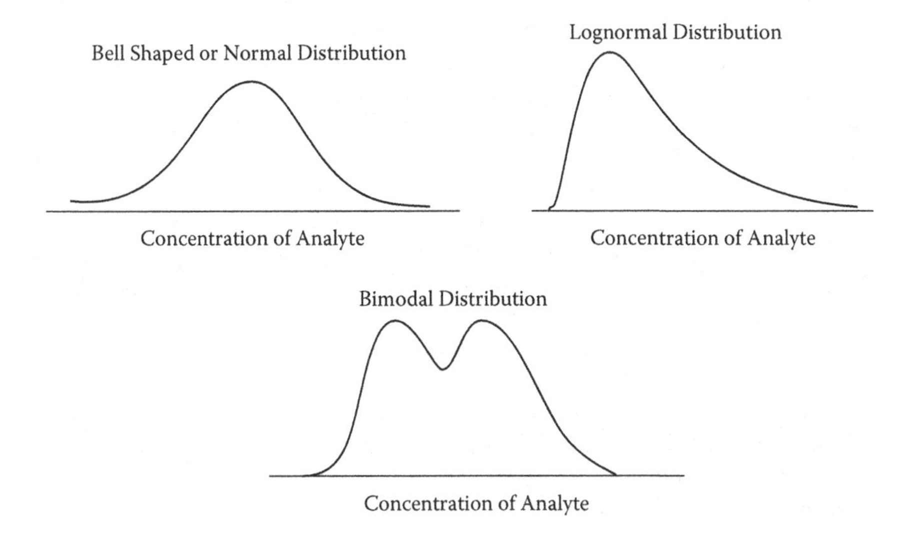

FIGURE 3.11 Distributions commonly observed in environmental studies. (From Byrnes, 2001. *Sampling and Surveying Radiological Environments,* Lewis Publishers, Boca Raton, FL.)

the measured concentrations cluster. This is roughly equivalent to the mean. Finally, a distribution provides a sense of how tightly clustered or how widely spread out the measured concentrations are. This feature of a distribution is called variability and can be summarized by calculating the standard deviation or variance of all the measured concentrations. The concept of variance has important implications for sampling, as will be discussed in the following text.

Three distributions commonly observed when working with environmental data sets are illustrated in Figure 3.11. These are theoretical or idealized distributions that can be thought of as hypothetical models for the site. The population distribution of real environmental data would never take the shape of one of these theoretical distributions exactly. However, the population distribution is often "close enough" to one of these theoretical distributions that it can provide a good approximation. This has tremendous benefits for data analysis.

The first distribution is often referred to as a *bell-shaped* or *normal* distribution. Its salient features are that it is symmetrical and unimodal – that is, it has only one "center" or mode around which a large percentage of measured concentrations cluster.

The second distribution is called a *lognormal* distribution. Note that it is very much like the normal, except that it contains a small number of extremely high measured concentrations. This causes the distribution to become asymmetrical, although it is still unimodal. A lognormal distribution may indicate the presence of hot spots, which create the upper tail of the distribution.

The third distribution is called a *bimodal* distribution. Note that there are two "centers" or modes around which measured concentrations tend to cluster. A bimodal distribution can be either symmetrical or asymmetrical. In either case, it often indicates that the site should be subdivided into subareas (strata) (see Figures 3.4). For example, suppose that half of a site was never contaminated, and the other half was heavily contaminated. If each half (or subarea) was evaluated separately, the distribution for each subarea would likely be unimodal (either normal or lognormal). However, if the site was not broken into subareas but rather evaluated as a whole, a bimodal distribution would result. The data clustered around the lower mode represent the uncontaminated subarea, whereas the data clustered around the higher mode represent the contaminated subarea. If a bimodal distribution is encountered, it is a good idea to review historical information to determine whether or not the site can be subdivided into subareas.

As mentioned earlier, distributions provide a good illustration of how tightly clustered or how widely spread out the measured concentrations in the population are. In statistical terminology, this is referred to as the *level of dispersion* or *variance* of a distribution. Numerically, the variance is summarized by calculating the standard deviation of the measured concentrations. In other words, the standard deviation provides one number that presents the variance of the measured concentrations in the population. The standard deviation plays an important role in calculating the number of samples that are required. Figures 3.12 and 3.13 illustrate different variances for a normal and lognormal distribution.

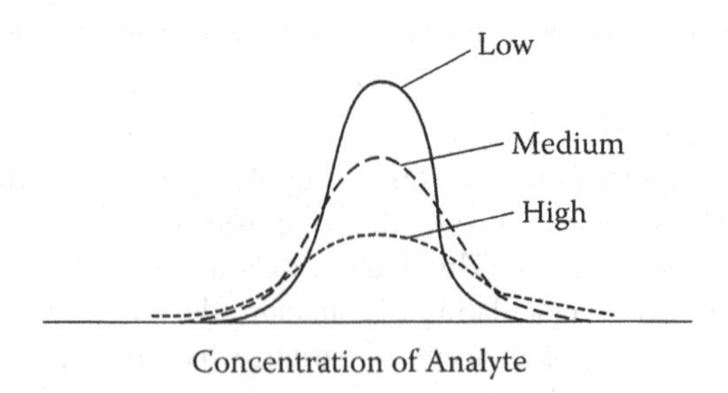

FIGURE 3.12 Differing levels of variability for normal distributions. (From Byrnes, 2001. *Sampling and Surveying Radiological Environments,* Lewis Publishers, Boca Raton, FL.)

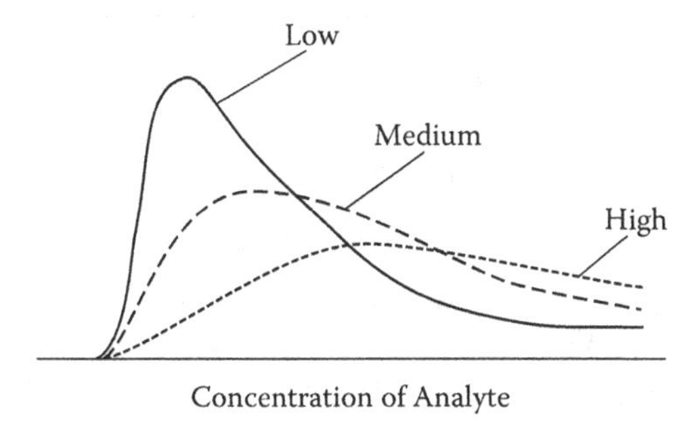

FIGURE 3.13 Differing levels of variability for lognormal distributions. (From Byrnes, 2001. *Sampling and Surveying Radiological Environments,* Lewis Publishers, Boca Raton, FL.)

To recap, although it is often useful to characterize a site in terms of a single number such as the mean or standard deviation, it is more appropriate to think of the true concentration of the site as a distribution of values.

How well can the mean of a subset of all possible samples represent the true mean of the site as a whole? – Unless all possible samples are taken, the true mean concentration of lead for the site cannot be known. The only reasonable alternative is to take a subset of all possible samples and to calculate the mean of this subset. The mean of the subset of samples is, then, an *estimate* of the mean of the population. Should the mean of just a few samples be trusted to estimate the true mean of the site? The answer to this question depends primarily (although not solely) on how much variance there is in the concentration of lead for the site as a whole. If the measured concentrations of lead in the population are tightly clustered (have low variance), then the mean of a few samples does a good job representing the site. However, if the measured concentrations of lead in the population are widely spread out, then the mean of a few samples is less likely to give an accurate representation of the site.

In either case, the mean of a few samples will never be exactly the same as the true mean of the population. So, when the mean of a few samples is used to make a decision about the true condition of the site, decision errors have some real (nonzero) probability of occurring. For example, the mean of the subset of samples may lead one to the conclusion that the true mean concentration of lead is below an action level, when in fact it is actually above the action level. The reverse may also happen where the mean of the subset of samples leads to the conclusion that the site is above the action level for lead, when in reality it is below the action level for lead. For a more detailed discussion of decision errors, see Section 3.2.5.6.3.

To summarize, sampling and analyzing a subset of all possible sampling units allows one to make an educated guess or inference about the population of interest. This can be done by calculating the mean of the samples taken and using this as an estimate of the true population mean. In addition, a distribution of the measured concentrations can be used to make inferences about the shape and variance of the population. The following sections discuss several commonly used methods for selecting the subset of all possible samples. Keep in mind that any sampling approach

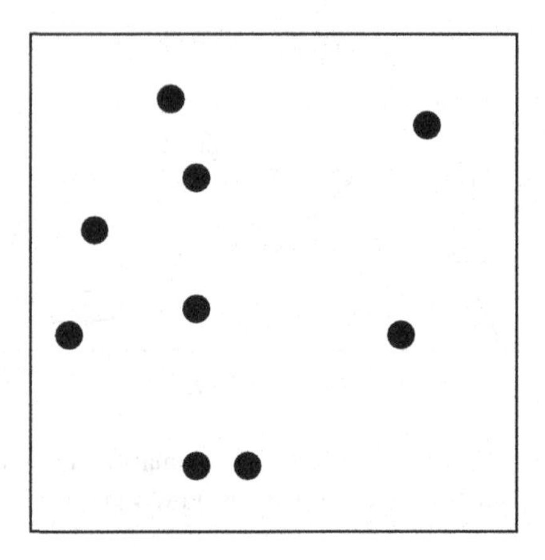

FIGURE 3.14 Simple random sampling design.

that collects only a subset of all possible samples has some probability of leading to a decision error.

3.2.5.7.2.2 Simple Random Sampling When little historical information about the site exists, simple random sampling is a good choice (Figure 3.14). Simple random sampling is implemented by dividing the site into a multitude of possible sampling units (i.e., the area that will be represented by a single sample). A site may be divided into sampling units by overlaying a grid with spacing determined by the minimum amount of area required to collect a sample with the equipment being considered. For example, the smallest area that may be excavated by a backhoe when collecting a sample may be 10 ft^2.

Each sampling unit within the population is initially assigned a number. A subset of sampling units is then chosen by drawing the assigned numbers at random. Each of the sampling units assigned a number has an equal probability of being chosen for sampling. If some sampling units are inaccessible, they are not assigned numbers and therefore have zero probability of being selected. Because of this, they cannot be considered as part of the population, and the results of any statistical hypothesis testing do not apply to these sampling units.

The biggest advantage of using a simple random sampling scheme is that it can be used when little or no historical information is available and it can provide data for almost any statistical hypothesis test, such as comparing a mean concentration to an action level. The greatest disadvantage is that the number of samples needed may be larger than that of other statistical sampling strategies.

3.2.5.7.2.3 Stratified Sampling In stratified sampling, the target population is separated into nonoverlapping subareas (strata), or subpopulations that are known or thought to be more homogeneous, so that there tends to be less variation among sampling units in the same subarea than among sampling units in different subareas. Subareas may be chosen on the basis of spatial or temporal proximity of the units,

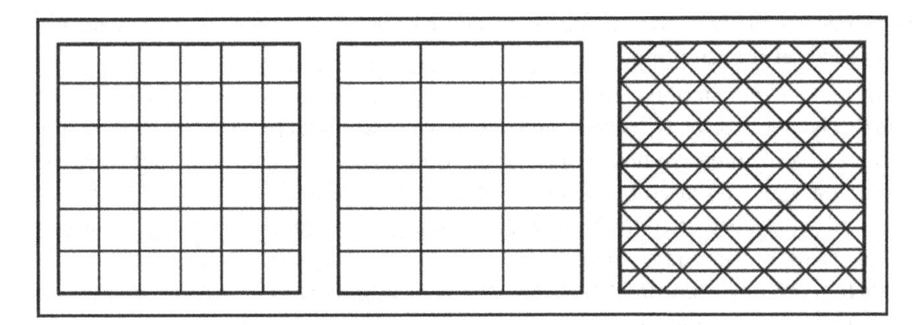

FIGURE 3.15 Systematic sampling design options.

or on the basis of preexisting information or professional judgment about the site or process (see Figure 3.4). This design is useful for estimating a parameter when the target population is heterogeneous and the area can be subdivided based on expected contamination levels. Advantages of this sampling design are that it has potential for achieving greater precision in estimates of the mean and variance. Greater precision can be obtained if the measurement of interest is strongly correlated with the variable used to make the subarea (EPA 2002).

3.2.5.7.2.4 Systematic Sampling Systematic sampling designs may be used when the objective is to search for leaks or spills, to determine the boundaries of a contaminated area, or to determine the spatial characteristics of a site. Basically, samples are collected from an evenly spaced grid where the starting point is randomly chosen.

To create a systematic sampling design, random coordinates from within the area are chosen for the first sample location. This establishes the initial location or reference point from which the grid is built. Once the grid is established, samples are systematically taken from the nodes or cross lines of the grid. The grid may be square, rectangular, or triangular in shape (Figure 3.15).

Systematic sampling may introduce a certain type of sampling bias. Because sampling occurs at the nodes, small areas of contamination may be missed if they are entirely within the grid. This could result in underestimating the contamination of the site. Conversely, if the spread of the contamination is very similar to the grid pattern, overestimation of the contamination could occur. Because of these factors, care must be taken in choosing both the size and type of sampling grid to be used. Sampling on a triangular grid pattern is often preferred because it reduces the possibility of sampling bias.

3.2.5.7.2.5 Ranked Set Sampling Ranked set sampling is an innovative design that can be highly useful and cost-efficient in obtaining better estimates of mean concentration levels in soil and other environmental media by explicitly incorporating the professional judgment of a field investigator or a field screening measurement method to pick specific sampling locations in the field. Ranked set sampling uses a two-phase sampling design that identifies sets of field locations, utilizes inexpensive analytical measurement methods to rank locations within each set, and then selects one location from each set for sampling.

In ranked set sampling, field locations are identified using simple random sampling. The locations are ranked independently within each set using professional judgment or inexpensive field screening measurements. One sampling unit from each set is then selected (based on the observed ranks) for subsequent measurement, using standard laboratory analytical methods for the contaminant of interest. Relative to simple random sampling, this design results in more representative samples and so leads to more precise estimates of the parameter of interest.

Ranked set sampling is useful when the cost of locating and ranking locations in the field is low compared to laboratory analytical costs. It is also appropriate when expert knowledge is available to rank population units with respect to the variable of interest (EPA 2002).

3.2.5.7.2.6 Adaptive Cluster Sampling In adaptive cluster sampling, *n* samples are taken using simple random sampling, and additional samples are taken at locations where measurements exceed some threshold value. Several additional rounds of sampling and analysis may be needed. Adaptive cluster sampling tracks the selection probabilities for later phases of sampling so that an unbiased estimate of the population mean can be calculated despite oversampling of certain areas. An example application of adaptive cluster sampling is delineating the borders of a plume of contamination.

Initial and final adaptive sampling designs are shown in Figure 3.16. Initial measurements are made at randomly selected primary sampling units using simple random sampling (designated by squares in Figure 3.16). Whenever a sampling unit is found to show a characteristic of interest (e.g., contaminant concentrations above

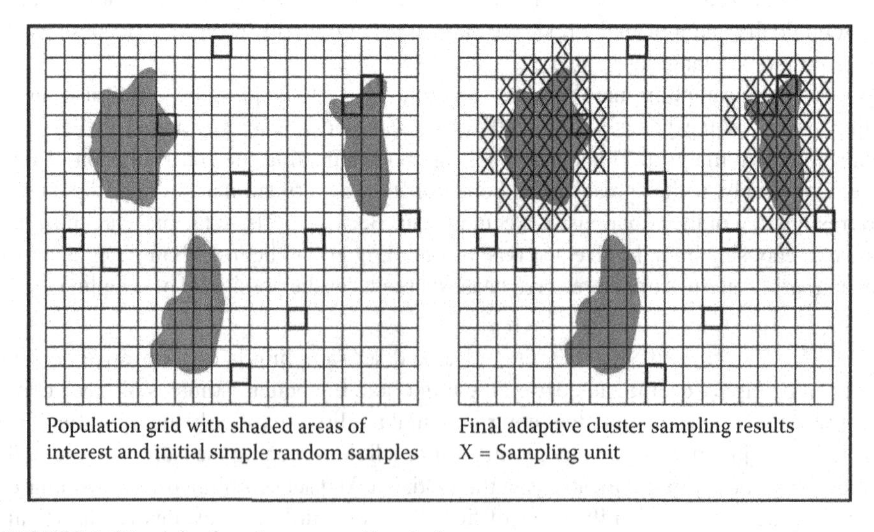

FIGURE 3.16 Adaptive cluster sampling design.

human health risk levels), additional sampling units adjacent to the original unit are selected and measurements are made.

Adaptive sampling is useful for estimating or searching for rare characteristics in a population and is appropriate for inexpensive quick turnaround field measurements. It enables delineating the boundaries of hot spots, while also using all data collected with appropriate weighting to give unbiased estimates of the population mean (EPA 2002).

3.2.5.7.2.7 Composite Sampling In composite sampling, volumes of material from several sampling locations are physically combined and mixed in an effort to form a single homogeneous sample, which is then analyzed. Compositing can be very cost-effective because it reduces the number of samples that need to be analyzed. Composite sampling is most cost-effective when analysis costs are large relative to sampling costs. Compositing is often used when the goal is to estimate the population mean and when information on spatial or temporal variability is not needed. It is extremely important to note that composite sampling should never be used when samples are to be analyzed for volatile organic compounds because the compositing procedure will cause volatiles to be lost to the atmosphere, and therefore the data resulting from the sample will not be representative.

3.2.5.7.2.8 Sequential Sampling When the site is expected to be either minimally or maximally contaminated, sequential sampling can often dramatically reduce the number of samples required. Unlike the other sampling approaches discussed earlier, a sequential sampling design does not define the required sample size in advance. Instead, after a few samples are collected, a decision is made to reject the null hypothesis (e.g., decide the site is clean), to fail to reject the null hypothesis (e.g., decide the site is contaminated), or to continue sampling.

Sequential sampling involves performing a statistical hypothesis test as results become available, rather than waiting until all the sampling results are in before running the test. The statistical hypothesis test is used to determine if the collection of additional samples is required to support the decision that the site meets or does not meet the cleanup standard. This sampling method is useful when using fast turnaround or field analytical methods in which results can be quantified very quickly. However, it should not be used in situations where the collection of an additional sample or samples requires remobilization. A more thorough discussion of sequential sampling is provided in EPA (1992) and Bowen and Bennett (1988).

3.2.5.7.2.9 Other Recommended Sampling/Surveying Designs In the process of characterizing a site or building for chemical or radioactive contaminants, EPA (2000) recommends a sampling approach that combines the use of low-cost screening surveys (using techniques such as the Global Positioning Environmental Radiological Surveyor [see Chapter 4, Section 4.1.4] and In Situ Gamma Spectroscopy [see Chapter 4, Section 4.1.5]) with a statistical sampling design. A similar approach should also be considered for chemical sites where low-cost field-screening methods (such as soil gas surveying [see Chapter 4, Section 4.2.2]) could be used to survey large portions of the site followed up by statistical sampling.

TABLE 3.10
EPA Recommended Land Area Sampling Designs

Survey Unit Classification		Statistical Test Required	Sampling Design	Surface Scans
Impacted	Class 1	Yes	Systematic	Coverage: 100%
	Class 2	Yes	Systematic	Coverage: 10–100% (Systematic and Judgmental)
	Class 3	Yes	Random	Judgmental
Nonimpacted		No	No	None

Source: From EPA, 2000. *Multi-Agency Radiation Survey and Site Investigation Manual (MARSSIM)*, EPA 402-R-97-016, Rev. 1.

Table 3.10 identifies a modified version of EPA's (EPA 2000) guidance on a recommended characterization sampling/survey design. Note that the term *screening* in Table 3.10 refers to the use of low-cost survey methods that can be used to cover large portions of the site at a relatively small cost. The EPA (2000) guidance uses the terms Class 1, 2, and 3 to refer to the degree to which areas of a site are contaminated. Class 1 refers to those areas that were known to have had contaminant concentrations exceeding action levels. Class 2 areas of a site refer to those areas that were known to have been impacted but contaminant concentrations were not expected to exceed action levels. Class 3 areas of a site refer to potentially impacted areas of the site that had a very low probability of showing elevated contaminant concentrations.

3.2.5.7.2.10 Statistical Hypothesis Testing Once the appropriate sampling design has been selected, the next step is to identify the *preferred* statistical hypothesis test to test the null hypothesis. Running a statistical hypothesis test to determine whether a site meets the appropriate action levels is the general approach taken by EPA (2000), EPA (2006a), and EPA (2006b).

The interpretation of the results of any statistical hypothesis test is essentially the same, even though the details of conducting the test may vary. The goal is to reject the null hypothesis (which assumes the site is contaminated until shown to be clean) and have a high degree of confidence that the site is not contaminated. To make this determination, an "observed" statistical parameter of interest (e.g., mean, median) is calculated from the samples that were collected. This observed statistical parameter of interest is compared with the action level. If this observed parameter of interest is significantly less than the action level, the null hypothesis is rejected, and the sample data support the conclusion that the site is not contaminated. If this observed statistical parameter of interest is significantly greater than the action level, then the null hypothesis cannot be rejected, and the sample data support the conclusion that the site is contaminated.

Note that, in either case, the true state of the site has not been determined with absolute certainty. It has not been "proven" that the population mean is above or below the action level. A statistical hypothesis test can only provide evidence that either supports or fails to support the null hypothesis. Nothing is ever proven with hypothesis testing because there is always some probability of making a false-rejection or false-acceptance decision error.

Parametric and Nonparametric Statistical Hypothesis Tests – There are two basic kinds of statistical hypothesis tests: parametric and nonparametric. Both types of tests have assumptions that must be met before the results of the test can be meaningfully interpreted. However, the assumptions of a nonparametric test are often less stringent than those of the corresponding parametric test. Because the choice between a parametric and a nonparametric test is based, in part, on assessing whether certain statistical assumptions have been met, it is wise to seek the advice of a statistician when making this choice. A detailed discussion of the assumptions underlying various parametric and nonparametric tests is beyond the scope of this book. However, it is important to verify any applicable assumptions before proceeding with the chosen statistical hypothesis test.

When choosing the most appropriate statistical hypothesis test to assess whether certain statistical elements have been met, seek guidance from the following EPA guidance manual:

- EPA, 2006c, Data Quality Assessment: Statistical Methods for Practitioners, EPA QA/G-9S, EPA/240/B-06/003, February

EPA (2006c) notes that parametric tests typically concern the population mean or quantile, use the actual data values, and assume data values follow a specific probability distribution. Nonparametric tests typically concern the population mean or median, use data ranks, and don't assume a specific probability distribution. A parametric test will have more power than a nonparametric counterpart if the assumptions are met. However, the distributional assumptions are often strict or undesirable for the parametric tests and deviations can lead to misleading results.

Table 3.11 provides details from a figure in Section 3.2 of EPA (2006c) which lists statistical tests that should be considered when evaluating environmental data.

3.2.5.7.3 Statistical Sampling Design Software

The following section provides a description, general guidance, and limitations on the use of statistical sampling design software such as Visual Sample Plan and Decision Error Feasibility Trials (DEFT).

3.2.5.7.3.1 Visual Sample Plan Software Visual Sample Plan (Figure 3.17) is a publicly available software tool sponsored by the U.S. Department of Energy, U.S. Environmental Protection Agency, and U.S. Department of Defense. It is designed to provide the user with a simple and defensible tool that defines a technically defensible sampling scheme that can be used to support site characterization or postremediation site-closeout sampling activities. This software helps the user determine the optimal number and location of samples using defensible statistical sampling design methods, and displays those locations on maps or aerial photos. Visual Sample Plan is applicable for any two-dimensional (2D) sampling plan, including surface soil, sediment, building surfaces, water bodies, or other similar applications.

TABLE 3.11
EPA 2006c Recommended Parametric and Nonparametric Testing Methods

	Parametric/ Nonparametric	Independent/ Paired	Test	Population Parameter	Distribution Assumption
One Sample Methods	Parametric	-	*t*-Test and Confidence Interval (CI)	Mean	Normal
			Stratified *t*-Test	Mean	Normal
			Chen Test	Mean	Right-Skewed
			Land's CI Method	Mean	Lognormal
			Test for a Proportion and CI	Proportion	-
	Nonparametric	-	Sign Test	Median	None
			Wilcoxon Signed Rank Test	Median/Mean	Symmetric
Two Sample Methods	Parametric	Independent	*t*-Test and CI (equal variances)	Difference in Means	Normal
			t-Test and CI (unequal variances)	Difference in Means	Normal
			Test for Proportions and CI	Difference in Proportions	-
		Paired	Paired *t*-Test	Difference in Means	Normal
	Nonparametric	Independent	Wilcoxon Rank Sum Test	Difference in Means	Same Variance
			Quantile Test	Right-Tail	-
			Slippage Test	Right-Tail	-
		Paired	Sign Test	Median	None
			Wilcoxon Signed Rank Test	Median/Mean	Symmetric

A copy of the Visual Sample Plan Version 7.15 software can be downloaded at no cost from the following Internet address: http://vsp.pnnl.gov/VSP_download/downloads.aspx.

One of the advantages in using the Visual Sample Plan software is that it is very visual. As shown in Figure 3.17, it can simultaneously show the following views:

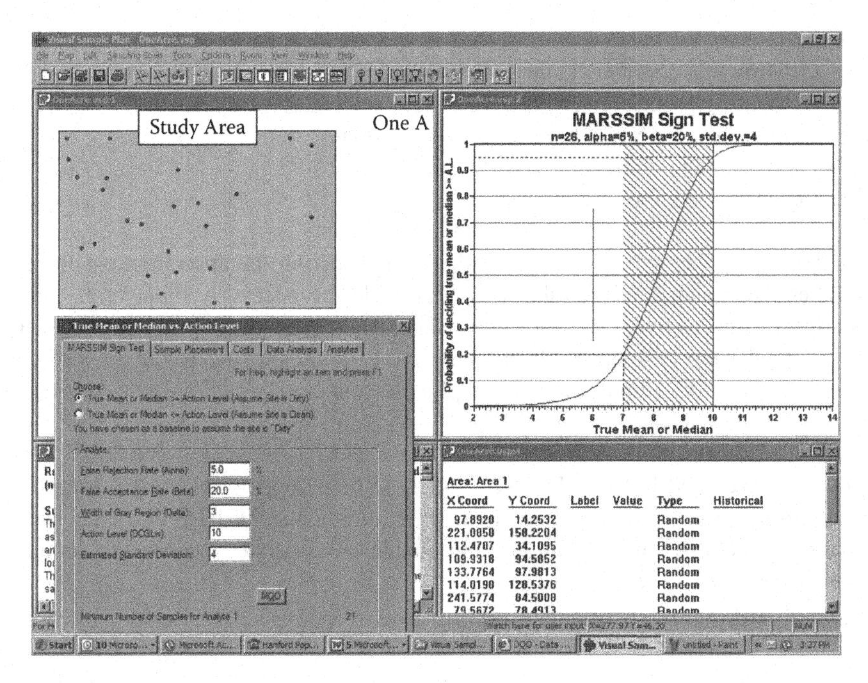

FIGURE 3.17 Visual sample plan statistical software.

- *Map View:* Shows the site map, selected sampling areas, and defined sampling locations
- *Cost View:* Presents statistics related to the number of samples, cost, probability, etc.
- *Graph View:* Presents a graph of the selected function
- *Plan View:* Displays the (X,Y) coordinates of the sampling locations

Being able to see all four of these views simultaneously is particularly beneficial when negotiating error tolerance requirements with regulatory agencies. For example, by projecting the Visual Sample Plan software up on a wall screen, regulators can see the impact that their stringent false-rejection decision error (also known as false-positive error or α error), false-acceptance error (also known as false-negative error or β error), and width of the gray-region requirements have on sampling costs. With Visual Sample Plan, error tolerances can be adjusted repeatedly until one comes up with a sample design that balances cost and uncertainty.

Visual Sample Plan allows the user to either sketch a simple outline of the study area in the shape of a square, rectangle, circle, ellipse, etc., or import an electronic site map. Once the site map has been established, the specific areas where samples are to be collected are then defined with the assistance of the computer mouse. Once the sampling areas have been identified, the Visual Sample Plan software allows the user to select between many design options, several of which include:

- Comparing average to fixed threshold
- Comparing average to reference average

- Estimating the mean
- Constructing confidence interval on mean
- Comparing proportion to fixed threshold
- Comparing proportion to reference proportion
- Estimating the proportion
- Locating a hot spot

Visual Sample Plan software assists the user in selecting the appropriate statistical hypothesis test and then asks the user to provide the necessary input (e.g., false-rejection [false-positive] rate, false-acceptance [false-negative] rate, width of gray region, action level, standard deviation, etc.) to complete the calculation of the required number of samples and the sampling locations.

3.2.5.7.3.2 Decision Error Feasibility Trials (DEFT) Software The DEFT software was developed by the EPA (EPA 2001) to help support selecting a sampling design that meets project-specific quality requirements. DEFT allows decision-makers and members of the planning team to quickly generate cost information for a variety of simple sampling designs.

DEFT is designed to calculate sample sizes to address the following types of questions:

- Is the population mean greater than or less than a fixed standard?
- Is a population proportion/percentile greater than or less than a fixed standard?
- Is the difference between two population means significant?
- Is the difference between two population proportions/percentiles significant?

DEFT is not as sophisticated as the Visual Sample Plan software (Section 3.2.5.7.3.1), and the EPA recommends that DEFT be used only to evaluate the feasibility of the data quality requirements specified through DQO Step 6 (see Section 3.2.5.6). Figure 3.18 shows an example of a DEFT software screen that has calculated that a total of 19 samples are required to be collected from a site with the following input assumptions:

- The population mean being the statistical parameter of interest
- Dealing with one population
- Minimum value detected based on historical data: 5 μg/L
- Maximum value detected based on historical data: 300 μg/L
- Action level: 30 μg/L
- Gray region: 20–30 μg/L
- Standard deviation of population: 16.67 μg/L
- Laboratory cost per sample: $1,000
- Field sampling cost per sample: $50
- False-rejection limit: 0.05 (or 5%)
- False-acceptance limit: 0.20 (or 20%)

For additional details on the DEFT software, see the EPA QA/G-4D (EPA 2001) guidance manual.

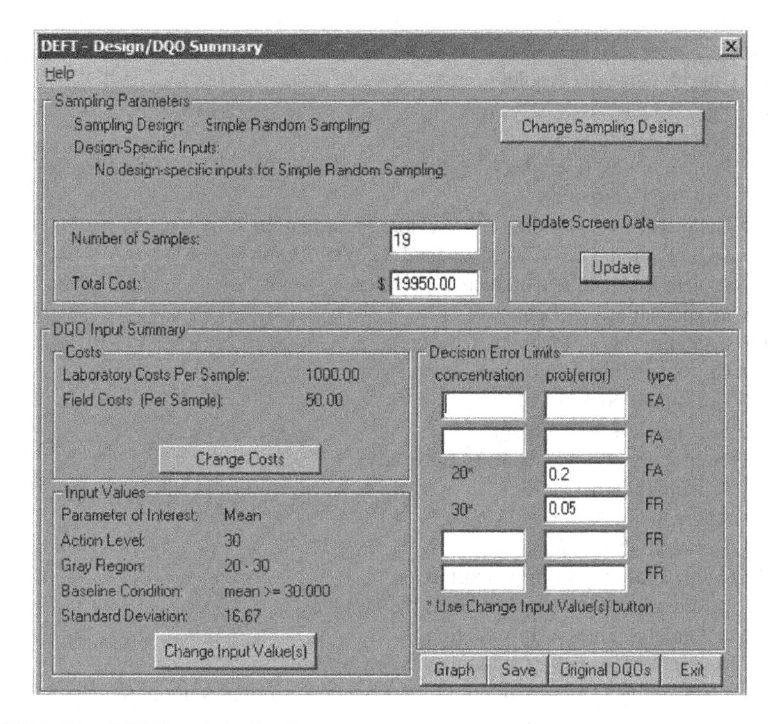

FIGURE 3.18 DEFT statistical software.

3.2.5.7.3.3 Other Available Statistical Sample Design Software Other statistical sample design software packages that are available at no cost to support environmental investigations include:

- MARSSIM Power 2000 (www.cvg.homestead.com/marssimpower2000.html)
- ProUCL (www.epa.gov/land-research/proucl-software)
- COMPASS (www.marssim.com/Tools.htm)
- SADA (www.sadaproject.net)
- GEOPACK (www.epa.gov/water-research/geostatistical-software-package-geopack)

3.2.6 Preparing a DQO Summary Report

The results from the seven-step DQO process can be documented in a DQO Summary Report. Figure 3.19 presents a recommended outline for this report.

3.2.7 Preparing a Sampling and Analysis Plan

A Sampling and Analysis Plan is prepared to provide the field sampling team with direction on how to implement the sampling design identified in Step 7 of the DQO process (Section 3.2.5.7). It is composed of a summary of the DQOs, a Quality Assurance Project Plan, and a Field Sampling Plan. Figure 3.20 provides a recommended outline for a Sampling and Analysis Plan.

Introduction
1.01 DQO Step 1: State the Problem
- 1.1 Project Objectives
- 1.2 Project Assumptions
- 1.3 Project Issues. (includes Global Issues and Task-Specific Technical Issues)
- 1.4 Existing References
- 1.5 Historical Background for Site Under Investigation
- 1.6 DQO Planning Team Members and Key Decision-Makers.
- 1.7 Project Budget and Contractual Vehicles
- 1.8 Project Milestone Dates
- 1.9 Contaminants of Potential. Concern
- 1.10 Current and Potential Future Land Use
- 1.11 Conceptual Site Model
- 1.12 Statement of the Problem

2.0 DQO Step 2: Identify the Goals of the Study
- 2.1 Principal Study Questions
- 2.2 Alternatives. Actions
- 2.3 Decision Statements

3.0 DQO Step 3: Identify Information Inputs
- 3.1 Information Required to Address the Problem
- 3.2 Potential Sources of Information
- 3.3 Action for Levels for Contaminants of Concern
- 3.4 Level of Data Quality Required. (Analytical Performance Requirements)
- 3.5 Appropriateness of Existing Data
- 3.6 Verifying Appropriate Sampling/Analysis Methods Exist for New Data

4.0 DQO Step 4: Define the Boundaries of the Study
- 4.1 Target Population of Interest
- 4.2 Spatial Boundaries of the Study Area
- 4.3 Temporal Boundaries of the Study Area
- 4.4 Conditions Most Favorable for Collecting Data
- 4.5 Time Frame over Which Each Decision Applies
- 4.6 Practical Constraints Affecting Data Collection
- 4.7 Scale of Decision-Making

5.0 DQO Step 5: Develop the Analytic Approach
- 5.1 Statistical Parameter of Interest for Decision-Making (e.g., mean, median, percentile)
- 5.2 Decision Rules

6.0 DQO Step 6: Specify Performance of Acceptance Criteria
- 6.1 Selection between Statistical vs. Nonstatistical Approach
- 6.2 Possible Range of Parameter of Interest (for Statistical Approaches Only)
- 6.3 Null Hypothesis (for Statistical Approaches Only)
- 6.4 Decision Errors (for Statistical Approaches Only)
- 6.5 Boundaries of Gray Region (for Statistical Approaches Only)
- 6.6 Tolerable Limits on Decision. Error (for Statistical Approaches Only)

7.0 DLO Step7: Develop the Plan for Obtaining Data
- 7.1 Nonstatistical (Judgmental) Designs
- 7.2 Statistical Designs

8.0 References

FIGURE 3.19 Recommended outline for DQO summary report.

Contents

1 Introduction
- 1.1 Project Scope and Objective
- 1.2 Background
  - 1.2.1 Site Geology/Hydrology
- 1.3 Data Quality Objective Summary
  - 1.3.1 Statement of the Problem
  - 1.3.2 Project Task and Problem Definition
  - 1.3.3 Decision Statements/Decision Rules
- 1.4 Contaminants of Concern/COPCs/Target Analytes
- 1.5 Project Schedule

2 Quality Assurance Project Plan
- 2.1 Project Management
  - 2.1.1 Project/Task Organization
  - 2.1.2 Quality Objectives and Criteria
  - 2.1.3 Methods-Based Analysis
  - 2.1.4 Special Training/Certification
  - 2.1.5 Documents and Records
- 2.2 Data Generation and Acquisition
  - 2.2.1 Analytical Methods Requirements/Field Analytical Methods
  - 2.2.2 Quality Control
  - 2.2.3 Measurement Equipment
  - 2.2.4 Equipment Testing, Inspection, Calibration and Maintenance
  - 2.2.5 Inspection/Acceptance of Supplies and Consumables
  - 2.2.6 Data Management
- 2.3 Assessment and Oversight
- 2.4 Data Review and Usability
  - 2.4.1 Data Review/Verification/Validation

3 Field Sampling Plan
- 3.1 Sampling Objectives/Design
- 3.2 Sample Location, Frequency, and Constituents to be Monitored
- 3.3 Sampling Methods
  - 3.3.1 Decontamination of Sampling Equipment
- 3.4 Documentation of Field Activities
  - 3.4.1 Corrective Actions and Deviations for Sampling Activities
- 3.5 Calibration of Field Equipment
- 3.6 Sample Handling
  - 3.6.1 Containers
  - 3.6.2 Container Labeling
  - 3.6.3 Sample Custody
  - 3.6.4 Sample Transportation

4 Management of Waste

5 Health and Safety

6 References

FIGURE 3.20 Recommended outline for a sampling and analysis plan.

The Quality Assurance Project Plan portion of the Sampling and Analysis Plan provides guidance on analytical performance requirements. Specifically, the Quality Assurance Project Plan provides details on:

- Special training requirements
- Analyses to be performed
- Analytical method requirements (e.g., detection limits, precision, accuracy)
- Quality control requirements (e.g., blanks, duplicates, spikes)
- Measurement requirements
- Equipment calibration and maintenance requirements
- Inspection and acceptance of supplies
- Data management requirements
- Assessment and oversight requirements
- Data review requirements
- Data verification/validation requirements

The Quality Assurance Project Plan should be prepared with input from a chemist, radiochemist, or data quality specialist who was involved in the preparation of the DQO Summary Report, in consultation with representatives from the laboratory performing the analyses.

The Field Sampling Plan portion of the Sampling and Analysis Plan identifies the:

- Sampling objectives
- Sampling locations and sampling frequency
- Sampling methods
- Sampling equipment decontamination procedures
- Field documentation requirements
- Calibration requirements for field equipment
- Procedures for sample handling
- Waste management handling requirements
- Health and safety requirements

3.2.8 Groundwater Modeling to Support Site Characterization and Remediation

Matt Tonkin

Mashrur A. Chowdhury

Groundwater flow and contaminant transport modeling is an important and necessary component of site remediation. Models play a range of roles throughout the various stages of site characterization and remediation. This chapter provides an overview of these various roles, organized by commonly recognized stages of the remediation lifecycle, with the goal of promoting a general understanding of the fundamental aspects of modeling in support of site characterization and remediation.

Guidance Documents

There are numerous guidance documents available from federal and state agencies that discuss the development and application of groundwater flow and contaminant transport models – i.e., F&T models – in the context of remediating contaminated sites. Some key guidance documents are listed below in order of publication year:

- NC DEQ, 2020, *Revised Technical Guidance for Risk-Based Environmental Remediation of Sites.*
- GA DNR, 2016, *Guidance: Groundwater Contaminant Fate and Transport Modeling.*
- CA EPA, 2012, *Guidance for Planning and Implementing Groundwater Characterization of Contaminated Sites.*
- OH EPA, 2007, *Technical Manual for Ground Water Investigations - Chapter 14: Ground Water Flow and Fate and Transport Modeling.*
- Zhang and Heathcote, 2001, *Guidance for Numerical Modeling in Tier 3 Assessments and Other Corrective Actions.*
- EPA, 1996, *Documenting Ground-Water Modeling at Sites Contaminated with Radioactive Substances*, EPA 540-R-96-003.
- EPA, 1994, *A Technical Guide to Ground-Water Model Selection at Site Contaminated with Radioactive Substances*, EPA 402-R-94-012.
- EPA, 1992, *Fundamentals of Ground-Water Modeling*, EPA 540-S-92-005.
- EPA, 1988, *Guidance on Remedial Actions for Contaminated Ground Water at Superfund Sites*, EPA 540-G-88-003.
- EPA, 1985, *Modeling Remedial Actions at Uncontrolled Hazardous Waste Sites*, EPA 540-2-85-001.

Because no two sites are the same, no two model applications are the same. The extent and appropriateness of modeling efforts vary enormously between sites, depending on the geographic and vertical scales of the site; the number and types of constituents; the impacted media; the technologies under consideration or in operation; and the anticipated duration of the remedy. For this reason, it is advisable at the outset of a project, and at various appropriate points during the project life cycle, to evaluate current and anticipated modeling needs so that model development and application proceed in a timely manner.

Common Questions Posed of Groundwater Models

Groundwater flow and contaminant transport models are developed to gain an understanding of subsurface conditions via hypothesis testing, to reconstruct past conditions through quantitative calibration or more qualitative analyses, and to predict or forecast conditions either in the future or at unsampled locations. As detailed in this chapter, models range from simple hand calculations to complete numerical models solved only on supercomputers: given this spectrum, the question arises – if a model is not to be used as the primary method to assimilate and evaluate field data and make predictions, what reasonable alternative is there? At the very least, the process of organizing data and information in the construction of a groundwater model forces the hydrogeologist to develop a thorough conceptualization of the site and confront many of the questions at hand. As stated by Anderson (1983) "*...applying a model is an exercise in thinking about the way a system works*". As such the mere process of model development is one that primes the hydrogeologist to make well-informed decisions.

Consequently, the question is usually not *whether to model*, rather *what is the appropriate complexity of modeling to undertake*. In any event, model development should not commence without a clearly defined purpose in mind; that purpose is usually established by the questions that are being posed. In simple terms, questions posed of groundwater models within the groundwater remediation context take the general form: "*how much, how fast, how far*?" More specifically, common questions include:

- How long will it take contaminants to migrate to potential receptors, and will they arrive at concentrations harmful to human health and the environment?
- What remedial options offer the best prospects to mitigate any potential impacts and achieve desirable cleanup objectives?
- Which of these remedial options provides the best overall balance of technical feasibility, likelihood of success, cost, and minimal impact (the latter including carbon footprint and other considerations)?

Although models often offer the best means for addressing these questions, blind reliance on them is a fool's errand. The quote "*all models are wrong, but some are useful*", often attributed to the British statistician George Box, serves as a reminder that as approximations of reality, models are only useful if they reflect reality to sufficient degree to reliably inform decision-making. The aphorism "*garbage in, garbage out*", signaling that flawed or nonsense model inputs produce erroneous and unreliable model outputs, tells us that models constructed on the basis of a flawed conceptual understanding of site conditions or processes are unlikely to provide any meaningful utility.

Types of Models and Their Applicability

Conventionally, F&T modeling refers to the use of process-based models. In these models, physical and chemical processes which control groundwater flow and subsurface contaminant transport are represented by governing equations that approximate those processes on appropriate scales. Solutions to these equations may be obtained analytically, thereby yielding an analytical model; numerically, by discretizing the problem in space and potentially in time; or through a hybrid, such as the analytic element method (AEM) (Strack 1989; Haitjema 1995).

The range of uses and applicability of different modeling approaches are detailed in several of the guidance documents listed above. In general, process-based models are considered appropriate "*if a robust conceptual model has been developed that can adequately be described by the mathematical relationships*" (McMahon et al. 2001). That is, the physical and chemical processes that the model is intended to represent are adequately understood and characterized; and also, that the model is a sufficiently reliable analog of the real-world system. The foregoing requirements reinforce the notion that if F&T models are to be developed and used to guide remediation efforts, particularly if their outputs will be used in a quantitative rather than qualitative capacity, their reliability is inexorably tied to the conceptual site model (CSM).

The CSM and Relevant Features, Events, and Processes (FEPs)

The CSM is principally a vehicle for communicating available data and information. It summarizes current knowledge of how, where, and at what rate(s) contaminants are expected to move and the impacts such movement may have. Consequently, CSMs evolve: if new data change or improve understanding of critical elements of the site, the CSM may need to be revised and any calculations developed on the basis of the CSM revisited accordingly. Several components of a CSM identified by the EPA (2019) are illustrated schematically in Figure 3.21.

CSMs are often qualitative narrative descriptors but can be supported in some instances by quantified elements. While narrative CSMs are valuable communication tools, selection of the approximate modeling method(s) for a site and a suitable simulation code or suite of codes to implement the necessary calculations require more explicit definition of modeling needs. In the context of radionuclide transport, for example, the Nuclear Energy Agency Organization for Economic Co-operation and Development (NEA 2000) developed an approach to define relevant scenarios for simulation studies based on FEPs, which are defined as follows (Sandia National Laboratories 2008):

- Feature – An object, structure, or condition that has a potential to affect repository system performance.
- Event – A natural or human-caused phenomenon that has a potential to affect repository system performance and that occurs during an interval that is short compared to the period of performance.
- Process – A natural or human-caused phenomenon that has the potential to affect repository system performance and that occurs during all or a significant part of the period of performance.

This approach identifies and prioritizes FEPs that potentially affect contaminant fate and transport, and then develops and models alternate scenarios, each of

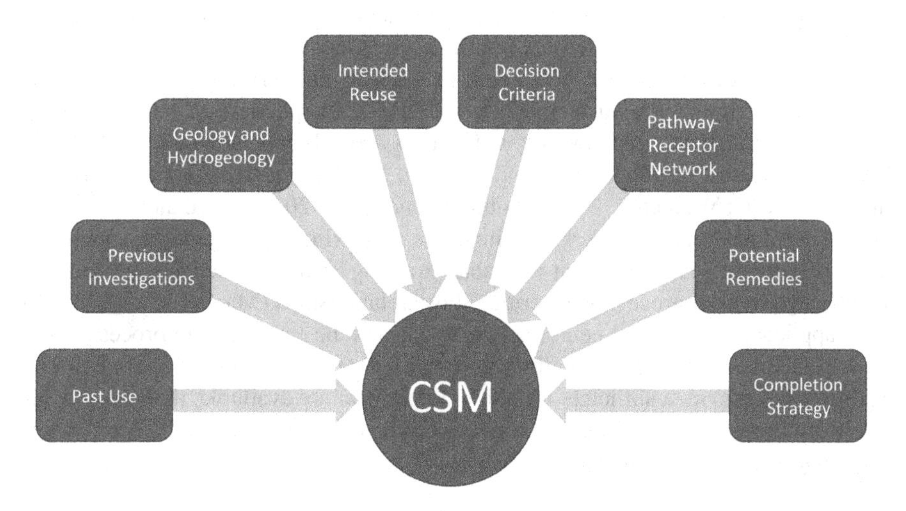

FIGURE 3.21 Components of a Conceptual Site Model (CSM) (Modified from EPA, 2019).

which consists of a well-defined, connected, sequence of FEPs. In the general context (as opposed to a nuclear waste repository), FEPs pertain to site conditions and the planned or operating remediation system. Therefore, one way to determine the modeling needs at a site is to explicitly identify FEPs that will require representation through modeling; and then identify appropriate methods and simulation codes possessing the capabilities necessary to represent those FEPs. In this sense, the definition of FEPs can be viewed as a refinement and quantification of a narrative CSM (e.g., Neptune and Company Inc. 2016).

Because modeling approximates complex, imperfectly known conditions, practitioners should have sound knowledge and understanding of the applicability and limitations of assumptions invoked when using process-based models. For example, the incorrect assumption of a linear sorption isotherm at high concentrations under conditions that exhibit nonlinear processes can underpredict contaminant travel rates, concentrations, and persistence (EPA 1994). Similarly, the simulation of contaminant transport using advective-dispersive methods when rate-limiting or scale-dependent processes are dominant has been shown to have implications for the design and performance of groundwater remedies (Hadley and Newell 2014). In recognition of this, methods such as dual domain (or dual porosity) representations of partitioning between solids, immobile porewater, and active groundwater have been demonstrated to better represent field processes (Brusseau 1992; Baeumer et al. 2001; Liu et al. 2007; Payne et al. 2008).

Other Classes of Models

Other classes of models besides the process-based models described above include geostatistical models, which are used to describe spatial and/or spatiotemporal phenomena using statistical methods rather than process-based governing equations, and hybrid methods that combine geostatistical and process-based theories and methods. Although geostatistical and hybrid methods are beyond the scope of this chapter, an important aspect of their application is that the use of such methods to predict values – such as concentrations – is accompanied by a measure of the uncertainty associated with that prediction. This accompaniment of the prediction with a measure of its uncertainty is a desirable quality of these methods that is not always available with the use of deterministic process-based modeling methods.

Finally, machine learning methods have also been applied at contaminated sites in data-driven analyses and coupled with process-based models (Rogers and Dowla 1994; Rogers et al. 1995; Johnson and Rogers 1995; Yan and Minsker 2006; Besaw and Rizzo 2007; Yoon et al. 2007; Majumder and Eldho 2020). Used alone, methods such as artificial neural networks facilitate the approximation of complex nonlinear relationships by "learning" from available "training" data without representing the underlying physical processes governing those relationships (Maskey et al. 2000). In other applications, artificial neural networks can also be coupled with process-based models and used to guide their calibration (training) to field data (Thiros et al. 2021). Additionally, provided sufficient reliable training data are available, the use of artificial neural networks can, on some occasions, obviate the need to develop a conventional fate and transport model.

Process-Based Models: Typical Applications

Recognizing that model development and application is unique to each site, Figure 3.22 presents an overview of general steps in the development and application of site-specific F&T models. Modeling at contaminated sites should be an iterative process, whereby field data are first used to inform the development of an initial CSM and accompanying computational model; and then the modeling analyses guide further field data collection, leading to revision and update of the CSM, and so on. This iterative procedure continues until sufficient confidence is built in the decision-making process, and/or until site objectives are met.

Because process-based models are an efficient method for integrating relevant FEPs of the CSM, they are often the primary tool used to make quantitative predictions

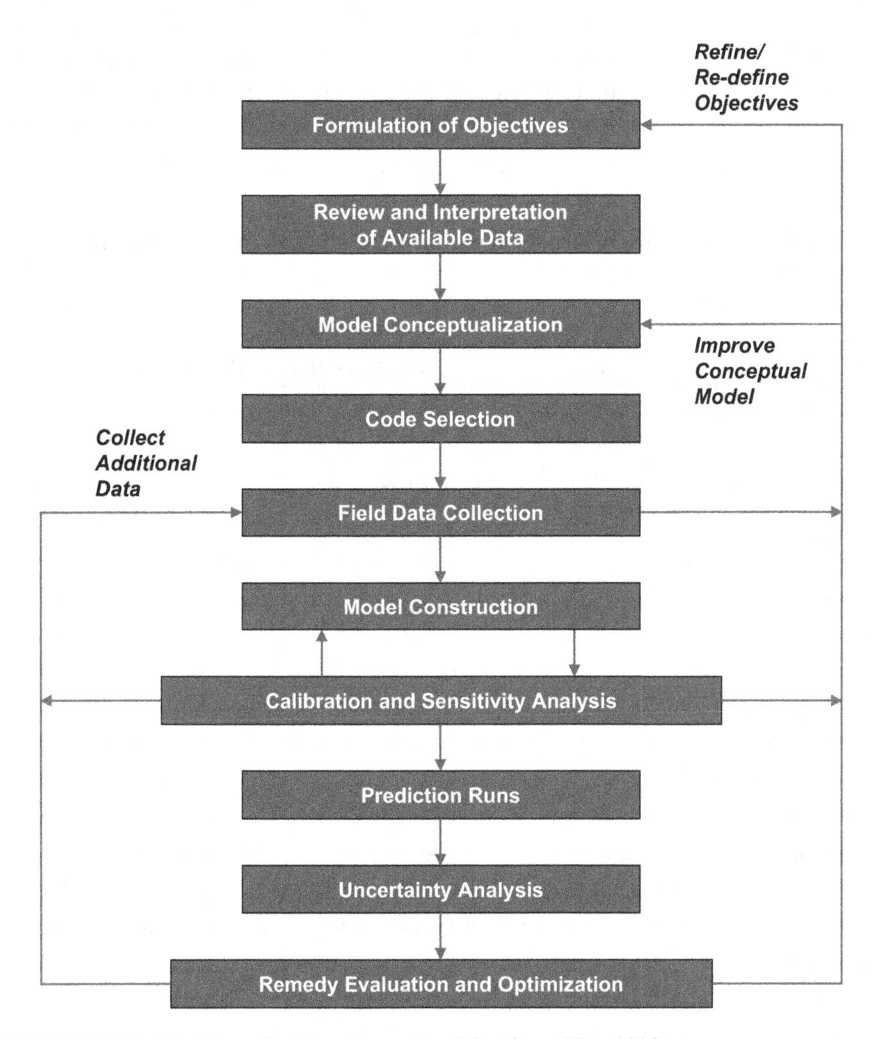

FIGURE 3.22 Model application process (modified from EPA, 1992).

to support decision-making. Additionally, it is desirable that model development and application provide value by reducing characterization and remediation timeframes and costs by an amount that exceeds the effort and cost of model development and application (Compernolle et al. 2013). To accomplish this goal, the modeling effort must be commensurate with the objectives identified for each phase of the remedy process (EPA 1994). Table 3.12 presents an overview of how modeling approach and applicability can change with different stages of the remedy process.

Common Remediation Technologies

As is evident from Figure 3.22 and Table 3.12, the type, complexity, geographic and temporal extents, and even the purpose of modeling can vary substantially by the project phase. It can also vary by the types and combination of technologies under consideration or selected for the remedy. The potential importance of site-specific FEPs can be closely related to the remedial technology (or combination thereof) under consideration or selected for the site. Therefore, the underlying CSM and the design, construction, and implementation, of the F&T model should

TABLE 3.12
Modeling Approach at Different Phases of Remediation (Modified from EPA, 1994)

Model Attribute(s)	Project Phase		
	Scoping	***Characterization***	***Remediation***
Accuracy	Conservative Approximations	Site-Specific Approximations	Remedial Action Specific
Assumptions Regarding Flow and Transport Processes	Simplified Flow and Transport Processes	Complex Flow and Transport Processes	Specialized Flow and Transport Processes
Temporal Representation of Flow Processes	Steady-State	Steady-State	Transient
Temporal Representation of Transport Processes	Steady-State	Transient	Transient
Boundary Conditions	Simplified	Non-Transient	Transient
Initial Conditions	Uniform	Nonuniform	Nonuniform
Dimensionality	1-Dimensional	1-/2-Dimensional/ Quasi 3-Dimensional	Quasi 3-Dimensional/ Fully 3-Dimensional
Lithology	Homogeneous/ Isotropic	Heterogeneous/ Anisotropic	Heterogeneous/ Anisotropic
Methodology	Analytical	Semi-Analytical/ Numerical	Numerical
Data Requirements	Limited	Moderate	Extensive

prioritize and incorporate the FEPs that are likely to dominate flow and transport conditions in the context of the considered or selected remedy. Examples include (EPA 1996):

- Vertical hydraulic gradients/three-dimensional F&T (e.g., pump and treat [P&T], extraction trenches, barrier walls)
- Aquifer de-saturation/re-saturation (e.g., P&T, multi-phase extraction)
- Matrix diffusion processes (e.g., P&T, soil-vapor extraction)
- Microbial processes (e.g., bio-sparging, monitored natural attenuation [MNA])
- Coupling with complex geochemical models to simulate surface complexation reactions (e.g., in-situ remediation technologies for metals)

In most applications, the full suite of modeling needs cannot be anticipated at the outset of the project; rather, they develop over time as the site proceeds through the remedy lifecycle. In recognition of this, sections that follow provide a guide to the model development process throughout the typical remedy life cycle.

A Note on Model Calibration

A critical aspect in the development of F&T models for site-specific application is model calibration. That is, in most cases, for a process-based model to be used in a quantitative role at a site, it must be demonstrated that the model reasonably reproduces real-world behavior. In simple terms, model calibration seeks to identify parameter values that result in acceptable correspondence between model outputs and measured data. Model calibration is a vast field of study that extends beyond the scope of this chapter. The following articles and books provide a wealth of information on many aspects of model calibration:

- Schilling et al., 2019, *How Diverse Observations Improve Groundwater Models.*
- Anderson et al., 2015, *Applied Groundwater Modeling: Simulation of Flow and Advective Transport.*
- Doherty and Hunt, 2010, *Approaches to Highly Parameterized Inversion: A Guide to Using PEST for Groundwater-Model Calibration.*
- Hill and Tiedeman, 2007, *Effective Groundwater Model Calibration: With Analysis of Data, Sensitivities, Predictions, and Uncertainty.*
- Reilly and Harbaugh, 2004, *Guidelines for Evaluating Ground-Water Flow Models.*
- Franssen et al., 2003, *Coupled Inverse Modelling of Groundwater Flow and Mass Transport and the Worth of Concentration Data.*

Throughout the rest of this section, it will be presumed that appropriate efforts at model calibration are undertaken throughout the various stages in the remedy life cycle.

3.2.8.1 Site Characterization

The primary objective during site discovery and characterization is to evaluate existing site data and knowledge, develop a conceptual understanding of the site, and, if possible, obtain additional data to further that knowledge (EPA 1988). This constitutes the Remedial Investigation (RI) phase of the CERCLA RI/Feasibility Study (FS) process. The principal roles of modeling at this stage are to facilitate an initial evaluation of the nature and extent of contamination and support screening-level assessments of the threat posed by contamination to potential receptors. Depending on the available data and information, and following the preparation of an initial CSM, modeling may be used to construct potentiometric surface maps to establish approximate groundwater flow directions and rates. These estimates can then be used to estimate plausible migration rates and extents for contaminants; to estimate peak concentrations at site boundaries, potential receptors, or other locations; and to strategically place monitoring wells or other characterization methods to corroborate or update these initial estimates.

Because data are limited in the characterization phase, F&T modeling may be comprised of hand calculations – such as applying Darcy's Law to estimate groundwater velocities and advective transport rates. As explained by Haitjema (2006), though often underappreciated, simple analytical calculations can help determine whether more sophisticated methods are warranted and if so, guide the development of more complex models. For example, a simple water balance calculation can be used to estimate the capture zone for an interim P&T remedy to be constructed while a longer-term remedy is identified. Analytical calculations may also be used to estimate the partitioning of mass from free-phase or soil sources into groundwater, indicating the likely persistence of contamination. Another benefit of rapid calculations using analytical methods is that they can be quickly revisited and revised as site-specific data become available. This can permit steady refinement of the initial CSM and even testing of hypotheses regarding the relative importance and sensitivity of key FEPs to site decisions. With sound judgment and sometimes purposefully conservative assumptions, such methods can cost-effectively facilitate the initial stages of site investigation and evaluation (EPA 1992). Calculations conducted in these early phases provide a foundation for potentially more sophisticated calculations required later in the remedy lifecycle.

3.2.8.2 Initial Remedy Selection and Design

The initial remedy selection and design phase often entails the development of remedial action objectives (RAOs) and comparative evaluations of potential remedies. This constitutes the Feasibility Study phase of the CERCLA RI/FS process. RAOs specify cleanup targets, the area of cleanup (area of attainment), and the anticipated time required to achieve cleanup (i.e., the restoration time frame) (EPA 1988). The nature and extent of contamination dictate general classes of potential remedies: for example, site conditions and the chemical compound(s) present may determine appropriate treatment technologies (i.e., physical/chemical/biological) and the three-dimensional (3D) distribution of the contamination will guide the applicability of in-situ versus ex-situ technologies. However, modeling is valuable for determining the specifics of any remedies suited to achieving the RAOs and comparing alternate designs. The objective is not typically to develop a final design basis suitable for construction and implementation, but to provide a preliminary design basis suitable for initial costing, sizing, and feasibility evaluation. Ideally, this may form the basis of an interim remedy design with the intent that the interim remedy constitutes a core component of the final remedy.

With more data available, modeling often transitions from analytical to numerical methods so that characteristics such as nonuniform aquifer properties, boundaries, or contaminant distributions, can be represented. Nevertheless, modeling objectives are unlikely to be as ambitious as those required for final remedy selection, design, and implementation. For example, simulations may consider non-reactive (e.g., conservative) rather than reactive transport, and 2D rather than 3D conditions. In addition, contaminants that exhibit similar characteristics may be grouped together to limit the number of simulations. The analyses may emphasize relative rather than absolute results – striving to determine which remedial technologies will attain the RAOs most expeditiously, and how best to implement those technologies, rather than predicting the absolute time to achieve RAOs. Despite this, modeling needs to reliably guide the remedy selection process to identify, scale, and approximately cost the combination of technologies that will form the core of the selected remedy.

3.2.8.3 Remedy Design and Implementation

Once the components of the remedy have been selected, the final design and implementation phase commences. During this phase, details of the remedy design are finalized to a sufficient degree that construction can commence, and operations begin. Although the utility of modeling varies depending on the scale of the remedy and the technologies selected, it usually performs at least two roles at this stage: providing quantitative inputs to the remedy design and setting expectations for performance monitoring. Modeling is often used to develop a "base case" design that provides the basis for the core remedy components and is then used to evaluate to what extent imperfect site knowledge causes uncertainty regarding the likely success of the core design, so that contingencies can be planned. The capability of simulating alternate scenarios, and gaining perspective on the likely success of the remedy over a range of conditions, is difficult to accomplish by any other means. The following examples illustrate the foregoing:

- P&T of a single contaminant – modeling may estimate the extent of hydraulic containment (capture) developed by pumping, the required number and location of extraction wells, and the degree to which imperfect knowledge of the extent of contamination, heterogeneity, or other factors impact successful hydraulic containment.
- P&T of multiple contaminants – in addition to the above, modeling may be used to estimate the time-varying influx of each contaminant to the system so that appropriate technologies can be combined and scaled for cost-effective long-term treatment.
- In Situ Chemical Reduction/In Situ Chemical Oxidation (ISCR / ISCO) – modeling may evaluate the mass, location, and delivery schedule of the amendment substrate together with anticipated concentration changes in response to substrate delivery.

- MNA – modeling may represent dominant attenuation mechanisms to predict anticipated concentration changes at point-of-compliance wells, together with related quantities such as changes in overall mass and mass flux throughout the extent of contamination.

In the above examples, modeled concentrations and mass recovery rates are used primarily to support treatment system design. However, other modeled quantities, such as changes in water levels for P&T remedies, and concentration changes in monitoring wells which are applicable to all technologies, provide expectations to which performance monitoring data can be compared. As such, it is usually during the design and implementation phase that expectations are set, and metrics are established, to guide remedy performance monitoring. For example, if modeling was used to estimate the length of time required to achieve an aquifer restoration RAO, then presumably the same model simulations provide metrics against which to compare field data and assess remedy progress – such as pumping-induced water-level changes or mass recovery.

Figure 3.23 depicts an image from sophisticated F&T modeling associated with the remedy design and implementation phase for a P&T remedy. The model in this instance includes a 3D heterogeneous representation of the subsurface, with transient simulation of groundwater elevations (depicted by the contour lines) and contaminant migration (depicted by the colored plume) in response to groundwater extraction, treatment, and re-injection, at multiple extraction and injection wells. For the project depicted in Figure 3.23, the treatment of multiple contaminants is simulated using the Contaminant Treatment System (CTS) package provided with the MT3D-USGS code (Bedekar et al. 2016), which represents contaminant treatment so that predictions accurately represent contaminant concentrations in re-injected water.

FIGURE 3.23 Example application of three-dimensional, transient flow and transport models to simulate remediation of a contaminated site via a pump-and-treat system.

3.2.8.4 Remedy Evaluation, Optimization, and Uncertainty Analysis

During remediation, data are obtained to measure progress toward RAO achievement. For remedies that are anticipated to operate for a relatively short period (such as a few years), performance evaluation and optimization may rely more upon empirical data interpretation than modeling. However, as the complexity and duration of the remedy increase – particularly if the remedy timeframe is decades or greater – the role of modeling for guiding performance evaluation and optimization increases. This is not to suggest that models are highly accurate when predicting conditions many years into the future. However, a well-considered model incorporating the salient FEPs of the CSM is the best available technology for establishing near-term remediation targets that are consistent with RAO attainment and that can be evaluated using monitoring data. For example, the RAO may be attainment of MCLs in groundwater; however, due to the long-anticipated remedy duration, concentrations sampled from monitoring wells during the early years of operation may not be a reliable indicator of progress. In this case, modeling can be used to establish the conditions required to achieve the RAO so that monitoring data can be compared with metrics established from the modeling for early remedy operations – such as mass recovery rates – to establish consistency.

As remedy operations and performance evaluation proceed, the necessity for and scope of optimization activities can be determined. Optimization may take many forms; for example, if monitoring data suggest that the remedy is performing as expected, optimization may not be warranted, or it may emphasize reducing costs while achieving comparable progress. In contrast, if monitoring data suggest the remedy is *not* performing as expected – i.e., that the RAOs are unlikely to be achieved within anticipated timeframes – then more comprehensive optimization efforts are likely warranted. In tandem with this determination, if it becomes evident that critical FEPs underpinning the remedy design are accompanied by greater uncertainty than anticipated, then uncertainty analysis may be warranted.

3.2.8.4.1 Modeling to Support Remedy Optimization

There is widespread literature on the use of modeling for remedy optimization. Model-based remedy optimization is comprised of two approaches, and in many instances, combines these two approaches with the goal of providing recommendations for ways in which the remedy may be modified to achieve the same (or appropriately revised) RAOs:

- Manual procedures, which are usually comprised of incorporating data via model reconstruction, calibration, or other means; and then revisiting the scenario-based simulations undertaken in the remedy design and implementation phase, to provide a revised design basis and updated remedy operational requirements.
- Automated procedures, which seek the same goal as manual optimization, but are usually facilitated by specialized methods and software that are capable of balancing multiple objectives and formally evaluating trade-offs between key elements of the remedy design and implementation – such as alternate

technologies, implementation scale and duration, and costs (Wagner 1995; Rizzo and Dougherty 1996; Zheng and Wang 2002; Maskey et al. 2002; Deschaine et al. 2013; Datta and Kourakos 2015).

Manual optimization efforts are undertaken first to identify major aspects of the remedy that warrant more detailed analysis, and to provide initial approximate solutions to the optimization problem, before initiating automated procedures that are facilitated by formal optimization software. The following list provides some illustrative examples of the foregoing approach:

- P&T of a single contaminant – initial efforts may identify the total system capacity needed to achieve the RAOs, whereas subsequent automated optimization may identify optimal well locations and rates to flush contaminants and speed the progress toward attainment.
- P&T of multiple contaminants – in addition to the above, automated optimization may be used to analyze and adjust the time-varying influx of each contaminant to the system so that treatment technologies can be modified and scaled to meet new expectations.
- ISCR/ISCO – initial efforts may revisit the mass-balance and substrate delivery requirements, such as changing the substrate to one of greater strength or increasing the mass of added substrate. Subsequent automated optimization may revisit the specific timing and locations of substrate delivery.
- MNA – initial analyses may re-assess the relative contributions of potentially applicable attenuation mechanisms – such as degradation, sorption, dispersion – to re-estimate the time required to achieve RAOs. If the remedy is then deemed to be failing, model-based optimization may be used to guide the implementation of contingency measures.

Model-based optimization of groundwater remedies often assumes that the major FEPs are sufficiently well known such that optimization therefore simply requires identifying the preferred solution that balances constraints. However, the failure of remedies to achieve their goals often results from imperfect knowledge about one or more critical FEPs. Common among these are imperfect knowledge of the true extent of contamination; the presence, strength, and persistence of continuing sources; the role of subsurface complexity, including random and non-random (but still unknown) heterogeneities; poorly understood contaminant migration and fate characteristics; and other challenges that hinder successful remedy design or implementation. Such imperfect knowledge can be evaluated through model-based uncertainty analysis.

3.2.8.4.2 Uncertainty Analysis and Multi-Model Methods

In the context of site remediation, it is desired that predictive modeling provides a reliable basis for making remedy decisions. However, the reliability of process-based models is compromised by imperfect knowledge. Acknowledgment that models are imperfect approximations of reality whose outputs are accompanied by uncertainty provides a framework for model development guided by project requirements and available data rather than a desire to perfectly represent the real world (e.g., Doherty and Moore 2019; Guthke 2017). The uncertainty associated with model outputs can be evaluated using a variety of methods and codes (Wu and Zeng, 2013). Broadly speaking, however, there are two major classes of methods to manage and incorporate conceptual model and parameterization uncertainties during remedy evaluation and optimization:

- Stochastic methods: to the extent that uncertainties can be characterized using the theory of random variables, their effects can be evaluated within a probabilistic framework (Sudicky and Huyakorn 1991; Liang et al. 2009).
- Multiple-model methods: multi-model methods are typically used when uncertainties are starkly different and cannot readily be represented using the stochastic continuum methods described above (Timani and Peralta 2015; Elshall et al. 2020).

The most common theoretical framework for stochastic uncertainty analysis is Bayes theorem (Wu and Zeng 2013), which ascribes probabilities to model inputs and computes conditional probabilistic outputs. Because of the computational burden of rigorous Bayesian methods, approximate methods have been developed. In either case, model inputs are generally provided in the form of probability distributions, and model outputs are generally obtained in the form of posterior probability distributions.

Multi-model analysis is used to evaluate complex boundary conditions, geology, transport processes, and other characteristics that require different representations – including binary "present-or-absent" states – that cannot be represented as a probabilistic continuum (Poeter and Hill 2007). These alternative representations lead to structurally different models rather than to alternative parameterizations of the same model structure. Because of the flexibility that it offers, the multi-model approach is conceptually appealing. However, multi-model analysis presents challenges to the decision maker when presented with several substantially different models that do not lie on a probabilistic continuum.

Lastly, at very challenging sites – particularly those that are candidates for very costly remedies due to the types of contaminants, presence of continuing sources, subsurface complexity, or a combination of these and other factors – optimization may need to be undertaken while representing non-trivial uncertainties. In such circumstances, the goal is to narrow prediction intervals for key model outputs – such as the time required to achieve RAOs. The theoretically and computationally challenging task of optimization under uncertainty is outside the scope of this chapter but addressed by several authors (such as Wagner and Gorelick 1989; Morgan 1993; Aly and Peralta 1999; Kunstmann et al. 2002; Ricciardi 2004; Mantoglou and Kourakos 2007; Singh and Minsker 2008; Baú and Mayer 2008; and He et al. 2009).

3.2.8.5 Long-Term Monitoring and Monitoring Network Optimization

Successful completion of the full remedy life cycle depends on obtaining data from a suitable monitoring network. During site characterization, monitoring network design is predominantly based on empirical observation and the initial CSM. However, if remedy design, operation, and optimization have been informed by process-based modeling, those same models offer great value in identifying the number and location of wells (or other sampling devices); the frequency of sampling; and the constituents to be sampled for monitoring purposes. Thoughtfully constructed process-based models provide useful predictions of spatio-temporal changes in contamination extents as remediation progresses. Among other things, modeling can be used to identify points of compliance, and the growth and reduction in monitoring requirements, including varying sampling frequencies for different locations and constituents throughout the monitoring network. Modeling can help identify both redundancies and gaps within existing monitoring networks and programs, and is therefore well suited to monitoring network optimization.

In general terms, if monitoring indicates that the remedy is working as anticipated and in accordance with model projections, then the primary objective of monitoring network optimization is to ensure that success continues while potentially reducing the costs of monitoring. However, if monitoring indicates that the remedy is not working as anticipated – particularly if this results from imperfect knowledge of key FEPs – modeling can be used to guide data collection to improve knowledge of those FEPs, so that the models themselves can be updated and improved upon to facilitate remedy re-design in light of the new information. Many publications describe the use of process-based models to support monitoring network design and optimization, including Pham and Tsai (2016), Wu et al. (2005), Hassan (2006), and Kollat et al. (2011), among others. Singh and Katpatal (2020) provide a general overview of methods to design and evaluate monitoring networks, while Kikuchi (2017) and others discuss the concept of "data worth" to guide data acquisition efforts. These methods and codes compute the relative value of different data collection strategies using process-based models as the means for determining data worth.

3.2.8.6 Attainment Demonstration and Rebound Studies

After a sufficient period of operation of the selected remedy, the site should progress to the attainment demonstration phase. While US EPA guidance is clear that attainment demonstration is empirical, based upon intra-well evaluation of contaminant concentrations, there can nonetheless be a role for process-based modeling to improve understanding of what appear to be the late stages of remediation. This is because long-term monitoring of contaminant concentrations often reveals the presence of contaminant concentration tailing and rebound phenomena. Tailing refers to the progressively slower rate of concentration declines that are observed with continued operation of the remedy. Tailing can be observed whether the remedy is active, such as P&T, or more passive, such as MNA. It is also often observed that contaminant concentrations may rebound following the cessation of operations of an active remedy component; for example, if pumping at a P&T remedy is discontinued after temporarily attaining a cleanup standard. Rebound phenomena are typically only observed at sites that included one or more active remedial technologies. Both tailing and rebound phenomena can greatly inhibit the attainment of cleanup goals. According to EPA (1997):

> Diagnosis of the cause of tailing and rebound, therefore, requires careful consideration of site conditions and usually cannot be made by examining

concentration-versus time data alone. Quantitative development of the conceptual model using analytical or numerical methods may help estimate the relative significance of different processes that cause tailing and rebound.

Studies suggest that a combination of physical and chemical processes might account for tailing and rebound phenomena. Consequently, the evaluation and optimization of groundwater remedies can be enhanced through the execution of planned rebound studies, or the opportunistic acquisition of data obtained through unplanned rebound events, and the analysis and evaluation of the rebound responses using process-based models.

3.2.8.7 Groundwater Modeling Summary

Groundwater flow and contaminant transport modeling is often a necessary component of site remediation. The foregoing sections discuss the various roles that models can play throughout the commonly recognized stages of the remediation life cycle. Although often envisioned as a one-time development and implementation activity, the use of process-based models has, appropriately, developed into an ongoing and iterative exercise. The initial model scoping, design, and development activities that take place during site discovery and characterization should, ideally, provide a firm conceptual basis for the gradual refinement and improvement of the modeling tools throughout later stages.

At no stage in the remedy life cycle should modeling be the only line of evidence used to inform decisions. Process-based models are valuable tools that integrate many important characteristics of a site, but that are of necessity and by design approximations of true site conditions. Consequently, empirical (monitoring and characterization) data continue to be invaluable and cannot be wholly supplanted by a model, no matter how sophisticated. It is usually unwise to rely upon models in an entirely literal sense, and the value provided by models is usually greater – and the inference drawn from models more robust – when used in "what-if" scenario analyses and to evaluate potential benefits of alternate actions in a relative rather than absolute manner. Indeed, the relationship between process-based models and field data should always be viewed as an iterative one, whereby data are used to inform the CSM, parameterize FEPs, and construct the process-based model, which is in turn used to make predictions and to guide further data collection to improve upon the CSM, FEPs, and process-based model itself. Rajabi et al. (2018) describe this iterative process as model–data interaction (MDI) and in this context, the maintenance and upkeep of process-based models throughout the remedy life cycle should emphasize continual updates and improvements to the CSM, FEPs, and their modeled representation, to increase confidence in the decision-making process.

REFERENCES

Aly, A.H., and R.C. Peralta, 1999, Optimal design of aquifer clean up systems under uncertainty using a neural network and a genetic algorithm. *Water Resources Research* 35, no. 8:2523–2532.

Anderson, M.P., 1983, Groundwater modeling – The emperor has no clothes. *Ground Water* 21, no. 6:666–669.

Anderson, M., W. Woessner, and R. Hunt, 2015, *Applied Groundwater Modeling: Simulation of Flow and Advective Transport* (second edition). Academic Press, Inc.

Baeumer, B., D.A. Benson, M.M. Meerschaert, and S.W. Wheatcraft, 2001, Subordinated advection-dispersion equation for contaminant transport. *Water Resources Research* 37, no. 6:1543–1550.

Baú, D.A., and A.S. Mayer, 2008, Optimal design of pump-and-treat systems under uncertain hydraulic conductivity and plume distribution. *Journal of Contaminant Hydrology* 100, no. 1:30–46.

Bedekar, V., E.D. Morway, C.D. Langevin, and M. Tonkin, 2016, MT3D-USGS Version 1: A U.S. Geological Survey release of MT3DMS updated with new and expanded transport capabilities for use with MODFLOW. U.S. Geological Survey *Techniques and Methods 6-A53*.

Besaw, L.E., and D.M. Rizzo, 2007, Stochastic simulation and spatial estimation with multiple data types using artificial neural networks. *Water Resources Research* 43, W11409:1–14.

Bowen and Bennett, 1988, *Statistical Methods for Nuclear Material Management,* Pacific Northwest Laboratory, Prepared for Office of Nuclear Regulatory Research, U.S. Nuclear Regulatory Commission, Washington, D.C., NRC FIN B2420, NUREG/CR-4604, PNL-5849, TK9152.S72, ISBN 0-87079-588-0.

Byrnes, M.E., 2001, *Sampling and Surveying Radiological Environments.* Boca Raton, FL: Lewis Publishers.

Brusseau, M.L., 1992, Transport of rate-limited sorbing solutes in heterogeneous porous media: application of a one-dimensional multifactor nonideality model to field data. *Water Resources Research* 28, no. 9:2485–2497.

California Environmental Protection Agency (EPA), 2012, *Guidance for Planning and Implementing Groundwater Characterization of Contaminated Sites*. June.

Compernolle, T., S.V. Passel, and L. Lebbe, 2013, The value of groundwater modeling to support a pump and treat design. *Groundwater Monitoring and Remediation* 33, no. 3:1–194.

Datta, B., and G. Kourakos, 2015, Preface: Optimization for groundwater characterization and management. *Hydrogeology Journal* 23:1043–1049.

Deschaine, L.M., T.P. Lillys, and J.D. Pintér, 2013, Groundwater remediation design using physics-based flow, transport, and optimization technologies. *Environmental Systems Research* 2, no. 6.

Doherty, J., and R.J. Hunt, 2010, *Approaches to Highly Parameterized Inversion: A Guide to Using PEST for Groundwater-Model Calibration*. U.S. Geological Survey Scientific Investigations Report 2010-5169. Available at: https://permanent.fdlp.gov/gpo2281/GWPEST_sir2010-5169.pdf.

Doherty, J., and C. Moore, 2019, Decision support modelling: Data assimilation, uncertainty quantification, and strategic abstraction. *Groundwater* 58, no. 3:327–337.

Elshall, A.S., M. Ye, and M. Finkel, 2020, Evaluating two multi-model simulation-optimization approaches for managing groundwater contaminant plumes. *Journal of Hydrology* 590:125427.

EPA, 2019, *Smart Scoping for Environmental Investigations Technical Guide.* EPA 542-G-18-004. November.

EPA, 2006a, *Guidance on Systematic Planning Using the Data Quality Objectives Process.* EPA QA/G-4, EPA/240/B-06/001, February.

EPA, 2006b, *Data Quality Assessment: A Reviewer's Guide.* EPA QA/G-9R, February.

EPA, 2006c, *Data Quality Assessment: Statistical Methods for Practitioners.* EPA QA/G-9S, EPA/240/B-06/003, February.

EPA, 2002, *Guidance on Choosing a Sampling Design for Environmental Data Collection.* EPA QA/G-5S, December.

EPA, 2001, *Data Quality Objectives Decision Error Feasibility Trials Software (DEFT)—User Guide.* EPA/240/B-01/007, EPA QA/G-4D.

EPA, 2000, *Multi-Agency Radiation Survey and Site Investigation Manual (MARSSIM).* EPA 402-R-97-016, Rev. 1.

EPA, 1997, *Design Guidelines for Conventional Pump-and-Treat Systems.* EPA 540-S-97-504. September.

EPA, 1996, *Soil Screening Guidance: User Guide.* EPA/540/R-96/018, July.

EPA, 1996, *Documenting Ground-Water Modeling at Sites Contaminated with Radioactive Substances.* EPA 540-R-96-003. January.

EPA, 1994, *A Technical Guide to Ground-Water Model Selection at Sites Contaminated with Radioactive Substances.* EPA 402-R-94-012. June.

EPA, 1992, *Statistical Methods for Evaluating the Attainment of Cleanup Standards,* Vol. 3: Reference-Based Standards for Soils and Solid Media, EPA 230-R-94-004, PB 94-176831, December.

EPA, 1992, *Fundamentals of Ground-Water Modeling.* EPA 540-S-92-005. April.

EPA, 1988, *Guidance on Remedial Actions for Contaminated Ground Water at Superfund Sites.* EPA 540-G-88-003. December.

EPA, 1985, *Modeling Remedial Actions at Uncontrolled Hazardous Waste Sites.* EPA 540-2-85-001. April.

Franssen, H.J., J. Gomez-Hernandez, and A. Sahuquillo, 2003, Coupled inverse modelling of groundwater flow and mass transport and the worth of concentration data. *Journal of Hydrology* 281, no. 4:281–295.

Georgia Department of Natural Resources (GA DNR), 2016, *Guidance: Groundwater Contaminant Fate and Transport Modeling.* October.

Guthke, A., 2017, Defensible model complexity: A call for data-based and goal-oriented model choice. *Groundwater* 55, no. 5:646–650.

Hadley, P.W., and C. Newell, 2014, The new potential for understanding groundwater contaminant transport. *Groundwater* 52, no. 2.

Haitjema, H., 1995, *Analytic Element Modeling of Groundwater Flow (first edition).* Academic Press.

Haitjema, H., 2006, The role of hand calculations in ground water flow modeling. *Groundwater* 44, no. 6:1–6.

Hassan, A.E., 2006, Developing a long-term monitoring network under uncertain flowpaths. *Groundwater* 44, no. 5:710–722.

He, L., G.H. Huang, and H.W. Lu, 2009, A coupled simulation-optimization approach for groundwater remediation design under uncertainty: An application to a petroleum-contaminated site. *Environmental Pollution* 157, no.8–9:2485–2492.

Hill, M.C., and C.R., Tiedeman, 2007, *Effective Groundwater Model Calibration: with Analysis of Data, Sensitivities, Predictions, and Uncertainty.* John Wiley & Sons, Inc., New Jersey.

Johnson, V.M., and L.L. Rogers, 1995, Location analysis in ground-water remediation using neural networks. *Groundwater* 33, no. 5:10.

Kikuchi, C., 2017, Toward increased use of data worth analyses in groundwater studies. *Groundwater* 55, no. 5:670–673.

Kollat, J.B., P.M. Reed, and R.M. Maxwell, 2011, Many-objective groundwater monitoring network design using bias-aware ensemble Kalman filtering, evolutionary optimization, and visual analytics. *Water Resources Research* 47, no. 2.

Kunstmann, H., W. Kinzelbach, and T. Siegfried, 2002, Conditional first-order second moment method and its application to the quantification of uncertainty in groundwater modeling. *Water Resources Research* 38, no. 4:6-1–6-14.

Liang, J., G. Zeng, S. Guo, and J. Li, 2009, Uncertainty analysis of stochastic solute transport in a heterogenous aquifer. *Environmental Engineering Science* 26, no. 2:359–368.

Liu, G., C. Zheng, and S.M. Gorelick, 2007, Evaluation of the applicability of the dual-domain mass transfer model in porous media containing connected high-conductivity channels. *Water Resources Research* 43:12.

Majumder, P., and T.I. Eldho, 2020, Artificial neural network and grey wolf optimizer based surrogate simulation-optimization model for groundwater remediation. *Water Resources Management* 34:763–783.

Mantoglou, A., and G. Kourakos, 2007, Optimal groundwater remediation under uncertainty using multi-objective optimization. *Water Resources Management* 21, no. 5:835–847.

Maskey, S., Y. Dibike, A. Jonoski, and D. Solomatine, 2000, Groundwater model approximation with artificial neural network for selecting optimum pumping strategy for plume removal. *Hydraulic and Environmental Engineering* 11.

Maskey, S., A. Jonoski, and D.P. Solomatine, 2002, Groundwater remediation strategy using global optimization algorithms. *Journal of Water Resources Planning and Management* 10.

McMahon, A., M. Carey, J. Heathcote, and A. Erskine, 2001, *Guidance on the Assessment and Interrogation of Subsurface Analytical Contaminant Fate and Transport Models.* Environment Agency.

Morgan, D.R., 1993, Aquifer remediation design under uncertainty using a new chance constrained programming technique. *Water Resources Research* 29, no. 3:551–561.

Neptune and Company Inc., 2016, *FEPS Analysis for the West Valley Site.* NAC-0071_R2. December.

North Carolina Department of Environmental Quality (NC DEQ), 2020, *Revised Technical Guidance for Risk-Based Environmental Remediation of Sites.* April.

Nuclear Energy Agency (NEA), 2000, *Features, Events and Processes (FEPs) for Geologic Disposal of Radioactive Waste.*

Oak Ridge National Laboratory, 1994, ELIPGRID-PC Program for Calculating Hot Spot Probabilities. ORNL/TM-12774, Oak Ridge, TN.

Oak Ridge National Laboratory, 1995, Monte Carlo Tests of the ELIPGRID-PC Algorithm. ORNL/TM-12899, Oak Ridge, TN.

Ohio Environmental Protection Agency (OH EPA), 2007, *Technical Manual for Ground Water Investigations - Chapter 14: Ground Water Flow and Fate and Transport Modeling.*

Payne, F.C., J.A. Quinnan, and S.T. Potter, 2008, *Remediation Hydraulics.* Boca Raton, FL: CRC Press/Taylor & Francis.

Pham, H.V., and F.T.-C., Tsai, 2016, Optimal observation network design for conceptual model discrimination and uncertainty reduction. *Water Resources Research* 52, no. 2:1245–1264.

Poeter, E.P., and M.C. Hill, 2007, *MMA, A Computer Code for Multi-Model Analysis.* U.S. Geological Survey Techniques and Methods 6-E3.

Rajabi, M.M., B. Ataie-Ashtiani, and C.T. Simmons, 2018, Model-data interaction in groundwater studies: Review of methods, applications and future decisions. *Journal of Hydrology* 567:457–477.

Reilly, T.E., and A.W. Harbaugh, 2004, Guidelines for evaluating ground-water flow models. U.S. Geological Survey Scientific Investigations Report 2004-5038. Available at: https://pubs.usgs.gov/sir/2004/5038/PDF/SIR20045038_ver1.01.pdf.

Ricciardi, K., 2004, Optimal groundwater remediation design subject to uncertainty in hydraulic conductivity with regional variations. *Development in Water Science* 55, no. 2:1215–1225.

Rizzo, D.M., and D.E. Dougherty, 1996, Design optimization for multiple management period groundwater remediation. *Water Resources Research* 32, no. 8:2549–2561.

Rogers, L.L., and F.U. Dowla, 1994, Optimization of groundwater remediation using artificial neural networks with parallel solute transport modeling. *Water Resources Research* 30, no. 2:457–481.

Rogers, L.L., F.U. Dowla, and V.M. Johnson, 1995, Optimal field-scale groundwater remediation using neural networks and the genetic algorithm. *Environmental Science Technology* 29, no. 5:1145–1155.

Sandia National Laboratories (SNL), 2008, Features, events, and processes for the total system performance assessment: Methods. ANL-WIS-MD-000026 REV 00, Sandia National Laboratories, Las Vegas, NV.

Schilling, O.S., P. Cook, and P. Brunner, 2019, How diverse observations improve groundwater models. *Eos*. 100, https://doi.org/10.1029/2019EO126933.

Singh, A., and B. Minsker, 2008, Uncertainty-based multiobjective optimization of groundwater remediation design. *Water Resources Research* 44, no. 2.

Singh, C.K., and Y. Katpatal, 2020, A review of the historical background, needs, design approaches and future challenges in groundwater level monitoring networks. *Journal of Engineering Science and Technology Review* 13, no. 2:135–153.

State of California Environmental Protection Agency (CA EPA), 2012, *Guidelines for Planning and Implementing Groundwater Characterization of Contaminated Sites*.

Strack, O.D.L., 1989, *Groundwater Mechanics*. Englewood Cliffs: Prentice Hall.

Sudicky, E.A., and P.S. Huyakorn, 1991, Contaminant migration in imperfectly known heterogenous groundwater systems. *Review of Geophysics* 19, no. S1:240–253.

Thiros, N.E., W.P. Gardner, M.P. Maneta, and D.J. Brinkerhoff, 2021, Quantifying subsurface parameter and transport uncertainty using surrogate modeling and environmental tracers. *Authorea*. Available at: www.authorea.com/doi/full/10.22541/au.163255068.89308766.

Timani, B., and R. Peralta. 2015. Multi-model groundwater-management optimization: Reconciling disparate conceptual models. *Hydrogeology Journal* 23:1067–1087.

Wagner, B.J., 1995, Recent advances in simulation-optimization groundwater management modeling. *Reviews of Geophysics* 33, no. S2:1021–1028.

Wagner, B.J., and S.M. Gorelick, 1989, Reliable aquifer remediation in the presence of spatially variable hydraulic conductivity: From data to design. *Water Resources Research* 25, no. 10:2211–2225.

Wu, J., C. Zheng, and C.C. Chien, 2005, Cost-effective sampling network design for contaminant plume monitoring under general hydrogeological conditions. *Journal of Contaminant Hydrology* 77, no. 1–2:41–65.

Wu, J., and X. Zeng, 2013, Review of the uncertainty analysis of groundwater numerical simulation. *Chinese Science Bulletin* 58, no. 25:3044–3052.

Yan, S., and B. Minsker, 2006, Optimal groundwater remediation design using an adaptive neural network genetic algorithm. *Water Resources Management* 42, no. 5 (W05407):1–14.

Yoon, H., Y. Hyun, and K.-K. Lee, 2007, Forecasting solute breakthrough curves through the unsaturated zone using artificial neural networks. *Journal of Hydrology* 335, no. 1:68–77.

Zhang, Y. -K., and R. Heathcote, 2001, *Final report - Guidelines for Numerical Modeling in Tier 3 Assessments and Other Corrective Actions*. July. Available at: www.iowadnr.gov/portals/idnr/uploads/ust/tier3nummod1.pdf.

Zheng, C., and P.P. Wang, 2002, A field demonstration of the simulation optimization approach for remediation system design. *Groundwater* 40, no. 3:258–265.

4 Field Investigation Methods

The following section provides guidance on which sampling methods are most effective for a multitude of sampling objectives. When performing Step 7 of the Data Quality Objectives (DQO) process (Chapter 3, Section 3.2.5.7), one should take advantage of guidance provided in this chapter to ensure that one is selecting the most appropriate sampling method to collect the data needed to resolve each of the principal study questions (Chapter 3, Section 3.2.5.2.2).

Field sampling methods are generally categorized into two major groups: nonintrusive and intrusive methods. As the name implies, nonintrusive methods do not require physical penetration of the ground surface. Examples of both of these methods are as follows:

Nonintrusive methods:

- Aerial photography
- Surface geophysical surveying
- Airborne gamma spectrometry radiological surveys
- Global Positioning Environmental Radiological Surveyor (GPERS-II)
- In situ gamma spectroscopy
- RadScan 800 surveying
- Laser-Assisted Ranging and Data System (LARADS) surveying
- Air sampling supporting environmental investigations and worker safety

Intrusive methods:

- Soil gas surveying
- Downhole screening methods
- Shallow and deep soil sampling
- Sediment sampling
- Surface water and liquid waste sampling
- Groundwater sampling
- Tracer testing
- Slug testing
- Drum and waste container sampling
- Building material sampling
- Pipe surveying
- Remote surveying

DOI: 10.1201/9781003284000-4

Because most sampling methods work effectively under unique environmental or sampling conditions, it is important to take these into consideration when selecting the most appropriate method for the job. For example, because ground-penetrating radar surveying works most effectively in sandy soil with low moisture content, one should consider an alternative geophysical or other method when these conditions are not present. Some of the more critical factors that should be considered when selecting the optimum sampling method include:

- Contaminants of concern
- Analyses to be performed on samples
- Type of sample being collected (grab, composite, or integrated)
- Sampling/surveying depth
- Clay content of the soil
- Moisture content of the soil
- Approximate depth to groundwater
- Aquifer characteristics

The following sections discuss many of the advantages and limitations of a multitude of sampling methods, and present standard operating procedures for many of those methods that do not require specialized academic training.

4.1 NONINTRUSIVE METHODS

For nonstatistical sampling designs, before collecting samples of environmental media for on-site or standard laboratory analysis, serious consideration should be given to performing one or more nonintrusive characterization surveys to assist the optimum positioning of the intrusive sampling points. Because nonintrusive characterization surveys are typically relatively inexpensive when compared to the cost of obtaining the same data through the collection and laboratory analysis of individual environmental samples, they can often save project cost in the long run. Some of the more effective nonintrusive soil characterization methods that should be considered include:

- Aerial photography
- Surface geophysical surveying
- Airborne gamma spectrometry
- GPERS-II
- In situ gamma spectroscopy

Some of the more effective nonintrusive building characterization methods that should be considered for radioactive sites include:

- RadScan 800
- Laser-Assisted Ranging and Data System
- In Situ Object Counting System

Air sampling is another nonintrusive method that can provide valuable data to help determine if a site is posing an immediate risk to human health, worker health, or the environment.

4.1.1 Aerial Photography

Through careful evaluation of historical or recent aerial photographs, it is often possible to identify the boundaries of fill areas, potential surface contaminant migration pathways and, in some cases, contaminant source areas. Fill area boundaries can sometimes be defined by identifying unnatural changes in topographic relief, or by identifying unnatural changes in vegetation patterns.

When both historical and recent aerial photographs are available, a comparison of the surface topography between the two sets of photographs will quickly reveal suspect areas. When looking for changes in vegetation types, color photographs are much more effective than black-and-white photographs.

Aerial photographs can be used to identify surface water drainages through which contaminants near the ground surface may be transported from the site. Identifying these pathways is critical to understanding the risk that a site poses to the surrounding environment, because they are routes for rapid contaminant migration. Historical aerial photographs may identify surface water drainages that are no longer present at the site. Because these historical drainages may have been pathways for contaminant migration in the past, one should consider collecting environmental samples from them as well as current drainages.

Source areas of contamination can sometimes be identified with the assistance of these photographs because vegetation near these areas often shows signs of stress. The effectiveness of aerial photography as a site characterization tool will be dependent on the quality and scale of the available photographs.

4.1.2 Surface Geophysical Surveying

Surface geophysical surveying methods can often provide useful data to assist in locating the presence of buried objects (e.g., drums, tanks, and pipes), defining the vertical and horizontal boundaries of fill areas, identifying the presence and continuity of low-permeability confining layers, highlighting preferential flow paths for contaminant migration, defining the boundaries of a contaminant plume, etc. The most appropriate and most effective geophysical survey methods for a particular site will be dependent on a number of factors including but not limited to:

- Objective of the survey
- Chemical composition of the soil and bedrock at the site
- Soil moisture content
- Soil clay content
- Electrical conductivity of the soil
- Depth of penetration required
- Types of surrounding interferences (e.g., electrical lines, underground piping, or tanks)

Some of the surface geophysical survey methods that have proved themselves to be the most effective in supporting the environmental industry include:

- Ground-penetrating radar
- Electromagnetic induction
- Magnetics
- Electrical resistivity
- Seismic surveying

These survey methods are described in the following subsections.

4.1.2.1 Ground-Penetrating Radar

Ground-penetrating radar radiates short pulses of high-frequency electromagnetic energy into the ground from a transmitting antenna to produce an underground cross-sectional image of the soil and subsurface features. Ground-penetrating radar utilizes a radar source and antenna, with the antenna being pulled along the ground by hand, mounted on a pushcart (Figure 4.1), or pulled behind a vehicle. The transmitted energy is reflected from various buried objects or earth layers. Figure 4.2 shows what buried pipes and tanks look like in a ground-penetrating radar survey. The antenna receives the reflected waves and stores them in a digital control unit for processing and display (GeoVision 2007).

The depth of penetration when using ground-penetrating radar is limited by the electrical conductivity of the ground and the transmitting frequency. As conductivity increases, the penetration depth decreases. This is because the electromagnetic energy is more quickly dissipated into heat energy, causing a loss in signal strength at depth. Higher frequencies do not penetrate as far as lower frequencies, but give better resolution. Optimal depth penetration is achieved in dry sandy soils or massive dry materials such as granite, limestone, and concrete, where the depth of penetration is up to 15 m. In moist or clay-rich soils and soils with high electrical conductivity, penetration is sometimes only a few centimeters (GeoVision 2007).

FIGURE 4.1 Ground-penetrating radar geophysical survey method.

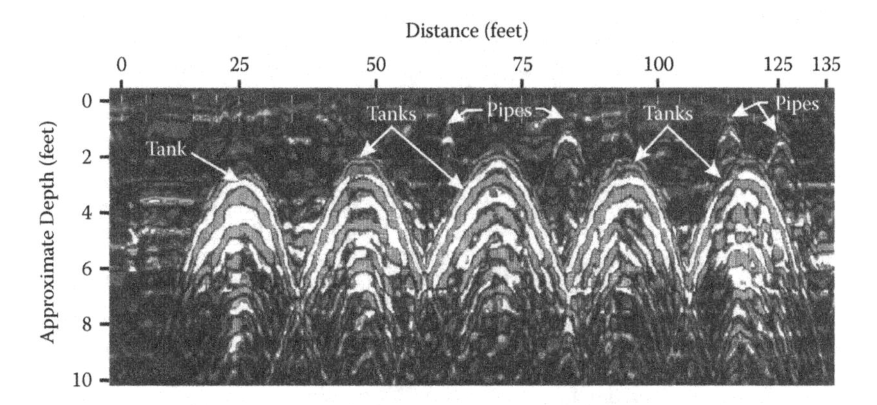

FIGURE 4.2 Example of ground-penetrating radar survey data plot.

Ground-penetrating radar surveys are typically conducted to:

- Locate metallic and nonmetallic pipes, utility cables, and underground storage tanks
- Delineate pits and trenches containing metallic and nonmetallic debris
- Define excavated and backfilled areas
- Identify shallow landfill boundaries, leach fields, and industrial cribs
- Map shallow groundwater tables
- Map shallow bedrock topography

4.1.2.2 Electromagnetic Induction

Electromagnetic induction surveys are often performed using an instrument such as the Geonics EM-31 or EM-34 terrain conductivity meters. Measurements are collected at regularly spaced stations along traverse lines. The spacing of the stations and number of lines used will vary depending on the objectives of the investigation and the size of the target one is looking for.

The EM-31 consists of a receiver coil mounted at one end and a transmitter coil mounted at the other end of a 3.7-m-long plastic boom (Figure 4.3). The EM-34 consists of a large transmitter and receiver coil connected by a reference cable. With these methods, electrical conductivity and in-phase component field strength are measured and stored in a digital data logger along with station and line numbers (GeoVision 2007).

Typical EM-31 anomalies over small, buried metallic objects consist of a negative response centered over the object and a lower-amplitude positive response to the sides of the object. When the instrument boom is oriented parallel to long linear conductors (such as pipelines), a strong positive response is observed. The EM-31 can explore to depths of about 6 m, but is most sensitive to materials about 1 m below ground surface. Single buried drums can typically be detected to depths of about 2 m. The EM-34 measures conductivity at coil separations of 10, 20, or 40 m and is used for exploration to depths of up to 60 m (GeoVision 2007).

FIGURE 4.3 Electromagnetic induction geophysical survey method. (Authorization provided by Mike Thompson [DOE-RL]).

In-phase component field strength measurements generally only respond to buried metallic objects. In contrast, conductivity measurements also respond to conductivity variations caused by changes in moisture (or salinity), soil type, and the presence of nonmetallic bulk wastes (GeoVision 2007). EM surveys are typically conducted to locate:

- Buried drums, tanks, and pipes
- Landfills, pits, and trenches containing metallic or nonmetallic debris
- Waste site boundaries
- Conductive soil and groundwater contamination
- Buried channel deposits
- Conductive fault and fracture zones

4.1.2.3 Magnetics

Magnetics is a geophysical survey method (Figure 4.4) that generally involves the measurement of the vertical gradient of the earth's magnetic field or the earth's magnetic field intensity. Anomalies in the earth's magnetic field are caused by induced or remanent magnetism. Induced magnetic anomalies are the result of secondary magnetization induced in a ferrous body by the earth's magnetic field (GeoVision 2007). The shape and amplitude of an induced magnetic anomaly is a function of the:

- Intensity and inclination of the earth's magnetic field at the survey location
- Orientation, geometry, size, depth, and magnetic susceptibility of the target

FIGURE 4.4 Magnetics geophysical survey method.

The magnetic geophysical survey method is an effective way to search for small metallic objects, such as buried ordnance and drums, because magnetic anomalies have spatial dimensions much larger than those of the objects themselves. Magnetic data are typically acquired in a grid pattern, with results being presented as color-enhanced contour maps. The approximate location and depth of magnetic objects can be calculated using geophysical survey support software (GeoVision 2007).

Typically, a single buried drum can be detected to a depth of about 10 ft. Larger metallic objects can often be located to greater depths. Induced magnetic anomalies over buried objects such as drums, pipes, tanks, and buried metallic debris generally exhibit an asymmetrical, south up/north down signature (positive response south of the object and negative response to the north; GeoVision 2007).

Magnetic surveys are typically conducted to locate:

- Buried drums, tanks, and pipes
- Landfills, pits, and trenches containing metallic debris
- Abandoned steel well casings
- Buried unexploded ordnance (UXO)
- New drilling locations

4.1.2.4 Electrical Resistivity

The electrical resistivity method (Figures 4.5 and 4.6) involves the measurement of the apparent resistivity of soil and rock as a function of depth or position. The resistivity of soils is a complicated function of porosity, permeability, ionic content of the pore fluids, and clay mineralization. The most common electrical resistivity methods

FIGURE 4.5 Laying electrical resistivity line.

FIGURE 4.6 Electrical resistivity instruments.

used in environmental investigations are vertical electrical soundings (resistivity soundings) and resistivity profiling (GeoVision 2007).

Resistivity surveys are performed by injecting a current into the ground through a pair of current electrodes, followed by measuring the potential difference between a pair of potential electrodes. The current and potential electrodes are typically arranged in a linear array. The apparent resistivity is the bulk average resistivity of all soils and rock influencing the current (GeoVision 2007).

In a resistivity sounding, the distance between the current electrodes and the potential electrodes is systematically increased, which yields information on subsurface resistivity from successively greater depths. The variation in resistivity with depth is modeled using forward and inverse modeling computer software.

Resistivity sounding surveys are typically conducted to:

- Define the extent of groundwater or soil contamination plumes
- Define the boundaries of landfill deposits
- Identify major subsurface geologic structures (e.g., paleochannels)
- Identify lithologic contacts between geologic units
- Determine the depth to groundwater and depth to bedrock

In resistivity profiling, the electrode spacing is fixed, and measurements are taken at successive intervals along a profile. Data are generally presented as profiles or contour maps and interpreted qualitatively (GeoVision 2007). Resistivity profiling is often used to

- Delineate the boundaries of disposal areas
- Define the lateral extent of conductive contaminant plumes
- Map heavy metal soil contamination
- Locate paleochannels
- Locate voids and karsts

4.1.2.5 Seismic Surveying

Seismic surveys are performed by inputting acoustic energy into the subsurface by an energy source (e.g., weight being dropped, air gun, explosive charge). The velocity at which acoustic waves propagate into the subsurface is dependent on the elastic properties of the material through which they travel. When the waves reach an interface where the density or velocity changes significantly, a portion of the energy is reflected back to the surface, and the remainder is transmitted into the lower layer. A portion of the energy is also critically refracted along the interface when the velocity of the lower layer is higher than that of the upper layer. Critically refracted waves travel along the interface at the velocity of the lower layer and continually refract energy back to the surface (GeoVision 2007). Seismic survey is effective in providing valuable subsurface geologic data hundreds to thousands of feet below the ground surface.

The incoming refracted and reflected waves are recorded by receivers (geophones) that are laid out in a linear array on the ground surface. The seismic refraction method involves analysis of the travel times of the first energy to arrive at the geophones. These first arrivals are from either the direct wave (at geophones close to the source) or critically refracted waves (at geophones further from the source). The seismic reflection method involves the analysis of reflected waves, which occur later in the seismic record (GeoVision 2007).

For environmental investigations, seismic refraction surveys are typically performed to:

- Estimate the depth to groundwater
- Map bedrock topography
- Map faults in bedrock
- Evaluate general bedrock properties

For environmental investigations, seismic reflection surveys are typically performed to:

- Map subsurface stratigraphy
- Define lateral continuity of geologic layers
- Identify the presence of buried paleochannels
- Map faults in sedimentary layers

4.1.2.6 Common Geophysical Techniques Used to Identify Unexploded Ordnance

Unexploded ordnance (UXO) potentially contaminate millions of acres of active military training ranges, as well as formerly used defense sites already in public hands. Even though UXO are unlikely to spontaneously detonate if disturbed (unlike land mines), one still needs to be very careful when performing remediation activities.

UXO may need to be detonated in place: Nearly all UXO are metal-cased, resulting in very high rates of detection with metal detectors. This is a key difference with land mine detection, where many land mines are made of plastic and are very difficult to detect. The two types of metal detectors used by the remediation industry include magnetometers and pulsed induction sensors. Magnetometers are passive sensors that detect ferrous objects (iron-bearing metal that a magnet will stick to). Inexpensive handheld magnetometers sold by the company Schonstedt as pipe locators are commonly used for human-portable UXO detection. Pulsed induction sensors are active sensors that create a current in any subsurface metal and sense the decay of that current. The handheld coin detectors frequently used by hobbyists for beach treasure hunting are pulsed induction sensors. Pulsed induction sensors outperform magnetometers for detection of small projectiles that have little or no steel in them, but magnetometers can outperform pulsed induction sensors for the detection of deep objects. Of the two sensors, magnetometers are more affected by iron-bearing soils that are a source of noise.

4.1.2.6.1 Techniques for Performing UXO Field Surveys

The following are common techniques that are used to perform UXO surveys:

"Mag and Flag": For many years, mag and flag (sometimes called "mag and dig" or "analog and dig") was the only technique used for UXO detection. In this technique, the survey, detection, and marking of potential UXO locations all occur in real time. Using this method an area is first divided into a grid. A surveyor then walks the grid while swinging a magnetometer back and forth. When the sensor makes an audible sound consistent with a signal from a buried UXO, the surveyor places a flag in the ground where the signal is strongest, indicating that a follow-up team should dig at this location. A highly skilled surveyor may be able to judge the different audible signals from whole UXO as opposed to ordnance fragments. However, often mag and flag surveys result in a large number of excavations that turn out to be fragments. Regardless of the skill of the surveyor, an inherent downside of mag and flag is that often no sensor data are recorded and therefore there is no record of where the surveyor has walked.

Integrating sensors and global positioning system (GPS): A more recent technique called Digital Geophysical Mapping separates the survey, the detection, and the marking into three separate processes. The survey employs larger pulsed induction sensors or more sensitive magnetometers than mag and flag, and includes a positioning system such as GPS and a recording computer. Both the sensor data and the position data are recorded during the survey. During processing, these data are postprocessed to generate a map that shows any metallic "hot spots." A preset detection threshold can then be used, enabling the location of all anomalies above this threshold to be quickly calculated. A second team can then return to the field with a list of these GPS locations to excavate.

In order to support the more powerful sensors and the GPS and the recording computer used for Digital Geophysical Mapping, the equipment is typically mounted on wheels and pushed by an operator. Arrays of sensors can be ganged together to create a wider sensor swath, which increases productivity. Because of the increased weight, these arrays rapidly become difficult for an operator to push, so they are towed by a small vehicle such as an all-terrain vehicle. This eliminates operator fatigue, which dramatically increases productivity but also restricts the topography the system can traverse. For treed or mountainous terrain, mag and flag may be the only practical choice.

Research to reliably discriminate UXO from clutter is steadily progressing. The most commonly employed technical approach is to carefully collect data with a pulsed induction sensor. This allows researchers to invert the data, extract the polarizability tensor, and get an estimate of the symmetry of the object. If the object has one clear semimajor axis and two clear semiminor axes, then it is oblong and likely to be an intact piece of ordnance as opposed to asymmetric scrap (which may be safely left in the ground). Researchers have recently shown that, using carefully collected data over an ordnance site known to contain a single ordnance type, they can discriminate whole intact UXO from clutter. However, the larger goal of rejecting all clutter on an arbitrary site with multiple UXO items remains very difficult to achieve.

4.1.3 Airborne Gamma Spectrometry Radiological Surveying

Airborne gamma spectrometry radiological surveys are typically performed as an initial screening method to support the radiological characterization of large areas impacted by fallout resulting from nuclear accidents (e.g., Chernobyl, Three Mile Island), nuclear weapons testing, or very large environmental sites (e.g., Hanford Nuclear Reservation). Airborne gamma spectrometry surveys may also be performed periodically during the remediation of radioactive sites to track progress, or following remedial activities to ensure no hot spots have been overlooked. These surveys are performed using either a helicopter or winged aircraft equipped with a multichannel spectrometer system and gamma ray sensors for measuring both terrestrial and atmospheric radiation.

Three naturally occurring radioactive elements that emit gamma radiation of sufficient intensity to be measured by airborne gamma ray spectrometry include potassium-40 (K-40), uranium-238 (U-238), and thorium-232 (Th-232). Cesium-137 (Cs-137) and several other anthropogenic isotopes also emit gamma radiation of sufficient intensity to be detected by the airborne gamma spectrometry. Cs-137, K-40, U-238, and Th-232 emit gamma rays at an energy level of 662, 1460, 1765, and 2614 keV, respectively.

The variations in the types of ground cover at the site should be carefully evaluated before interpreting the results from the airborne gamma spectrometry surveys because concrete, asphalt, building structure, and other types of ground cover will shield the activity of the underlying soil. The attenuation of gamma rays in most materials is proportional to the electron density of the material. For this reason, the absorption of gamma rays both in the ground and by the mass of the air between the surface and the aircraft must be taken into account (IAEA, 1991).

4.1.3.1 Instrumentation

Sodium iodide and some other crystalline substances give off scintillations (flashes of light) when struck by alpha or beta particles, or gamma rays. These scintillations are counted by a photomultiplier, thus giving an indication of the intensity of radiation in the vicinity. Gamma ray sensors are designed primarily for use in multichannel spectrometer systems. Each gamma ray sensor contains multiple detectors, each of which is composed of a rectangular, thallium-activated sodium iodide crystal (up to 4 L in size). A photomultiplier tube is attached to the end of each detector. The detectors are mounted on a shock-absorbing chassis, and enclosed in a well-insulated protective aluminum container (Figures 4.7 and 4.8).

Each second, the pulse results are examined and output as digital values of counts per second. As a general rule, one should use the largest volume of sodium iodide crystals as practical when performing a survey: 17 or 33 L for helicopter surveys and 33 or 50 L for fixed-wing surveys.

FIGURE 4.7 Sodium iodide detectors mounted on protective aluminum containers.

FIGURE 4.8 Aircraft often used to support airborne gamma spectrometry radiation surveys.

4.1.3.2 Flight Line Direction and Speed

For study areas where little or no information is available regarding distribution patterns of radiological contaminants, airborne gamma ray spectroscopy surveys are typically flown in either a north-south or east-west direction. On the other hand, if the results from historical investigations show radioactive contaminants to be distributed in an elongated pattern (i.e., fallout elongated along dominant wind direction at the time of release), consideration should be given to flying the airborne surveys parallel to the long axis of the elongation. This approach will minimize the total number of flight line passes needed to complete the survey, and will minimize the chances of missing hot spots.

While helicopter surveys are typically flown at speeds ranging from 30 to 60 mph, fixed-wing aircraft surveys are typically flown at speeds ranging from 80 to 120 mph. Because instrument readings are collected at a standard time interval (e.g., one reading per second), slower airspeeds are preferred as they reduce the horizontal distance between measurement positions.

When designing a sampling program that includes airborne radiological surveys, the preferred flight direction, flying speed, flying height, and horizontal distance between flying lines should be clearly defined in the Sampling and Analysis Plan (see Section 3.2.7) and should be based on the radiological target size, shape, and orientation.

4.1.3.3 Flying Heights and Spacing

In order for an airborne gamma ray survey to provide reliable results, it is essential that the airborne survey is performed at a constant height. Because gamma rays are attenuated by air in an exponential fashion, lower flying heights provide more reliable survey results as anomalies can more easily be distinguished from background. For terrains that have very little topographic relief, flying heights as low as 100 to 150 ft are used. Lower flying heights and slower speeds are particularly desirable when mapping sites where contamination is suspected to be only slightly elevated above background levels. However, it is important to recognize that lower flying heights have significantly greater safety risks, particularly in areas that have significant topographic relief. Generally, flying heights of 300 to 400 ft are used for mapping nuclear fallout (IAEA, 1991).

It is important to select a spacing between flight lines that will provide the data needed to address the decision statements identified in Step 2 of the DQO process (Chapter 3, Section 3.2.5.2.4). The selected spacing should take into consideration the size and shape (i.e., circular, elliptical) of the target one is looking for, the flying height, and the potential consequences (e.g., low, moderate, severe) of missing the target. If the potential consequences of missing the target are severe, the selected spacing between flight lines should be tight. On the other hand, if the potential consequences of missing the target are minor, a wider spacing may be more appropriate. When mapping a very large area that is known to be contaminated (e.g., nuclear fallout), one should consider initially using a wide line survey to identify the general areas where higher-level activity is concentrated. Then, a tight line survey would follow to define the boundaries of the high-activity areas more clearly.

IAEA (1991) has published additional details on implementing an airborne gamma spectrometry survey.

4.1.4 Global Positioning Environmental Radiological Surveyor

The Global Positioning Environmental Radiological Surveyor (GPERS-II) system was developed by Eberline Services to combine a GPS with standard radiation survey instrumentation. As shown in Figures 4.9 and 4.10, GPERS-II is most often constructed in a backpack configuration in which the backpack carries the GPS unit, power supply

FIGURE 4.9 Performing field survey using GPERS-II backpack unit.

Courtesy of Eberline Services (Mike Dillon).

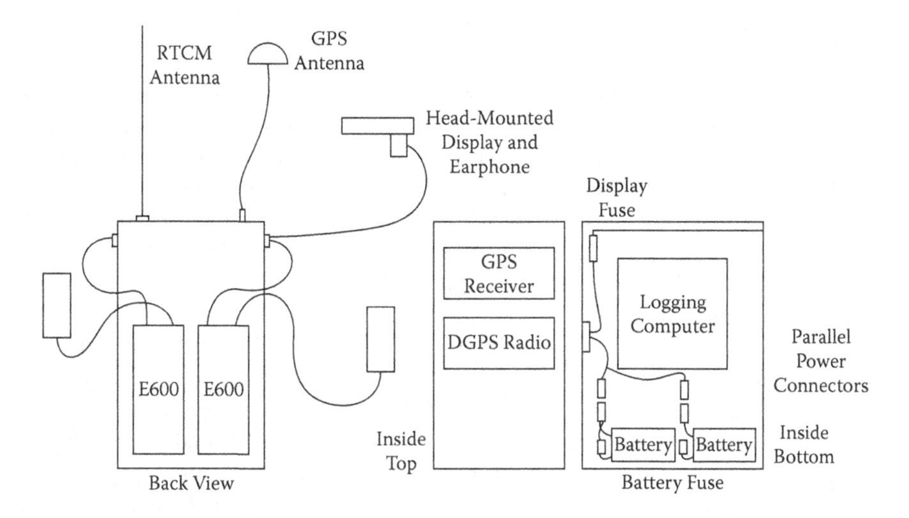

FIGURE 4.10 Components of GPERS-II backpack unit.

(From Byrnes, 2001. *Sampling and Surveying Radiological Environments.* Lewis Publishers, Boca Raton, FL.)

(batteries), and a small logging computer that records location coordinates every few seconds and ties these coordinates to gross radiation measurements. As many as two radiation survey detectors can be attached to the GPERS-II unit at the same time, and data from these detectors can be collected simultaneously. This eliminates the need

FIGURE 4.11 GPERS-II unit attached to remotely operated vehicle.

Courtesy of Eberline Services (Mike Dillon).

to perform two separate surveys with two separate detectors. When two detectors are used, one most often will use two separate gamma detectors measuring two separate energy ranges (e.g., low- and high-range gamma). The technician carrying the GPERS-II backpack typically walks across the study area along tightly spaced grid lines while swinging the survey detector from side-to-side. A small head-mounted display screen allows the technician to see the data being collected in real time overtop of a map of the study area. An earphone also allows the technician to hear the data as they are being collected.

For highly radioactive environments the GPERS-II unit can be mounted on a robotic or remotely operated vehicle (RADCART) such as that shown in Figure 4.11. One of the primary advantages of the GPERS-II system is that it is very versatile, in that it can be mounted on a motorized vehicle, cart, or wagon, in addition to being carried as a backpack unit. It also has the advantage of being able to receive input from two different detectors simultaneously. One of the limitations of the GPERS-II unit is that the performance of the system is limited by the 1-m accuracy of the positioning system. A second limitation is that the radio signals are unable to penetrate building walls, and are impacted by heavily forested areas, or steep canyon walls. For further details on the capabilities of this system, see www.eberlineservices.com.

4.1.5 In Situ Gamma Spectroscopy Surveying

Recent advances in computer technology have made it possible to develop laboratory-grade portable gamma spectroscopy systems that can be effectively utilized in the field to support the surveying of radioactive sites. An in situ gamma spectroscopy system consists of a high-purity germanium (HPGe) (or sodium iodide) detector, a tripod for holding the detector at the optimum height above the ground surface, a notebook computer with a multichannel analyzer board, and gamma spectra analysis software.

The detector is typically positioned 3 ft above the ground surface using a tripod. At this height an unshielded detector will collect data from the area on the ground surface represented by a circle around the detector that has a radius of approximately 33 ft (66 ft diameter). The depth below ground surface that the detector is reading is approximately 1 ft immediately below the detector and reduces to 0 ft at the radius of 33 ft.

To use this method, a grid is first established across the site to be investigated. The grid spacing selected should take into consideration the understanding that the detector will be collecting data from 33 ft on each side of each grid node. The detector is positioned at the first sampling point to "count" long enough to meet the detection limit requirements specified in DQO Step 3 (Chapter 3, Section 3.2.5.3.4). Count times of an hour or two are not uncommon. The detector is then moved to the next sampling point.

The advantage of using in situ gamma spectroscopy is that it can provide quantitative soil concentration results without requiring the physical collection of a soil sample that is sent to the laboratory for analysis. As a result, this method can be used to reduce the need for soil sampling and standard laboratory analyses. A sodium iodide detector may also be used to collect in situ gamma spectroscopy measurements, as opposed to an HPGe detector, if there are only a few isotopes of concern that are present in relatively high concentrations. This should be taken into consideration because an HPGe detector is much more expensive than a sodium iodide detector.

The Canberra In Situ Object Counting System (Figure 4.12) is a type of in situ gamma spectroscopy instrument.

4.1.6 Nonintrusive Building Characterization Methods

Two of the more effective nonintrusive building characterization methods that should be considered for radioactive sites include the RadScan 800 and the Laser-Assisted Ranging and Data System (LARADS). These two techniques are commonly used to help support building decontamination and decommissioning activities and are designed to identify radioactive hotspots. Once a hot spot has been identified, it can either be decontaminated in place, or it can be cut out and separated from the nonradioactive building materials.

4.1.6.1 RadScan 800 Surveying

The RadScan 800 Gamma Scanner (Figure 4.13), made by Cavendish Nuclear, is designed to pinpoint the origin of measured gamma radiation by overlaying radioactive data on a video image. The resulting plot shows rings of color (representing different ranges of gamma activity levels) superimposed on a still video image of the scene under investigation. Figure 4.14 shows an example of two radiological hotspots that were found on a plastic container using the RadScan 800.

FIGURE 4.12 Canberra In Situ Object Counting System.

(From Byrnes, 2001. *Sampling and Surveying Radiological Environments.* Lewis Publishers, Boca Raton, FL.)

FIGURE 4.13 RadScan 800 gamma scanner.

FIGURE 4.14 Radiological hotspots identified by RadScan 800.

The RadScan 800 is particularly useful in supporting building decontamination and decommissioning activities associated with nuclear reactors or other facilities associated with the nuclear industry. The RadScan 800 can be mounted on either a height-adjustable tripod or attached to a remotely operated vehicle. In an effort to minimize worker exposure, this instrument can be operated from a distance as great as 100 m from the study area, although it is most commonly operated at distances less than 10 m. The RadScan 800 can provide a panoramic view of an entire room in a single image, with all of the contamination areas highlighted. More detailed images can then be produced from those areas where contamination is most concentrated. A single object can often be imaged in just a few minutes, and an entire plant can often be imaged in a few days.

This instrument is often used to assist health physicist professionals perform dose rate surveys before allowing personnel into an area suspected of being highly contaminated. Often, images are collected using the RadScan 800 multiple times throughout a decommissioning effort to document progress and to highlight areas needing additional decontamination work. This instrument can also be used to help minimize the cost of decommissioning activities by assisting in the separation of waste material that needs to be disposed of in a radiological landfill from that which can be disposed of in a municipal landfill.

For further details on the capabilities of this monitor, refer to www.cavendishnucl ear.com.

4.1.6.2 Laser-Assisted Ranging and Data System Surveying

LARADS is a radiation screening method developed by Eberline Services that is very effective in supporting building radiological characterization and decontamination/

FIGURE 4.15 LARADS used to support radiological floor scanning surveys.

(From Byrnes, 2001. *Sampling and Surveying Radiological Environments*. Lewis Publishers, Boca Raton, FL.)

decommissioning activities. This method is effective for screening of radiological contamination on interior and exterior building surfaces.

LARADS is composed of a tracking system, radio or wired data communication device, a portable computer that can be operated on site or from a remote location, and customized software. LARADS is capable of interfacing with many different kinds of radiological detectors for assisting in defining contaminant distribution, dose rates, etc. LARADS, similar to GPERS-II (see Section 4.1.4), is capable of logging data from two radiological instruments simultaneously.

The activity measurements, and x, y, and z positional coordinates, collected using LARADS are electronically recorded. These data can be superimposed on computer-aided design drawings or digital photographs of the walls and floor. The location of each reading is recorded using a laser-assisted mapping system (similar to that utilized by civil land surveyors for mapping geographical locations). Figure 4.15 shows how LARADS can be used to support facility decontamination and decommissioning operations.

4.1.6.3 In Situ Object Counting System

In a decontamination and decommissioning (D&D) project of a radioactive building, one needs a gamma-ray identification and assay system to identify radioactive isotopes and to qualitatively determine the amount of radioactive material that is present. This information is used for planning, for monitoring the effectiveness of a decontamination process, and for determining the subsequent disposition of material generated during the D&D process. Canberra, Inc. developed the In Situ Object Counting System (ISOCS) (see Figure 4.12) which is a portable, in-situ Germanium-based spectroscopy system that is designed to provide information on the type and amount of radioactive material that is present. This system is able to provide quantitative information in real time which reduces costly delays from offsite laboratory analyses.

The Canberra ISOCS consists of an ISOCS characterized Germanium detector with portable cryostat; a cart support for holding the detector, lead shielding and collimators; a portable spectroscopy analyzer; a portable computer; and calibration software. The ISOCS uses a Germanium detector. Steel-jacketed lead shielding can be mounted around the Germanium detector to provide 1 or 2 in. of shielding from background radiation, and to change the field of view between 30, 90, or 180°. The detector rotates on the cart for alignment with the target. The computer controls the analyzer and the software provides peak identification as well as data and error analysis.

Argonne National Laboratory and the U.S. Army Corps of Engineers performed a test of this instrument that is presented in the following document:

- DOE, 1999, In-Situ Object Counting System, Deactivation and Decommissioning Focus Area, DOE/EM-0477, September.
  This study concluded that the Canberra ISOCS system performed well during the demonstration by successfully obtaining data over a wide range of objects and surfaces. The conclusions from the study were that:
- There were no problems identified with the system during the three-day test.
- The high-resolution Germanium detector and spectroscopy system were found to be easy to use and the associated databases provided useful information on radionuclide gamma peak identification.
- The use of the ISOCS system to assay concrete or soil samples was determined to be considerably more efficient compared to collecting individual samples and having them analyzed by an off-site laboratory.
- The ISOCS system could be used to analyze core samples directly.
- Some training is required to operate the system and the use of the assaying software required considerable experience in modeling the source distribution.

4.1.7 Air Sampling Supporting Environmental Investigations

While air sampling is outside of the scope of this book, this section provides some key references to air sampling guidance manuals and procedures. Air sampling may need to be performed at an environmental site to determine if a site poses any health risk to the surrounding population. The data quality objectives process (see Section 3.2.5) and sampling and analysis plan (see Section 3.2.6) supporting the environmental investigation will specify the purpose of the air sampling, the number and type of air samples to be collected, the sampling devices to be used, the air sampling locations, and the analyses to be performed on the air samples. The sampling and analysis plan will also specify the quality control requirements, and will either provide enough detail to support the implementation of the air sampling or will cite EPA (or other) standard operating procedures.

Some of the more common types of air samplers include:

- *High-volume air samplers*: Samples collected using this method are typically analyzed for total suspended particulates, particulate matter with an aerodynamic diameter of <10μm (PM10) or <2.5 μm (PM2.5), semi-volatile organic compounds, pesticides, etc.
- *Polyurethane foam high-volume air sampler*: Samples collected using this method are typically analyzed for: semi-volatile organic compounds, polychlorinated biphenols, dioxins, furans, pesticides, etc.
- *Low-volume air sampler*: Samples collected using this method may be analyzed for the same constituents as the high-volume air sampler. This method simply samples a smaller volume of air.
- *SUMMA® electropolished stainless steel canisters*: Samples collected using this method are analyzed for volatile organic compounds (Figure 4.17).
- *Tedlar Bag*: Samples collected using this method are analyzed for volatile organic compounds.
- *Silcosteel® canisters*: Samples collected using this method are analyzed for volatile organic compounds.
- *Passive air samplers*: Samples collected using this method are analyzed for volatile organic compounds and semi-volatile organic compounds.

Below are the Code of Federal Regulations and other references that provide specific air sampling and analysis requirements for some of the more common contaminants of concern, along with procedures for implementation.

- Title 40 CFR, Part 50, Appendix A – Reference Method for the Determination of Sulfur Dioxide in the Atmosphere (Pararosaniline Method)
- Title 40 CFR, Part 50, Appendix B – Reference Method for the Determination of Suspended Particulate Matter in the Atmosphere (High-Volume Method)
- Title 40 CFR, Part 50, Appendix C – Measurement Principle and Calibration Procedure for the Measurement of Carbon Monoxide in the Atmosphere (Non-Dispersive Infrared Photometry)
- Title 40 CFR, Part 50, Appendix D – Measurement Principle and Calibration Procedure for the Measurement of Ozone in the Atmosphere
- Title 40 CFR, Part 50, Appendix F – Measurement Principle and Calibration Procedure for the Measurement of Nitrogen Dioxide in the Atmosphere (Gas Phase Chemiluminescence)
- Title 40 CFR, Part 50, Appendix G – Reference Method for the Determination of Lead in Suspended Particulate Matter Collected From Ambient Air
- Title 40 CFR Part 50, Appendix J – Reference Method for the Determination of Particulate Matter as PM10 in the Atmosphere
- Title 40 CFR Part 50, Appendix L – Reference Method for the Determination of Fine Particulate Matter as PM2.5 in the Atmosphere

- EPA, 2019, Method TO-15A, Determination of Volatile Organic Compounds (VOCs) in Air Collected in Specially Prepared Canisters and Analyzed by Gas Chromatography-Mass Spectrometry (GC-MS), September
- EPA, 2016, Quality Assurance Guidance Document 2.12, Monitoring $PM_{2.5}$ in Ambient Air Using Designated Reference or Class I Equivalent Methods, EPA-454/B-16-001, January
- EPA Region 4, 2016, Ambient Air Sampling, SESDPROC-303-R5, Athens, Georgia, March
- EPA Region 1, 2011, Canister Sampling Standard Operating Procedure, ECASOP-Canister Sampling SOP5, September
- US EPA. 2002. EPA Quality Assurance Guidance Document: Method Compendium, Field Standard Operating Procedures for the PM2.5 Performance Evaluation Program, United States Environmental Protection Agency Office of Air Quality Planning and Standards, Revision No. 2, March
- EPA, 1999, Compendium of Methods for the Determination of Inorganic Compounds in Ambient Air, EPA/625/R-96/010a, June
- EPA, 1999, Compendium of Methods for the Determination of Toxic Organic Compounds in Ambient Air – Second Edition, EPA/625/R-96/010b, January
- EPA, 1999, Compendium of Methods for the Determination of Toxic Organic Compounds in Ambient Air, Compendium Method TO-4A, Determination of Pesticides and Polychlorinated Biphenyls in Ambient Air Using High Volume Polyurethane Foam (PUF) Sampling Followed by Gas Chromatographic/Multi-Detector Detection (GC/MD), Second Edition, EPA/625/R-96/010b, January
- EPA, 1999, Compendium of Methods for the Determination of Toxic Organic Compounds in Ambient Air, Compendium Method TO-13A, Determination of Polycyclic Aromatic Hydrocarbons (PAHs) in Ambient Air Using Gas Chromatography/Mass Spectrometry (GC/MS), EPA/625/R-96/010b, January
- EPA, 1999, Compendium of Methods for the Determination of Toxic Organic Compounds in Ambient Air, Method TO-9A, TO-11A, and TO-15A
- EPA, 1998, Guidance for Using Continuous Monitors in $PM_{2.5}$ Monitoring Networks, EPA/R-98-012, May
- EPA, 1984, Method (TO-1) for the Determination of Volatile Organic Compounds in Ambient Air Using Tenax® Adsorption and Gas Chromatography/Mass Spectrometry (GC/MS).

4.1.8 Air Sampling Supporting Worker Safety

Various types of air sampling methods are used by industrial hygienists to define the personal protective equipment that is required to ensure worker safety. Permissible exposure limits to various contaminants of concern are available on the following Occupational Safety and Health WEB page: www.osha.gov/annotated-pels/table-z-1. The specific analytical method that is to be used to test for each of the contaminants of concern can be found on the following National Institute for Occupational Safety and Health WEB page: www.cdc.gov/niosh/nmam/.

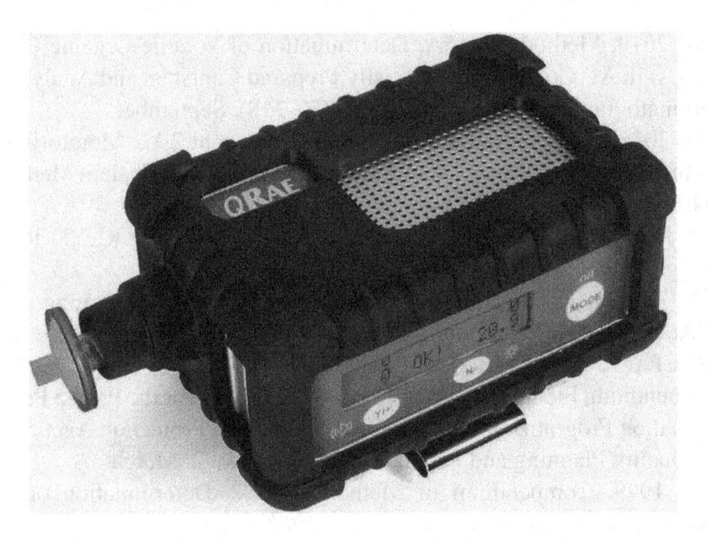

FIGURE 4.16 Example of multigas analyzer for monitoring air quality.

Air sampling can be performed in a number of different ways. One of the more common methods is to use a portable gas analyzer (Figure 4.16) that is specifically designed to detect the contaminants of concern for the site. There are a multitude of these analyzers on the market today that can screen for many contaminants of concern at the same time, a few of which include:

- EntryRAE Portable Gas Monitor with Photoionization Detector: This monitor detects oxygen, lower explosive limit, hydrogen sulfide, and carbon monoxide as well as over 300 volatile organic compounds that are found in fuels, oils, degreasers, and paints.
- MultiRAE PLUS and MultiRAE IR: These are multigas monitors with interchangeable plug-in sensors that can detect a variety of toxic and combustible gases including carbon monoxide, hydrogen sulfide, sulfur dioxide, nitric oxide, nitrogen dioxide, chlorine, hydrogen cyanide, ammonia, phosphine, and ethylene oxide.
- Q-RAE Four-Gas Monitor: This monitor detects lower explosive limit, oxygen, hydrogen sulfide, and carbon monoxide with plug-in smart sensors. It is an ideal monitor for confined-space applications.

Air sampling can also be performed by using an air sample pump to draw a specific volume of air into a sampling container (Figure 4.17; e.g., bag, canister, or tube), which is then sent to the laboratory for analysis. This method can provide a very accurate measurement of contaminant concentrations in air.

FIGURE 4.17 Summa air sampling canisters.

4.2 INTRUSIVE METHODS

Intrusive sampling methods are different from nonintrusive methods in that they require the penetration of the ground or water surface. Examples of these methods include:

- Soil gas surveying
- Downhole screening methods (e.g., Ribbon NAPL Sampler, gross gamma logging)
- Shallow and deep soil sampling
- Sediment sampling
- Surface water and liquid waste sampling
- Groundwater sampling/testing
- Drum and waste container sampling
- Building material sampling
- Pipe surveying
- Remote surveying

As discussed earlier in this chapter, serious consideration should be given to performing one or more nonintrusive surveys before the commencement of an intrusive sampling program for the purpose of more effectively selecting intrusive sampling locations. To ensure the success of a field sampling program, the following questions should be asked before the commencement of fieldwork:

- Have the proposed sampling locations been cleared by local utilities?
- Is work being performed on private property? If so, have access agreements been prepared and approved by the landowner?
- Has a secured area been established to store drums containing waste materials?
- Have the potential hazards at the site been thoroughly researched, and is a site-specific Health and Safety Plan in place?
- Is there a qualified Health and Safety Officer on-site with the appropriate monitoring equipment to ensure the health and safety of the workers?
- Have field quality assurance procedures been developed?
- Has all the sampling and monitoring equipment been properly calibrated?
- Has all sampling equipment been decontaminated?
- Has the laboratory provided the appropriate sample bottles, sample preservatives, and information on the holding time for each of the analyses to be performed?

It is important to contact local utilities before performing any intrusive activities. Most areas have a local "Digger's Hotline" telephone number (e.g., 811) where one can report the locations of proposed sampling points. The hotline representative will provide a Work Authorization Number and a date on which work can begin. If any sampling points are found to be near a utility line, a hotline representative will contact you before the work start date and require that the sampling point be moved. If a utility line is encountered during sampling, the Work Authorization Number protects the sampler from liability.

If work is to be performed on private property, it is a legal requirement to get an Access Agreement signed by the landowner before beginning work. This agreement should provide background information on the purpose of the study, information regarding the location and depth of the samples to be collected, equipment to be used, and outline the steps that will be taken to restore the site to its original condition.

If wastewater or soil will be generated from any of the sampling activity, a secured area is needed to store the waste material before final disposal. To ensure the health and safety of the field workers, it is essential that all the potential hazards at the site are thoroughly researched and a site-specific Health and Safety Plan (see Chapter 10) is developed to address each of these hazards. To further ensure the safety of the workers, a Health and Safety Officer must be present on-site with appropriate calibrated instruments to screen the working atmosphere before, and during, field activities. On the basis of results from an initial screening, the Health and Safety Officer must determine the appropriate level of protection to begin work. If conditions change in the process of sampling, the level of protection must likewise change. The Health and Safety Officer is responsible for seeing that all the workers have the appropriate training, are on an annual medical monitoring program, and follow good health and safety practices while performing fieldwork.

To ensure the effectiveness of the field sampling procedures, a Quality Assurance Project Plan must be in place at the time of sampling (Chapter 3, Section 3.2.7). The primary objective of this plan is to outline the frequency and method of collecting quality control samples (e.g., rinsate blanks, field blanks), sample documentation and laboratory analytical requirements, and equipment calibration and decontamination procedures.

Before beginning any field operations, all data-gathering instruments must be properly calibrated using standard calibration media. In addition, all sampling equipment must be decontaminated following the procedures outlined in Chapter 9. Instrument calibration and equipment decontamination are necessary to ensure the accuracy and integrity of the data being collected, and to prevent the spread of contamination from the site. After a piece of sampling equipment has been decontaminated, it should be carefully wrapped in aluminum foil, with the shiny side of the foil facing outward. This will prevent the equipment from being contaminated before the time of sample collection.

A contract with an analytical laboratory should be signed and in place before beginning sample collection. This laboratory should provide the appropriate sample bottles, sample preservatives, and information on analytical holding times. Most laboratories will provide sample bottle labels and chain-of-custody seals for sample bottles upon request.

Sampling tools used to collect different types of media samples for laboratory analysis should be constructed of non-reactive materials that will neither add nor alter the chemical or physical properties of the material that is being sampled (EPA 2020b). In other words, the sampler should not be constructed of a material that could cause a loss or gain in contaminant concentrations measured due to sorption, desorption, degradation, or corrosion (EPA 2002). Because stainless steel, Teflon®, and glass are inert substances, sampling equipment made of these materials should be preferentially selected over equipment made of other materials. USGS (2003) notes that fluorocarbon polymers (Teflon®, Kynar®, and Tefzel®), stainless steel 316-grade, and borosilicate glass (laboratory grade) are acceptable for both inorganic and organic analyses; polypropylene, polyethylene, polyvinyl chloride (PVC), silicone, and nylon are only acceptable for inorganic analyses; and stainless steel 304-grade and other metals (brass, iron, copper, aluminum, and galvanized and carbon steels) are only acceptable for organic analyses. Having a rigorous quality control sampling program (e.g., collecting rinsate blanks) in place will help ensure that the sample integrity is not being impacted by the sampling equipment (see Chapter 6).

It is recommended that field sampling should be delayed until all of the foregoing items have been thoroughly considered.

4.2.1 General Media Sampling

When collecting samples of soil, sediment, surface water, groundwater, concrete, paint, dust, or any other type of media, one must carefully select the type of "sample" to collect. The term *sample* in this case refers to the physical collection of representative material from a media for the purpose of analytical testing. Selecting the appropriate type of sample is important because it will influence the resulting analytical data. The four types of samples that are most frequently collected for environmental investigations include grab, composite, swipe, and integrated. These are discussed in the following subsections.

4.2.1.1 Sample Types

The four primary sample types when collecting samples for environmental investigations include:

- Grab samples
- Composite samples
- Swipe samples
- Integrated samples

The following subsections briefly describe each of these sample types and provide guidance on when they should be used.

4.2.1.1.1 Grab Samples

A grab sample refers to the physical collection of a media sample from a single location for analysis. When collecting a grab, the sample is transferred directly from the sampling tool into the sample jar or bottle. No mixing or compositing is performed when collecting this type of sample. Care should be taken when selecting the sampling tool, to be certain that the method of sample collection does not in itself composite the sample. For example, when collecting shallow soil grab samples, the scoop (Section 4.2.4.1.1) or core barrel (slide hammer) method (Section 4.2.4.1.3) is preferred over the hand auger method (Section 4.2.4.1.2) because the rotation of the hand auger blades composite the soil during the sample collection process. Generally, a grab sample is not collected over sampling intervals greater than 0.5 to 1.0 ft in length, or outside the perimeter of a 1-ft^2 area. Grab samples should be collected when one would like to know the range of concentration levels (minimum and maximum) for a contaminant present at a site. Grab sampling is effective in collecting samples for site characterization, waste characterization, risk assessment, feasibility study, remedial design, and postremediation confirmation sampling.

4.2.1.1.2 Composite Samples

A composite sample is collected by either taking multiple grab samples from different locations and homogenizing them together, or by collecting samples from multiple depths from the same sampling location and homogenizing these depth intervals together for analysis. Composite sampling is most frequently used to help reduce the cost of site characterization, and in some cases is used for waste characterization activities. However, it is important to recognize that the results only provide the mean concentration for the composited intervals, and will not provide a reliable estimate of the range of concentration levels present at a site. Composite sampling is less frequently used to collect data for human health or environmental risk assessment because it is important to know the full range of contaminant levels that a receptor may be exposed to. Composite sampling is generally not recommended for site-closeout sampling because it tends to "dilute" the analytical results. When collecting composite samples, it is important that the composited interval is not so large that the resulting data are diluted beyond the point of providing meaningful information.

4.2.1.1.3 *Swipe Samples*

Swipe samples are distinct from grab, composite, or integrated samples in that they are collected from the surfaces of walls, floors, ceiling, piping, ductwork, etc. Swipe samples are most often collected for the purpose of determining the amount of removable radioactivity from a surface. Once a swipe sample has been collected, it is typically analyzed for gross alpha or gross beta activity using a swipe counter. Swipe samples are most often collected to support building characterization, risk assessment, and building closeout. Section 4.2.9.1 identifies the procedure used for collecting a swipe sample.

4.2.1.1.4 *Integrated Samples*

Integrated sampling involves the collection of a sample from one location over an extended period of time (e.g., days, weeks, or months). Integrated sampling is most commonly performed when assessing surface water, groundwater, or air quality over time. It is different from composite sampling in that samples are collected from the same location over an extended period of time. For example, one integrated surface water sample may be collected from a location downstream from a chemical plant, where a small volume of water is collected every week throughout the three-month wet season. At the end of the three-month period, the integrated sample is analyzed to identify the mean concentration of contaminants in the surface water over that time period. Integrated sampling should not be used when the contaminants of concern are volatile organic compounds.

4.2.2 Soil Gas Surveying

Soil gas surveying is one of the investigative techniques that is commonly used to assist remedial investigations in which the contaminants of concern include volatile organic compounds. This method is relatively inexpensive when compared to the cost of obtaining similar data through the collection and chemical analysis of individual soil samples.

Soil gas in the form of methane occurs naturally in soils as a result of the decomposition of organic material. The concentration of this gas depends on a number of environmental factors, such as soil moisture content, buffer capacity, pH, nitrogen and phosphorus content, and the temperature of the surrounding environment. The optimum conditions for natural soil gas generation of methane are soils with high moisture content, neutral pH, high nitrogen and phosphorus content, and moderate temperatures.

In addition to methane gas, it is not uncommon for deep pockets of naturally occurring hydrocarbons to provide trace or, in some cases, stronger detections of various volatile organic compounds at the ground surface. In order to avoid misinterpreting these detections as contamination, it is necessary to collect background soil gas samples. Background locations should be selected from an undisturbed area upgradient from the study area, and should represent the same soil formation. When the background soil gas results are subtracted from the study site results, contamination will be revealed. Great care should be taken in selecting the background locations because, depending on the stratigraphy, soil gas can migrate upgradient.

The soil gas technique is most effective in mapping low-molecular-weight, halogenated solvent compounds and petroleum hydrocarbons possessing high vapor pressures and low aqueous solubilities. These compounds readily partition out of the groundwater and into the soil gas as a result of their high gas–liquid partitioning coefficients. Once in the soil gas, volatile organic compounds diffuse vertically and horizontally through the soil to the ground surface, where they dissipate into the atmosphere. Because the contaminants in the ground act as a source and the atmosphere above the ground surface acts as a sink, a concentration gradient typically develops between the two.

Some of the more effective soil gas sampling procedures include the field screening method, mobile gas chromatograph method, BESURE method, and Amplified Geochemical Imaging, LLC (AGI) sample module method. Each of these methods has its own advantages and disadvantages. The field screening method is an "active" soil gas sampling method (analytical results available almost immediately) that utilizes a portable flame ionization or photoionization detector in combination with a hand-driven stainless steel sampling rod. Using this method, the investigator is able to sample the relative concentration of organic soil vapors in the field as deep as several feet below the ground surface. The advantage of this method is that sampling results can be obtained quickly and very inexpensively. This method is less sophisticated than other soil gas methods because the instrument is not able to differentiate between organic compounds; rather, the instrument readings are a measure of relative equivalents to the gas which the instrument is calibrated to (typically, methane or pentane). Because this instrument cannot provide results below the 1 part-per-million (ppm) range, it is most commonly used as a "quick and dirty" method of identifying areas where organic contamination is most concentrated.

The mobile gas chromatograph method is another "active" soil gas sampling method (analytical results available an hour or so after sample collection) that utilizes a laboratory-quality gas chromatograph (GC) or a gas chromatograph/mass spectrometer (GC/MS) that is mounted on a mobile laboratory near the job site. In this method, soil gas sampling rods are most often hydraulically pushed into the ground to allow the collection of soil gas samples from selected sampling depths. Using this method, soil gas samples can be collected as deep as 100 ft below the ground surface. The depth of penetration is dependent on both the size and weight of the vehicle used to advance the sampling rods as well as the type of soil the rods are being pushed through. Soil gas samples are often collected using an air pump to pull a vacuum on either the soil gas sampling rod itself or a sampling tube that is lowered down the inside of the sampling rod. The soil gas sample is most often collected in a Tedlar bag before the sample is transferred to the GC or GC/MS for analysis. Another method of collecting soil gas samples involves lowering a negatively pressurized sample bottle down the inside of the soil gas sampling rod. When the bottle reaches the bottom of the sampling rod, a pin punctures the septum cap, which allows soil gas to flow into the sample bottle. The bottle is then retrieved for analysis using the GC or GC/MS.

To collect soil gas samples at depths greater than 100 ft below the ground surface using the mobile gas chromatograph method, a drill rig is needed to advance a borehole to the desired sampling depth. The mobile gas chromatograph method has the advantage of providing preliminary analytical results within an hour or so after sampling, with detection limits for specific contaminants in the parts-per-billion (ppb) range. This method supports the "Observational Approach" in which the data from the last sampling point is used to support the selection of the next sampling point. This method also provides the flexibility of collecting soil gas samples from multiple depth intervals from the same borehole. The primary disadvantage of the mobile gas chromatograph method is that sampling is restricted to areas that can be accessed by the vehicle used to reach the sampling depth; the method tends to be more expensive than other soil gas methods; and the investigator must select which analytes are to be screened for before analyzing the sample so that the instrument can be properly calibrated.

The BESURE soil gas sampling method is a "passive" method, meaning that a multitude of samplers are installed in the ground (typically in a grid pattern) for a period of time (e.g., days) before being retrieved and then analyzed. This method utilizes adsorbent samplers that are emplaced subsurface to adsorb compounds in soil gas without forcing the flow rate of soil gas (as active soil gas sampling does). The method analyzes samples using gas chromatography/mass spectroscopy (GC/MS) instrumentation and is generally able to identify with greater precision a broader range of compounds than active soil gas techniques. The analytical results for a passive soil gas method are presented in units of mass or concentration for comparison between other sampling locations from the same sampling event and risk assessment. This method has the advantage of being less expensive than the mobile gas chromatograph method for collecting the same volume of data. This method also works very effectively in environments in which contaminant levels are very low because the samplers can be left in the ground as long as needed to obtain positive results. Passive samplers that have validated uptake rates, such as the sampler provided by Beacon, allow the data to be reported as concentrations ($\mu g/m^3$). A disadvantage of the passive method is that data are not available for a week or more following sample tube installation.

The AGI sample module (previously known as the GORE Module) can be used as another "passive" soil gas sampling method. The AGI sample module is approximately 0.25 in. in diameter and 13 in. in length and consists of a tube of GORE-TEX® membrane. Housed inside the membrane tubing are several packets of hydrophobic sorbents that have an affinity for a broad range of volatile and semi-volatile organic compounds. The AGI sample modules are typically placed in a grid pattern across the study, similar to the BESURE soil gas survey method. The AGI sample modules are left in the ground for approximately 3–5 days before being retrieved for analysis. The modules are then shipped to the manufacturer's laboratory to be analyzed using a gas chromatograph/mass spectrometry. The results are then often plotted on a map and contoured. The advantages and disadvantages of this method are similar to the BESURE method described above.

4.2.2.1 Field Screening Soil Gas Surveying Method

The field screening method is an "active" soil gas sampling method (analytical results available almost immediately) that utilizes a portable flame ionization or photoionization detector in combination with a hand-driven sampling rod to measure the concentration of organic vapors several feet below the ground surface. The sampling tool is composed of a slide hammer, sampling rod, and probe (Figure 4.18). The probe has vapor holes for gas entry and a removable inner liner rod that prevents soil intrusion into the probe.

When using this method, the first sample should be collected near the suspected source areas of contamination and proceed radially outward until the boundaries of the soil gas plume have been defined (Figure 4.19). The advantage of this method is that sampling results can be obtained quickly and very inexpensively. This method is less sophisticated than other soil gas methods because the instrument is not able to differentiate between organic compounds; rather, the instrument readings are a measure of relative equivalents to the gas which the instrument is calibrated to (typically methane or pentane). Because this instrument cannot provide results below the 1-ppm range, it is most commonly used as a quick and dirty method for identifying areas where organic contamination is most concentrated.

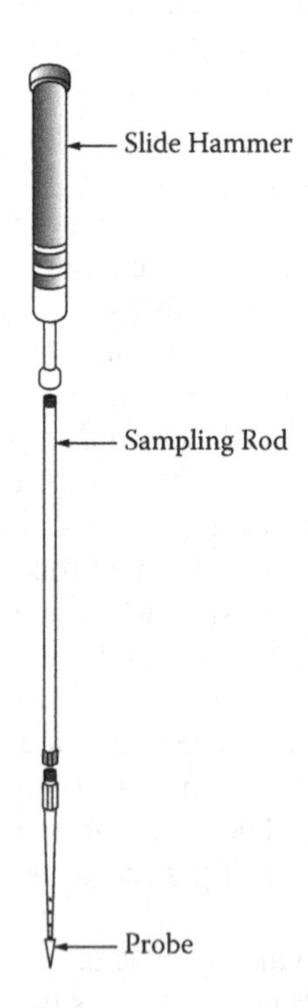

FIGURE 4.18 Sampling tool used to collect soil gas samples using field screening method.

(From Byrnes, 1994. *Field Sampling Methods for Remedial Investigations.* Lewis Publishers, Boca Raton, FL.)

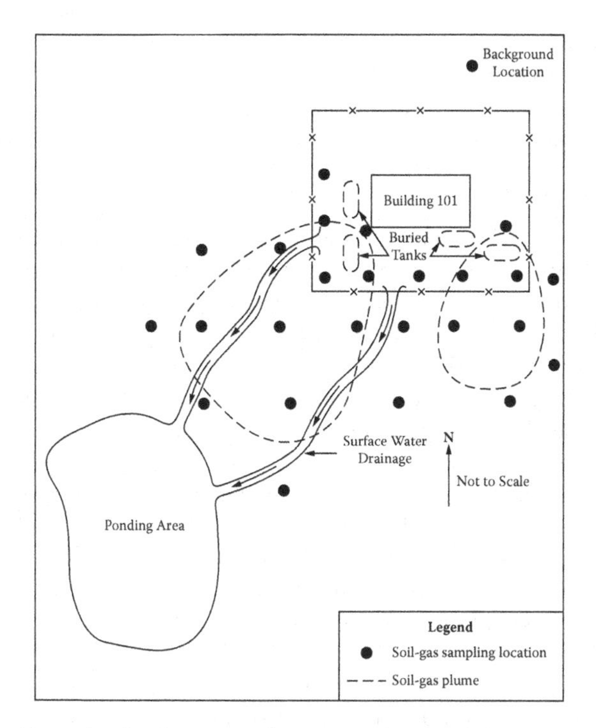

FIGURE 4.19 Example of soil gas sampling approach using field screening or mobile gas chromatograph methods.

(From Byrnes, 1994. *Field Sampling Methods for Remedial Investigations.* Lewis Publishers, Boca Raton, FL.)

If the objective of the soil gas survey is to precisely define the boundaries of an organic contaminant plume, the mobile gas chromatograph, BESURE, or AGI sample module soil gas sampling methods are more appropriate (see Sections 4.2.2.2, 4.2.2.3, and 4.2.2.4). This technique is more designed to determine whether or not there is a serious organic vapor problem at the site, and if so, generally where the source is.

For most sampling programs, four people are sufficient to implement this sampling procedure. Two are needed for driving and removing the sampling rod, collecting the soil gas measurement, and documenting the results. A third is needed for health and safety and quality control, and a fourth is needed for equipment decontamination and management of investigation-derived waste.

After laying out a sampling grid, the following equipment list and procedure are used to collect soil gas samples using this method:

1. Slide hammer
2. Sampling rod/liner and probe
3. Portable flame or photoionization detector
4. Instrument calibration gas
5. Sample logbook

6. Health and safety screening equipment
7. Health and safety clothing
8. DOT-approved 55-gal waste drum
9. Plastic sheeting

4.2.2.1.1 Sampling Procedure

1. In preparation for sampling, confirm that all necessary preparatory work has been completed, including obtaining property access agreements, meeting health and safety and equipment decontamination requirements, and checking the calibration of all health and safety and chemical and/or radiological field screening instruments.
2. After suiting up into the appropriate level of protective clothing, cut a center hole in the plastic sheeting several inches in diameter, and then lay it over the location to be sampled.
3. Using the slide hammer, beat the sampling rod to the desired sampling depth.
4. Remove the liner from the sampling rod, and quickly insert the probe of the ionization detector. Leave the probe in the sampling rod for 1 minute, and record the highest reading in a logbook.
5. Remove the sampling rod from the ground by rocking it back and forth several times before lifting upward.
6. Very little, if any, waste material is generated from this sampling procedure; however, whatever waste is generated should be containerized in a DOT-approved 55-gal drum. Before leaving the site, the waste drum should be sealed, labeled, and handled appropriately (see Chapter 11).
7. At least one point on the sampling grid should be surveyed by a professional surveyor to provide an exact sampling location. The remainder of the sampling grid can be tied into this one surveyed location.

4.2.2.2 Mobile Gas Chromatograph Soil Gas Surveying Method

The mobile gas chromatograph method is another "active" soil gas sampling method that utilizes a laboratory-quality GC or a GC/MS that is mounted on a mobile laboratory near the job site. With this method soil gas samples are analyzed in the field immediately following sample collection, and preliminary analytical results are commonly available within an hour or so after sampling. The number of analytes being screened for directly affects the analysis time and cost.

A mobile laboratory is commonly equipped with one or more laboratory-grade GCs or GC/MSs, a temperature-programmable oven, and various types of instrument detectors such as electron capture, flame ionization, photoionization, and thermoconductivity. It also contains a computer to record the instrument output, and a generator for self-contained operation. Vendors who provide this service send a chemist to operate the analytical equipment, and an environmental technician to collect the samples. The analytical equipment must be calibrated daily for each of the analytes to be screened for. The calibration is performed each morning before the collection of the first sample. The time required for calibration is dependent on the number of analytes being screened for. In general, the calibration procedure takes between 1 and 2 hours.

FIGURE 4.20 Collecting and analyzing soil gas samples using the mobile gas chromatograph method.

(From Byrnes, 1994. *Field Sampling Methods for Remedial Investigations.* Lewis Publishers, Boca Raton, FL.)

With this method, soil gas sampling rods are most often hydraulically pushed (Figure 4.20) into the ground to allow the collection of soil gas samples from selected sampling depths. A retractable tip is placed at the end of the sampling rod to prevent soil from entering the rods when it is advanced. When the tip is retracted, it allows soil gas to flow into the rods. Using this method, soil gas samples can be collected as deep as 100 ft below the ground surface. The depth of penetration is dependent on both the size and weight of the vehicle used to advance the sampling rods as well as the type of soil the rods are being pushed through. Soil gas samples are often collected using an air pump to pull a vacuum on either the soil gas sampling rod itself or a sampling tube that is lowered down the inside of the sampling rod. Three or more volumes of air are then purged from the sampling rod or tube before sample collection. The soil gas sample is most often collected in a Tedlar bag before the sample is extracted and injected to the GC or GC/MS for analysis. Another method of collecting soil gas samples involves lowering a negatively pressurized sample bottle down the inside of the soil gas sampling rod. When the bottle reaches the bottom of the sampling rod, a pin punctures the septum cap, which allows soil gas to flow into the sample bottle. The bottle is then retrieved for analysis using the GC or GC/MS.

To collect soil gas samples at depths greater than 100 ft below the ground surface using the mobile gas chromatograph method, a drill rig is needed to advance a borehole to the desired sampling depth. A sampling tube is then lowered down the inside of the drill casing to the bottom of the hole. A packer is inflated above the sampling interval to block off the upper portion of the drill casing. The method of sample collection and analysis is the same as described earlier for the hydraulic push method.

For measurement of the concentration of compounds in soil gas, samples should be collected at a flow rate not exceeding 200 mL/min so as not to disturb the equilibrium conditions for representative results. The sample can be drawn into a Summa canister (Figure 4.17) or over a tube packed with adsorbent before analysis of the sample. Refer to EPA Compendium Method TO-15 (EPA 1999a) and Method TO-17 (EPA 1999b) for specific details regarding how to properly collect air samples using a canister or packed tube, respectively.

Common laboratory quality assurance procedures, such as running laboratory blank, duplicate, and spike samples, are used to ensure the accuracy and precision of the analytical results. The soil gas detection limits are a function of the injection volume as well as the detector sensitivity for individual compounds. Generally, the larger the injection size, the greater the sensitivity. However, peaks for compounds of interest must be kept within the linear range of the detector. If any compound has a high concentration, it is necessary to use small injection volumes and, in some cases, dilute the sample to keep it within linear ranges. This may result in higher detection limits for the other compounds in the analysis.

When using this soil gas method, sampling should begin at areas known to be, or suspected to be, the most contaminated. Samples are then collected radially away from these areas until soil gas concentrations drop to background levels (Figure 4.19). Because soil gas samples can be collected at various depths, this technique can provide useful information regarding the depth of soil contamination.

A step-by-step procedure has not been provided for this method, because this operation must be performed by professionals specifically trained in operating the sampling tools and instruments. The mobile gas chromatograph method has the advantage of providing preliminary analytical results within an hour or so after sampling, with detection limits for specific contaminants in the ppb range. This method supports the "Observational Approach," which the data from the last sampling point is used to support the selection of the next sampling point. This method also provides the flexibility of collecting soil gas samples from multiple depth intervals from the same borehole. The primary disadvantage of the mobile gas chromatograph method is that sampling is restricted to areas that can be accessed by the vehicle used to reach the sampling depth; the method tends to be more expensive than other soil gas methods; and the investigator must select which analytes are to be screened for before analyzing the sample so that the instrument can be properly calibrated.

For general planning purposes when hydraulically pushing sampling rods into the ground, one can expect to collect and analyze approximately five to ten soil gas samples per day using this method. This number will be significantly less when collecting soil gas samples with the assistance of a drill rig.

4.2.2.3 BESURE Soil Gas Surveying Method

The BESURE passive soil gas method (developed by Beacon Environmental) also analyzes soil gas samples using GC/MS instrumentation and is able to identify a broader range of compounds than other soil gas techniques. The analytical results for a passive soil gas method are presented in units of mass or concentration for comparison between sample locations. The results from BESURE passive soil gas surveys are most often used to identify sources of volatile and semivolatile organic contamination, to delineate the lateral extent of volatile and semivolatile organic contamination (including migration pathways), to assist with monitoring the performance of remediation activities, and to assess vapor intrusion risks.

The BESURE passive soil gas method utilizes adsorbent samplers that are emplaced subsurface to adsorb compounds in soil gas without forcing the flow rate of soil gas (as active soil gas sampling does). Samplers are typically placed in a grid pattern (Figure 4.21) across the study area. The grid spacing used should

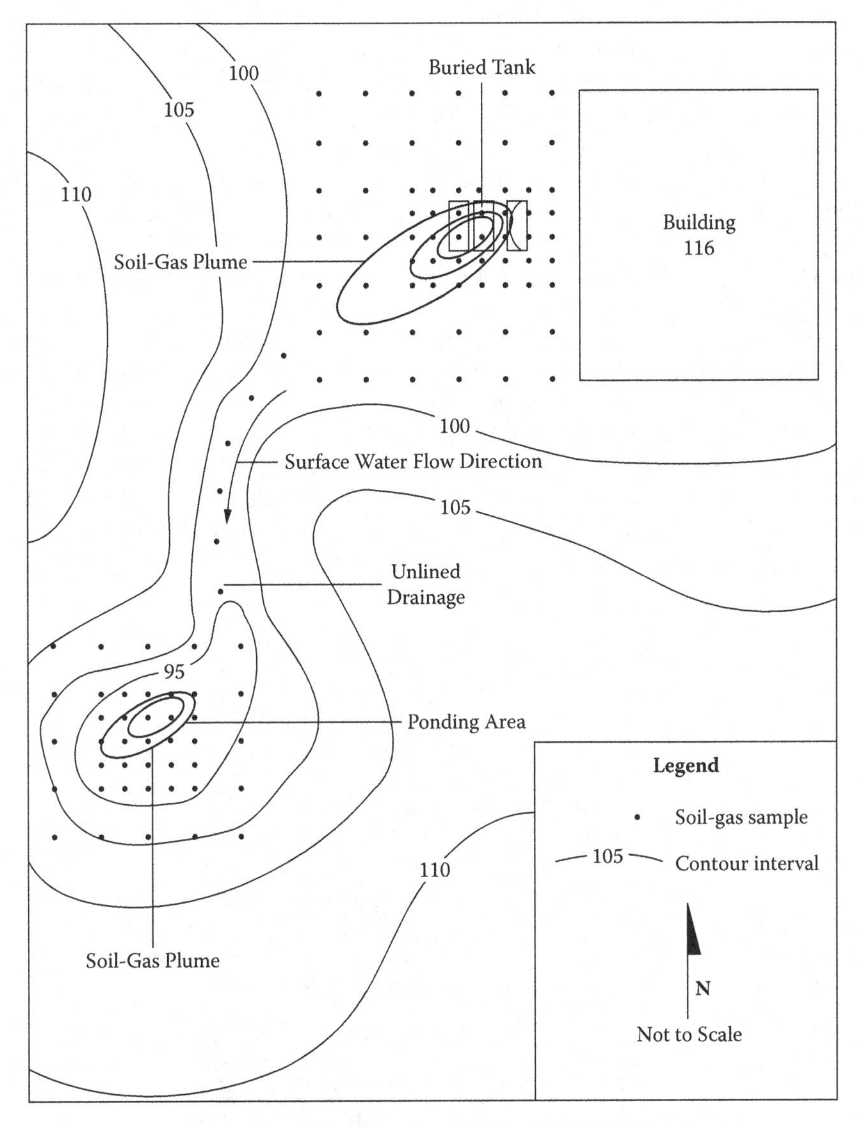

FIGURE 4.21 Example of a soil gas sampling grid for the BESURE method.

(From Byrnes, 1994. *Field Sampling Methods for Remedial Investigations.* Lewis Publishers, Boca Raton, FL.)

take into consideration the minimum size of the contaminant plume being targeted. By sampling all grid locations at the same time, the temporal variations in soil gas concentrations that are known to occur daily, and even hourly, are normalized. In addition, the spatial variability of contamination is often better defined with a passive soil gas survey because of the lower sampling and analytical costs (as compared to the mobile gas chromatograph method), which allow for more locations to be sampled. The BESURE passive soil gas method can target petroleum hydrocarbons (e.g., MTBE, BTEX, total petroleum hydrocarbons, PAHs, and complex mixtures), chlorinated hydrocarbons (e.g., vinyl chloride, carbon tetrachloride, trichloroethylene, and tetrachlorethylene), nitroaromatics, and chemical warfare agents and their breakdown products. Additional information concerning the application of passive soil gas methods can be found at the following web site: www.beacon-usa.com.

The BESURE passive soil gas sampler consists of a borosilicate glass vial prewrapped with wire (Figure 4.22) for installation and retrieval of the sampler. Each sampler contains two sets of adsorbent cartridges to adsorb compounds in soil gas. The adsorbent materials used are hydrophobic, with low affinity for water vapor that makes them effective even in water-saturated conditions.

BESURE passive soil gas samplers are most often installed in a 3/4-in.-diameter hole that is between 6 in. and 3 ft in depth. Note that whatever installation depth is selected, all of the samplers within the sampling grid should be installed at the same depth. Holes may be advanced to greater depths when necessary; however, the samplers need only be suspended in the upper portion of the hole because compounds in soil gas that enter the hole will migrate up to the sampler. After all the samplers in the grid have been installed, they are typically left in the ground between 3 and 14 days. The length of time the samplers are left in the ground is dependent on the site

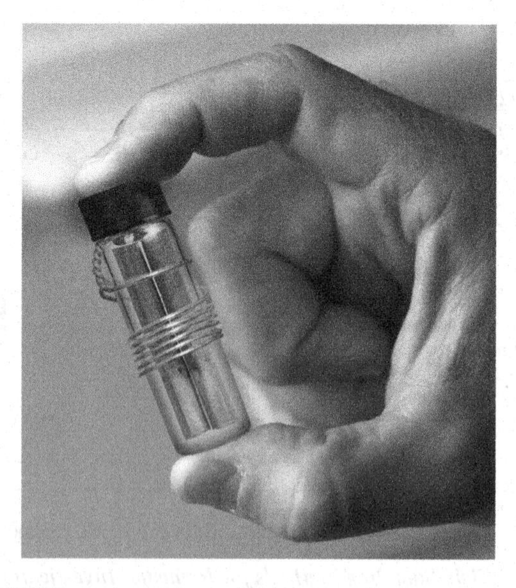

FIGURE 4.22 BESURE passive soil gas adsorbent sampler.

conditions, the depth at which contaminants are expected to be present, the expected contaminant concentration, and the overall objectives of the survey.

For planning purposes, one person can typically install 50–100 samplers per day with standard shallow installation when there is no asphalt or concrete present on the ground surface. When asphalt or concrete is present, a two-person team can install on average of 50 samplers per day using a hammer drill (with an approximately 1-in.-diameter drill bit). For retrieval of the samplers, one person can typically retrieve 100 samplers per day regardless of whether asphalt or concrete surfacing is present.

BESURE passive soil gas samples are analyzed by Beacon Environmental using GC/MS instruments following EPA Method 8260C or TO-17, modified for the introduction of samples by thermal desorption and to target a broad range of compounds, including VOCs and semi-volatile organic compounds (SVOCs). The results from a soil gas survey are then plotted on color contour maps (Figure 4.23) to assist with interpretation.

Passive soil gas results are based on a minimum of a five-point initial calibration, with the lowest point on the calibration curve at or below the practical quantitation limit of each compound. Internal standards and surrogates are included with each analysis – per EPA Method 8260C or TO-17 – to provide proof of performance that the system was operating properly for each sample and to provide consistent reference points for each analysis, which enables an accurate comparison of measured quantities. Trip blanks are analyzed with each batch of samples, and because two sets of adsorbent cartridges are provided in each sampler, duplicate or confirmatory analyses can be performed for any of the sample locations if requested.

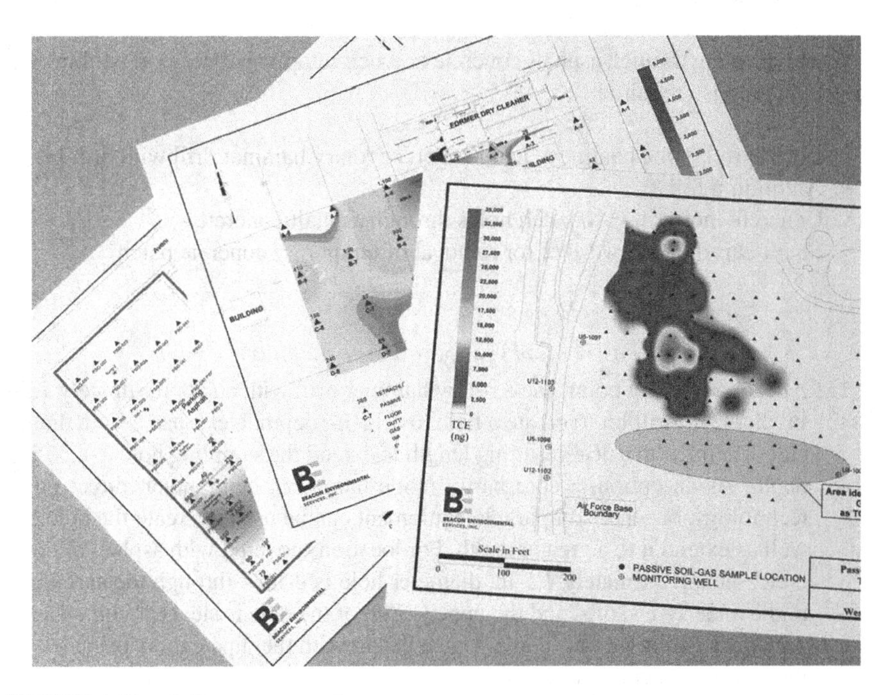

FIGURE 4.23 Soil gas survey color contour maps.

A BESURE Sample Collection Kit contains the following equipment and materials:

- BESURE passive soil gas samplers
- Sampling caps
- Cap storage container
- Trip blanks
- 3 × 4 in. plastic sampler bag
- Return shipment bags
- Small hand tools (e.g., wire cutters and pipe cutters)
- Metal pipe (when needed)
- Gauze cloths
- Chain-of-custody forms
- Tug-tight custody seals

Additional equipment required to install samplers include:

- Hammer
- Pin flags and flagging tape
- Clean towel
- Nitrile gloves
- Health and safety instruments
- Health and safety clothing
- Waste container (e.g., 55-gal drum), if needed
- Plastic waste bags

When sampling through asphalt/concrete or when sampling at depth, the following equipment are also required:

- Core barrel (slide hammer), hand auger, or rotary hammer drill with drill bit
- Aluminum foil
- Concrete mortar mix to patch holes through asphalt/concrete
- Small chisel or screwdriver for removal of temporary concrete patch

4.2.2.3.1 Sampling Procedures

4.2.2.3.1.1 BESURE Passive Soil Gas Sampler Installation

1. At each sampling point, use a rotary hammer drill with an approximately 1.5-in. diameter drill bit to create a hole to a 12-in. depth. Next, use a ½-in diameter drill bit with a 36-in. cutting length to extend the sampling hole to a 36-in. depth. As an option, a core barrel (slide hammer), hand auger, direct push technology, or other comparable equipment can be used to create the hole, as well as extend it to a greater depth. For locations covered with asphalt or concrete, an approximately 1.5-in. diameter hole is drilled through the surfacing to the underlying soils, and the upper 12 in. of the hole is sleeved with a metal pipe provided in the kit. The pipe can be cut with the pipe cutters provided in the sample collection kit, if necessary.

2. After the hole is created, remove a BESURE sampler (a borosilicate glass vial containing two sets of hydrophobic adsorbent cartridges) from the sample collection kit and unwind the retrieval wire wrapped around it. Holding the capped end of the vial in one hand, pull the wire tight (to straighten it) with the other hand. Remove the solid cap on the sampler vial and replace it with a sampling cap (a one-hole cap with a screen meshing insert). Store the solid cap in the cap storage container.
3. Using the retrieval wire, lower the sampler into the hole with the sampling cap pointing down. (If the hole was created to a greater depth, it is only necessary to suspend the sampler in the upper portion of the hole because compounds in soil gas that enter the hole will migrate up to the sampler.) With a portion of the retrieval wire extending above the hole, collapse the soil above the sampler. Coil the wire remaining above the hole, and lay it flat on the ground surface. When appropriate, stake some flagging tape near the sampler so that it can be seen easily. For those locations drilled through concrete or asphalt, lower the sampler through the metal pipe and plug the top of the hole with aluminum foil and a thin concrete patch to effectively seal the sampler in the ground.
4. On the Chain-of-Custody, record (a) sample-point number, (b) date and time of emplacement (to nearest minute), and (c) other relevant information (e.g., soil type, vegetation, and proximity to potential source areas). Mark the sample location and take detailed notes (i.e., compass bearings and distances from fixed reference points or GPS coordinates). Move to the next location.

4.2.2.3.1.2 BESURE Passive Soil Gas Sampler Retrieval

1. At each sample location open the sample collection kit, and place it and the wire cutters within easy reach. Remove a square of gauze cloth, and place it and a clean towel on the open kit. Remove a solid cap from the Cap Storage Container and place it on the kit.
2. Expose the sampler by pulling on the wire when in soils or using a chisel and hammer to chip the thin concrete patch away when in asphalt/concrete. Retrieve the sampler from its hole by pulling on the retrieval wire. Holding the sampler upright, clean the sides of the vial with the clean towel (especially close to the sampling cap). Remove the sampling cap, cut the wire from the vial with the wire cutters, and clean the vial threads completely with the gauze cloth.
3. Firmly screw the solid cap on the sampler vial and, with a ballpoint pen, record on the cap's label the sample number corresponding to the sample location.
4. On the Chain-of-Custody, record (a) date and time of retrieval (to nearest minute) and (b) any other relevant information.
5. Return the sampling cap to the cap storage container. Place the sealed and labeled sampler vial into a 3 in. × 4 in. plastic sampler bag. Then place the individually bagged and labeled sampler into the larger bag labeled "Return Shipment Bag." Each sampler is individually bagged and placed in a Return Shipment Bag that can hold as many as 40 samplers and one or more trip blanks.

6. After all samples have been retrieved, verify that the caps on each sampler are sealed tightly and that the seals on the sampler bags are closed. Verify that all samplers are stored in the return shipment bag, which contains an adsorbent pack. Seal the return shipment bag, place it in the upper tray of the kit, and place the provided tools and materials in the lower compartment of the kit.
7. Complete the Chain-of-Custody form for shipment of samplers. Seal the BESURE sample collection kit with the provided tug-tight custody seal, which has a unique identification number that is documented on the Chain-of-Custody. Place the kit and paperwork in a cardboard box, and ship via overnight delivery to Beacon Environmental for analysis of the samples.

4.2.2.4 AGI Sample Module Soil Gas Surveying Method

The AGI Sample Module (previously known as the GORE Module) was developed to sample soil gas and groundwater for a variety of VOC and SVOC compounds. This section is focused on how the AGI Sample Module can be used to support soil gas sampling. See Section 4.2.7.5.1.4 for using this sample module to support groundwater characterization.

The AGI Sample Module is approximately 0.25 in. in diameter and 13 in. in length and consists of a tube of GORE-TEX® membrane which is expanded polytetrafluoroethylene that is chemically inert, vapor-permeable, and waterproof (Figure 4.24). Housed inside the membrane tubing are several packets of hydrophobic sorbents that have an affinity for a broad range of volatile and semi-volatile organic compounds. Reportedly, the AGI Sample Module can be used to detect chlorinated solvents, fuel-related compounds, oxygenates, 1,4-dioxane, some explosives, chemical warfare agent breakdown compounds, pesticides, and polycyclic aromatic hydrocarbons (US Army Corps of Engineers 2014). When used to support soil gas investigations, AGI Sample Modules are typically placed in a grid pattern across the study area, similar to that shown in Figure 4.21 for the BESURE soil gas survey method.

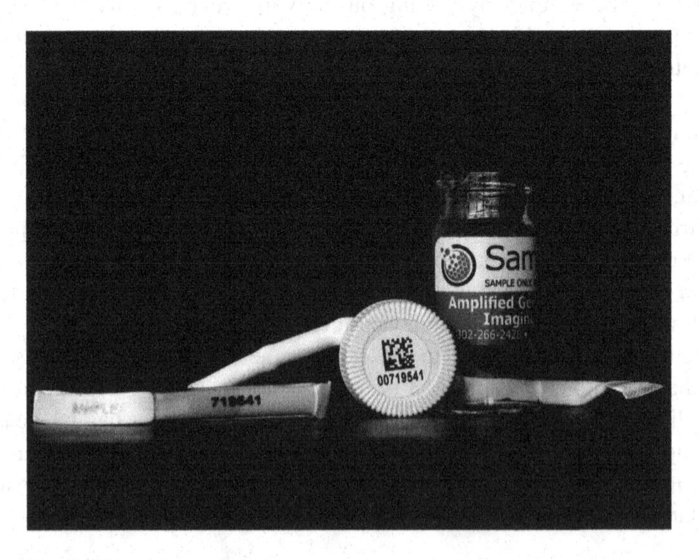

FIGURE 4.24 AGI sample module.

To collect a soil gas sample using the AGI Sample Module one typically drills a 0.75 in. (20 mm) diameter borehole in the soil that is 1.5 to 3 ft (0.5 to 0.9 m) deep. The sampling and analysis plan will identify the optimum depth for installation. Note that whatever installation depth is selected, all of the samplers within the sampling grid should be installed at that same depth. The procedure below identifies how the AGI sampler is to be installed. The AGI Sample Module is typically left in the borehole for 3–5 days before being retrieved for analysis, but it may be left in the ground longer if contaminant concentrations are expected to be low. For example, an EPA study (documented in EPA 1998a) left the sampler in the ground for 10 days. To ensure the AGI Sample Modules are left in the ground the optimum length of time, one should consider installing five to ten "test" AGI Sample Modules near the center of the sampling grid where soil gas concentrations are expected to be highest. One or two of these "test" AGI Sample Modules should be retrieved every other day and shipped to the manufacturer's laboratory for analysis. The analytical results from these "test" AGI Sample Modules should be used to determine the optimum time to retrieve all of the AGI Sample Modules for analysis.

Once retrieved from the borehole, the module is shipped to the manufacturer's laboratory. Analyses are performed by GC/MS. EPA 1998a reports that the guidelines used for on-site analysis are similar to SW-846 Method 5021 (Volatile Organic Compounds in Soils and Other Solid Matrices Using Equilibrium Headspace Analysis), modified to include high- and low-concentration procedures similar to those described in SW-846 Method 5035 (Closed-System Purge-and-Trap and Extraction for Volatile Organics in Soil and Waste Samples). See EPA 1998a for a study on how this soil gas sampling method compared to other soil gas sampling methods in several case studies.

Advantages of this sampling method include:

- Screens for a broad range of VOCs and SVOCs
- Can be used to sample multiple depths within a borehole
- Sample does not require low-temperature storage following sample collection
- Installation and retrieval of sample module are not difficult or time-consuming.

Disadvantages of this sampling method include:

- In highly contaminated environments, if AGI Sample Modules are left in the ground too long, sorbent saturation may occur (EPA 1998a).
- Clay layer(s) in soil can disrupt soil gas results measured at ground surface.
- The AGI Sample Module must be left in the borehole for multiple days prior to analysis.

Equipment

1. AGI Sample Modules contained in sealed glass vials
2. AGI-supplied strings and corks
3. AGI-supplied stainless steel insertion rod
4. Power drill with multiple 0.75 in. (20 mm) diameter drill bits. The length of the drill bits must reach the sample depth specified in the project-specific sampling and analysis plan

5. Generator, fuel, and power cord to support power drill
6. Measuring tape
7. Project-specific sampling and analysis plan (see Section 3.2.6)
8. Field logbook
9. Stainless steel scissors (or razor blade)
10. Bentonite grout
11. Brass survey marker
12. Sample labels
13. Cooler
14. Trip blank
15. Chain-of-custody forms
16. Chain-of-custody seals
17. Permanent ink marker
18. Health and safety instruments
19. Health and safety clothing (see Chapter 10)
20. Waste container (e.g., 55-gal drum)
21. Plastic waste bags

Procedure

1. In preparation for sampling, confirm that all necessary preparatory work has been completed, including obtaining property access agreements, meeting health and safety and equipment decontamination requirements, and checking the calibration of all health and safety instruments.
2. Use health and safety instruments (e.g., photoionization, flame-ionization detector) to screen air quality at the soil gas sampling location. Based on results, adjust personal protective equipment if needed in accordance with project health and safety plan.
3. Attach 0.75 in. (20 mm) diameter drill bit (long enough to reach sampling depth) to power drill.
4. Start generator. Connect drill power cord to generator.
5. Drill borehole to the depth specified in the project-specific sampling and analysis plan.
6. Remove the AGI Sample Module from the sealed glass vial. Tie one end of the string supplied by AGI to the loop end of the AGI Sample Module and tie the other end to the cork supplied by AGI.
7. Remove drill bit from borehole.
8. Insert the tip of the AGI-supplied stainless steel insertion rod into the pocket at the bottom of the AGI Sample Module, then use the insertion rod to push the AGI Sample Module down the borehole.
9. Remove the insertion rod leaving the AGI Sample Module in the borehole. Push the cork into the top of the borehole to prevent ambient air or water from entering. Label the cork with the sampling location and sample number.
10. Record in field logbook the date and time the AGI Sample Module was installed, along with soil gas sampling grid location, and sample number.

11. Repeat Steps 2 through 10 for each soil gas sampling grid location.
    a. As noted in the text above, it is recommended that five to ten "test" AGI Sample Modules be installed near the center of the sampling grid. One or two of these "test" AGI Sample Modules should be retrieved every other day and shipped to the manufacturer's laboratory for analysis.
12. Leave AGI Sample Modules in place until "test" AGI Sample Modules indicate it is the optimum time to retrieve all of the AGI Sample Modules.
13. Retrieve AGI Sample Module at first grid location by pulling straight upward on the cork to bring the AGI Sample Module to ground surface.
14. Record in field logbook the date and time the AGI Sample Module was retrieved, along with soil gas sampling grid location and sample number.
15. Use stainless steel scissors (or razor blade) to free the AGI Sample Module from string, transfer the module back into glass vial it came in, and screw on cap.
16. Attach a sample label and custody seal to the glass vial and immediately place into a sample cooler.
17. Repeat Steps 13 through 16 for each soil gas sampling grid location.
18. See Chapter 5 for details on preparing sample bottles and coolers for sample shipment.
19. Use bentonite grout to fill each soil gas sampling location. Install at least one brass survey marker at one of the soil gas sampling locations. Arrange for the coordinates of that one location to be surveyed in.
20. Containerize any waste in a waste container. Before leaving the site, all waste containers should be sealed, labeled, and handled appropriately (see Chapter 11).
21. Decontaminate sampling equipment in accordance with procedures outlined in Chapter 9.

4.2.3 Downhole Screening Methods

The following subsections present several screening methods that can be used downhole to either determine the presence or absence of nonaqueous phase liquid (NAPL), determine if there are depth intervals showing elevated radioactivity levels, or define other important features of the subsurface.

4.2.3.1 Ribbon NAPL Sampler

The Ribbon NAPL sampler (Figure 4.25) is a sampling device that allows detailed depth-discrete mapping of NAPLs in a borehole. This characterization method provides a "yes" or "no" answer to the presence of NAPLs and is used to complement and enhance other characterization techniques. Several cone penetrometer deployment methods are available for the Ribbon NAPL sampler, and methods for other drilling techniques are also under development.

This characterization technique uses the Flexible Liner Underground Technologies (FLUTe), Ltd., flexible liner, which is pressurized against the wall of the borehole. Attached to the flexible liner is a hydrophobic absorbent ribbon that absorbs and

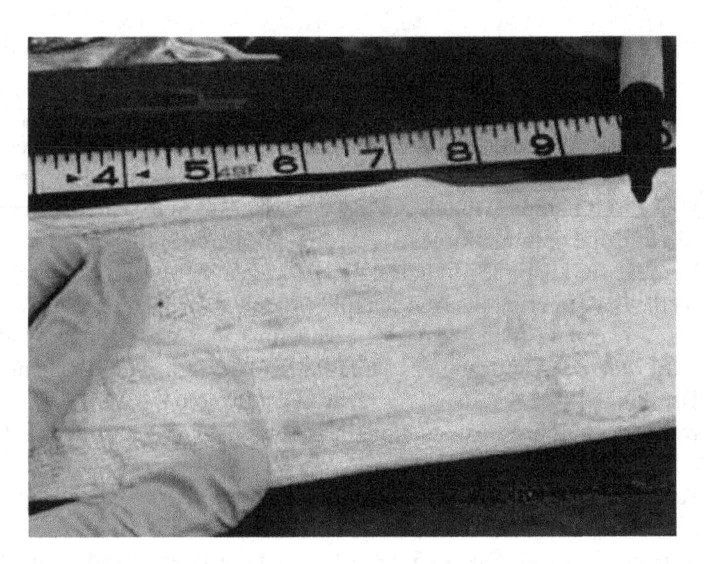

FIGURE 4.25 Ribbon NAPL sampler. (See color insert following page 42.)

stains when it comes in contact with NAPL. The Ribbon NAPL sampler is deployed with air or water pressure. Before use, the hydrophobic adsorbent ribbon is attached to the FLUTe flexible liner, which is then everted (turned inside out). This eversion method prevents the hydrophobic adsorbent ribbon from sliding along the borehole and smearing the NAPL on the ribbon. After the liner has been pressurized against the wall of the borehole, it is retrieved and then re-everted at the surface and inspected for the presence of NAPL. The reusable flexible liner is available in custom lengths and can use any length of the replaceable hydrophobic adsorbent ribbon. A 2-in.-diameter membrane is used in cone penetrometer boreholes, and other diameters are available.

For more information on the Ribbon NAPL sampler, see www.osti.gov/servlets/purl/15027.

4.2.3.2 Downhole Gross Gamma Logging

Downhole gross gamma logging is an effective method for quickly identifying borehole depth intervals showing elevated gamma activity levels. This method is most often used to support site characterization activities by assisting in the selection of soil sampling depth intervals to be selected for laboratory analysis.

With this method, once the maximum depth of a borehole has been reached, a sodium iodide detector is lowered to the bottom of the hole. Gross gamma activity counts are then collected at systematic time and depth intervals. For boreholes less than 100 ft in depth, gross gamma activity measurements are often collected at 1-ft or 2-ft depth intervals through the entire length of the borehole, using a short count time (e.g., few minutes) at each interval. For deeper boreholes, measurements are typically collected at a greater spacing (e.g., every 5 ft). Downhole gross gamma logging is typically performed inside the drill casing because the hole would likely collapse if the casing were removed before collecting measurements.

Once the entire depth of a borehole has been counted, the count-per-minute readings are then used to identify the depth intervals showing the highest gross gamma activity, as well as the first depth interval to show background activity. The advantage of using the downhole gross gamma logging method is that it is inexpensive to implement and can help minimize the number of samples that need to be sent to the laboratory for analysis. The sampling and analysis plan (see Section 3.2.7) will identify the specific depth intervals to collect samples for laboratory testing. For a site characterization study, typically a minimum of three soil sampling intervals are selected from a borehole for analytical testing. These include the interval from the top 10 ft of the borehole showing the highest gross gamma activity level (for human health and risk assessment purposes), the interval showing the highest gross gamma activity for the entire borehole, and the first interval below this that shows background gross gamma activity. The remaining sampling intervals should be archived in case they are needed at a later time.

The following equipment and procedure can be used to perform downhole gross gamma logging:

Equipment

1. Method for drilling 6-in.-diameter (or larger) borehole
2. Sodium iodide detector and handheld meter
3. Measuring tape
4. Stopwatch
5. Sample logbook
6. Health and safety instruments
7. Health and safety clothing
8. Waste containers (55-gal drums)
9. Plastic sheeting
10. Plastic waste bags

Sampling Procedure

1. In preparation for logging, confirm that all necessary preparatory work has been completed, including obtaining property access agreements, meeting health and safety and equipment decontamination requirements, and checking the calibration of all health and safety and chemical/radiological field screening instruments.
2. Cut a 1-ft-diameter hole in the center of the plastic sheeting, and center the hole over the drilling location. The purpose of this sheeting is to help prevent the spread of contamination.
3. Use a drill rig to advance a 6-in.-diameter (or larger) borehole to the desired depth (advancing drill casing to keep the borehole open if appropriate).
4. After removing drill string and drill bit from borehole, lower the sodium iodine detector to the bottom of the borehole and collect a gross gamma activity measurement for that depth over a selected count time defined in the sampling and analysis plan (see Section 3.2.6). Record the gross

gamma activity measurement and count time for that interval in the sample logbook.

5. Lift the detector up to the next measurement interval, and collect the next gross gamma activity measurement using the same count time. Record the gross gamma activity measurement and count time for that interval in the sample logbook.
6. Repeat Step 5 until the entire borehole has been logged.
7. Transfer any waste soil from the drilling and logging operation into waste containers. Before leaving the site, all waste containers should be sealed, labeled, and handled appropriately (see Chapter 11).
8. Transfer any other sampling-related wastes (e.g., gloves and foil) into a plastic waste bag.
9. Have a professional surveyor survey the coordinates of the borehole to preserve the exact location.

4.2.3.3 Downhole High-Purity Germanium Logging

Downhole high-purity Germanium logging is performed in the same manner as downhole gross gamma logging (see Section 4.2.3.2), except that a high-purity Germanium detector is lowered down the inside of the drill casing (or open borehole) instead of a sodium iodide detector. The advantage of using a high-purity Germanium detector over a sodium iodide detector is that specific activity levels for specific isotopes can be obtained that are of laboratory quality, as opposed to simply gross gamma activity levels. The length of the count time at each of the sampling intervals using the high-purity Germanium detector will be dependent on the detection limit requirements. Longer count times will provide lower detection limits. Typically, count times range from 20 to 60 minutes in length. It should be noted that a high-purity Germanium detector is much more expensive than a sodium iodide detector and, as a result, this method may not be cost-effective unless many field measurements are to be collected.

4.2.3.4 Portable X-Ray Fluorescence (XRF) Measurements

Portable XRF instruments may be used to run EPA SW-846 Method 6200 to screen soil or sediment samples for 26 metal analytes concentrations during field investigations to help reduce the number of samples that is needed to be sent to the laboratory for analysis. While the detection limits for Method 6200 are typically above the toxicity characteristic regulatory level for most Resource Conservation and Recovery Act (RCRA) analytes, the project-specific data quality objectives may still find this method useful for helping make some field decisions. See Table 1 in EPA 2007 for detection limits that can be achieved by Method 6200.

When running in situ analyses using this method, remove any large or non-representative debris from the soil surface. This debris includes rocks, pebbles, leaves, vegetation, and roots. Also, the soil surface must be as smooth as possible so that the probe window will have good contact with the surface. This may require some leveling of the surface using a stainless steel trowel. In situ measurements should only be collected when soil moisture is <20%. Method 6200 also suggests tapping the soil to increase soil density, which in turn will increase repeatability and representativeness. Source count times for in situ analysis usually range from 30 to 120 seconds

(EPA 2007) and will be defined by project-specific detection limit requirements (see Section 3.2.5.3).

For running intrusive analyses on surface soils or sediment, Method 6200 recommends that a sample be collected from a 4-in. by 4-in. square that is 1-in. deep, which is enough to fill an 8-ounce sample jar. The sample should be homogenized, dried, and ground with a mortar and pestle until at least 90% of the original sample passes through a 60 mesh sieve. An aliquot of the sieved sample should then be placed in a 31.0-mm polyethylene sample cup (or equivalent) for analysis. The sample cup should be one-half to three-quarters full at a minimum. The sample cup should be covered with a 2.5 μm Mylar (or equivalent) film for analysis (EPA 2007). The XRF instrument needs to be programmed for source count times that will meet the project-specific detection limit requirements.

Tests that were run and reported in EPA 2007 concluded that precision dramatically improved when running intrusive analyses over running in situ analyses. For that reason, the procedure below is for intrusive analyses only. While EPA 2007 only addresses surface soil or sediment sampling, portable XRF analyses can also be performed on shallow or deep soil samples that are brought to the ground surface. To verify the accuracy of the portable XRF field measurements, the field study should require a certain percentage (e.g., 5–10%) of the samples split and analyzed by both the portable XRF and an offsite laboratory. Ravansari (2020) provides guidance on a few important issues that one should be aware of when using this field screening method.

The following equipment and procedure should be used to run intrusive portable XRF analyses. This procedure applies to both grab and composite samples since both require homogenization prior to being dried, ground with a mortar and pestle, and then passed through a 60 mesh sieve.

Equipment

1. Calibrated portable XRF
2. Quality control materials to run systems check and field calibration check
3. Project-specific sampling and analysis plan (see Section 3.2.7)
4. Soil sample
5. Sample labels
6. Stainless steel bowl
7. Stainless steel spoon
8. Stainless steel No. 60 mesh sieve
9. Field logbook
10. Permanent ink marker
11. Health and safety clothing
12. Waste container (e.g., 55-gal drum)
13. Plastic waste bags

Sampling Procedure

1. In preparation for sampling, confirm that all necessary preparatory work has been completed, including obtaining property access agreements,

meeting health and safety and equipment decontamination requirements, and checking the calibration of all health and safety instruments.
2. Use health and safety instruments (e.g., photoionization, flame-ionization detector) to screen air quality as soil or sediment is collected and prepared for XRF analyses. Based on results, adjust personal protective equipment if needed in accordance with project health and safety plan.
3. Verify all equipment has been properly decontaminated (see Chapter 9).
4. In preparation for running portable XRF analyses, refer to the instrument user manual and run all required checks (e.g., systems check, field calibration check) to verify the instrument is properly calibrated and operating correctly. Recheck every 5 hours (EPA 2017b).
5. Grab or composite samples to be analyzed are first transferred into a stainless steel bowl and homogenized using a stainless steel spoon, dried, and then ground with a mortar and pestle until at least 90% of the original sample passes through a 60 mesh sieve.
6. An aliquot of the sieved sample is then placed in a 31.0-mm polyethylene sample cup (or equivalent) for analysis. The sample cup should be one-half to three-quarters full at a minimum. The sample cup is then covered with a 2.5 μm Mylar (or equivalent) film for analysis. The sample cup needs to be labeled with the sample number, sampling time/date, etc.
7. Program the XRF instrument for source count times that will meet the project-specific detection limit requirements (see Section 3.2.5.3).
8. Collect the first XRF instrument measurement on the sample and record the results in a field logbook.
9. Repeat Step 6 through Step 8 ten times using a clean sample cup and new Mylar film. Record the results in a field logbook.
10. Calculate the mean concentration, standard deviation, and 95% upper confidence level (UCL) for each metal being tested and record the results in the field logbook.
11. When using the data to make field decisions, EPA (2017b) recommends using the 95% UCL value be compared to the action level.
12. Decontaminate equipment (e.g., bowl, spoon, sieve, soil moisture meter) in accordance with the procedures outlined in Chapter 9 prior to analyzing the next sample.

4.2.4 Soil Sampling

The following section provides the reader with shallow and deep soil sampling methods that should be considered to support soil characterization and soil remediation studies. The criteria used in selecting the most appropriate method include the analyses to be performed on the sample, the type of sample being collected (grab or composite), and the sampling depth. Standard operating procedures have been provided for each of the methods to facilitate implementation.

At chemical and radiological sites, soil sampling is most often performed for the following purposes:

- Defining the nature and extent of contamination
- Assessing the risk that a site poses to human health and the surrounding environment
- Defining soil distribution coefficients and geotechnical properties for the purpose of supporting transport modeling studies and the evaluation of potential remedial alternatives
- Performing treatability testing and other engineering evaluations
- Determining whether or not remedial action objectives have been met

The seven-step DQO process (Chapter 3, Section 3.2) should be used to define the sampling approach, required number of samples, analyses to be performed, and analytical performance requirements. When defining the nature and extent of contamination, it is typically most cost-effective to define the sources and horizontal and vertical extent of contamination using a combination of low-cost nonintrusive methods (e.g., aerial photography and surface geophysical surveying) and low-cost intrusive methods (e.g., soil gas surveying) combined with confirmation soil sampling and laboratory analysis. This is referred to as a judgmental sampling approach because soil samples are collected from locations that have the highest likelihood of showing contamination (Chapter 3, Section 3.2.5.7.1).

Soil sampling is often required to provide the data needed to assess the risk that a site poses to human health and the surrounding environment. Samples collected for this purpose are often collected from statistically selected locations to provide estimated average exposures to receptors (Chapter 3, Section 3.2.5.7.2). Samples to support risk assessments may also be collected from contamination source locations (judgmental locations) to provide an estimate of the worst-case exposure. Soil sampling performed to support engineering, modeling, or treatability studies may require samples to be collected from judgmental or statistical locations, depending on the focus of the study.

After a soil remediation effort is complete, soil samples are most often collected from statistically derived locations to determine whether or not the site can be declared clean (see Chapter 3, Section 3.2.5.7.2).

The following sections present the most effective shallow and deep soil sampling methods, along with detailed procedures on how to use them.

4.2.4.1 Shallow Soil Sampling

Soil samples collected from a depth of 5 ft or less are generally referred to as *shallow*. The most effective shallow soil sampling methods include the scoop, hand auger, core barrel (slide hammer), open-tube, split-tube or solid-tube, thin-walled tube (Shelby Tube), and En Core® methods. When preparing a sampling program, considerable thought should go into selecting appropriate sampling methods because the selected method can significantly influence the analytical results. For example, if the contaminants of concern at a site include volatile organic compounds, it would not be good practice to collect samples using the hand auger method, because this method churns up the soil, which facilitates volatilization. The core barrel (slide hammer) or split-tube or solid-tube sampler would be a more appropriate selection, because both of these tools can be used to remove a compacted but undisturbed core of soil.

The scoop sampler can be used to collect grab or composite samples of surface soil. To collect a grab sample, the surface soil from one location is scooped directly into a sample jar. A composite sample can be collected by scooping surface soil from the multiple locations to be composited into a stainless steel bowl, and homogenized before filling sample jars.

The hand auger is most effectively used to collect composite soil samples from sites where the contaminants of concern do not include volatile organic compounds. When using this tool, samples for radiological or chemical analysis generally are not collected over intervals greater than 1 ft because larger intervals tend to dilute the composition of the sample beyond the point of providing useful data.

The core barrel (slide hammer) and split-tube or solid-tube samplers can be used to collect either grab or composite samples. When collecting a grab sample, these tools are commonly loaded with sample liners that can be quickly removed and capped after sample collection. When composite samples are collected, the soil from the intervals (or locations) to be composited is transferred into a stainless steel bowl and homogenized with a stainless steel spoon before filling sample jars.

When shallow soil samples are needed for lithology description only, the open-tube sampler is an effective tool. The open-tube sampler has a sampling tube of small enough diameter so that it can easily be advanced several feet into the ground to provide a small-diameter soil core. Because the sampling tube is open on one side, the soil lithology can be described without removing the sample from the tube. When soil samples are needed for geotechnical analysis, the thin-walled tube (Shelby tube) sampler hydraulically pushed into the ground is the preferred sampling method.

The En Core® sampler can be used to collect a small core of soil for volatile organic analysis from a core of soil brought to the ground surface using either a core barrel (slide hammer), or split-tube or solid-tube sampler.

Table 4.1 summarizes the effectiveness of each of the seven recommended sampling methods. A number "1" in the table indicates that a particular procedure is most effective in collecting samples for a particular laboratory analysis, sample type, or sampling depth. A number "2" indicates that the procedure is acceptable, but less preferred, whereas an empty cell indicates that the procedure is not recommended. For example, Table 4.1 indicates that the core barrel (slide hammer), split-tube/solid-tube, and En Core® methods are most effective in collecting soil samples for volatile organic analysis. The scoop method is considered acceptable when collecting samples for this analysis, whereas the hand auger, open-tube, and thin-walled tube (Shelby tube) methods are not recommended. The following sections provide further details and standard operating procedures for each of the recommended soil sampling methods.

4.2.4.1.1 Scoop Method

The scoop is a handheld sampling tool that is effective in collecting samples of the top 0.5 ft of soil. Figure 4.26 presents a variety of AMS Inc. (AMS) scoop samplers.

TABLE 4.1
Rating Table for Shallow Soil Sampling Methods

	Laboratory Analyses								Sample Type			Depth		
	Radio-nuclides	Volatiles	Semi-volatiles	Metals	Pesti-cides	PCBs	TPH	Geotech-nical	Grab	Composite (Vertical)	Composite (Areal)	Surface (0.0–0.5 ft)	Shallow (0.0–5.0 ft)	Lithology Description
Scoop	1	2	2	1	1	1	1		1		1	1		1
Hand auger	1		2	1	1	1	1			1	1	1	1	2
Core barrel (slide hammer)	1	1	1	1	1	1	1		1	1	2	1	1	2
Open-tube									1			1	1	1
Split tube/ solid tube	1/1	1/1	1/1	1/1	1/1	1/1	1/1		1/1	1/2	2/2		1	1/2
Thin-walled tube (Shelby tube)								1	1				1	
En Core® sampler		1							1			1[a]	1[a]	2

Note: 1 = preferred method; 2 = acceptable method; empty cell = method not recommended.

[a] The En Core® sampler may be used to collect a soil core sample from any depth at which a soil core is brought to the ground surface using either the core barrel (slide hammer) or split tube/solid tube sampler.

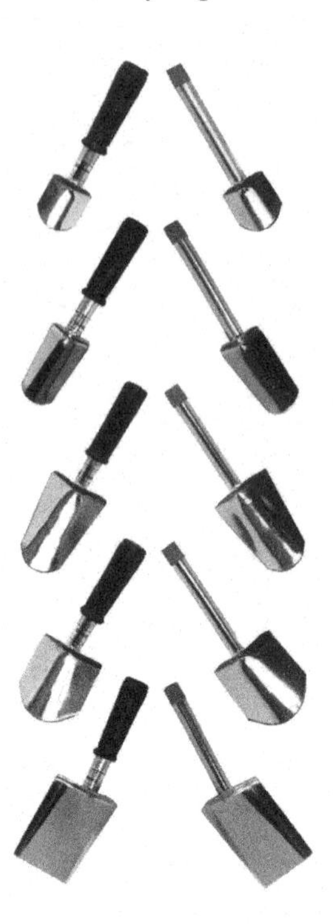

FIGURE 4.26 AMS scoop samplers.

This method is commonly used to collect samples of discolored soil observed at the ground surface, or to collect samples from areas where, for some reason, deeper sampling is not possible or is not necessary. Grab or areal composite samples can be collected using this method by either spooning soil from one location directly into a sample jar or by compositing soil from more than one location in a stainless steel bowl before filling a sample jar (Figure 4.27). For additional information on scoop samplers, see www.ams-samplers.com.

For most sampling programs, four people are sufficient for this procedure. Two are needed for sample collection, lithology description, labeling, and documentation; a third is needed for health and safety; and a fourth is needed for miscellaneous tasks such as waste management and equipment decontamination.

The following equipment and procedure can be used to collect shallow soil samples using the scoop method for chemical or radiological analysis:

Equipment

1. Scoop
2. Stainless steel bowl
3. Stainless steel spoon
4. Sample jars
5. Sample labels

FIGURE 4.27 Compositing soil sample.

6. Cooler packed with Blue Ice® (Blue Ice is not required for radiological analysis.)
7. Trip blank (only required for volatile organic analyses)
8. Coolant blank (not required for radiological analysis)
9. Sample logbook
10. Chain-of-custody forms
11. Chain-of-custody seals
12. Permanent ink marker
13. Health and safety instruments
14. Chemical or radiological field screening instruments
15. Health and safety clothing
16. Waste container (e.g., 55-gal drum)
17. Sampling table
18. Plastic sheeting
19. Plastic waste bags

Sampling Procedure

1. In preparation for sampling, confirm that all necessary preparatory work has been completed, including obtaining property access agreements, meeting health and safety and equipment decontamination requirements, and checking the calibration of all health and safety and chemical and/or radiological field screening instruments.
2. Cut a 1-ft-diameter hole in the center of the plastic sheeting, and center the hole over the sampling point. The purpose of this sheeting is to help prevent the spread of contamination.

3. Begin collecting the sample by applying downward pressure on the scoop until the desired sampling depth is reached, and then lift. If a grab sample is being collected, transfer the soil from the scoop directly into a sample jar. If an areal composite sample is being collected, transfer the soil from each location to be composited into a stainless steel compositing bowl and homogenize with a stainless steel spoon before filling a sample jar.
4. Scan the hole where the sample was collected using chemical and/or radiological field screening instruments, and record the results in a bound logbook.
5. After the jar is capped, attach a sample label and custody seal to the jar and immediately place it into a sample cooler. Samples for chemical analysis should be packed in Blue Ice®.
6. See Chapter 5 for details on preparing sample jars and coolers for sample shipment.
7. Transfer any soil left over from the sampling into a waste container. Before leaving the site, all waste containers should be sealed, labeled, and handled appropriately (see Chapter 11).
8. Transfer any other sampling-related wastes (e.g., gloves and foil) into a plastic waste bag.
9. Have a professional surveyor survey the coordinates of the sampling point to preserve the exact sampling location.

4.2.4.1.2 Hand Auger Method

The hand auger is an effective shallow soil sampling tool when the contaminants of concern do not include volatile organic compounds, because the augering motion facilitates volatilization. This tool is composed of a bucket auger, which comes in various shapes and sizes, a shaft, and a T-bar handle (Figure 4.28). Extensions for the shaft are available to allow sampling at deeper intervals. However, in most soils this tool is only effective in collecting samples to a depth of 5 ft, because the sample hole typically begins to collapse by this depth. Because the auger rotation automatically homogenizes the sampling interval, this method is used to collect composite samples.

For most sampling programs, four people are sufficient for this sampling procedure. Two are needed for sample collection, lithology description, labeling, and documentation; a third is needed for health and safety; and a fourth is needed for miscellaneous tasks such as waste management and equipment decontamination.

The following equipment and procedure can be used to collect shallow soil samples for chemical or radiological analysis:

Equipment

1. Hand auger
2. Stainless steel bowl
3. Stainless steel spoon
4. Sample jars
5. Sample labels

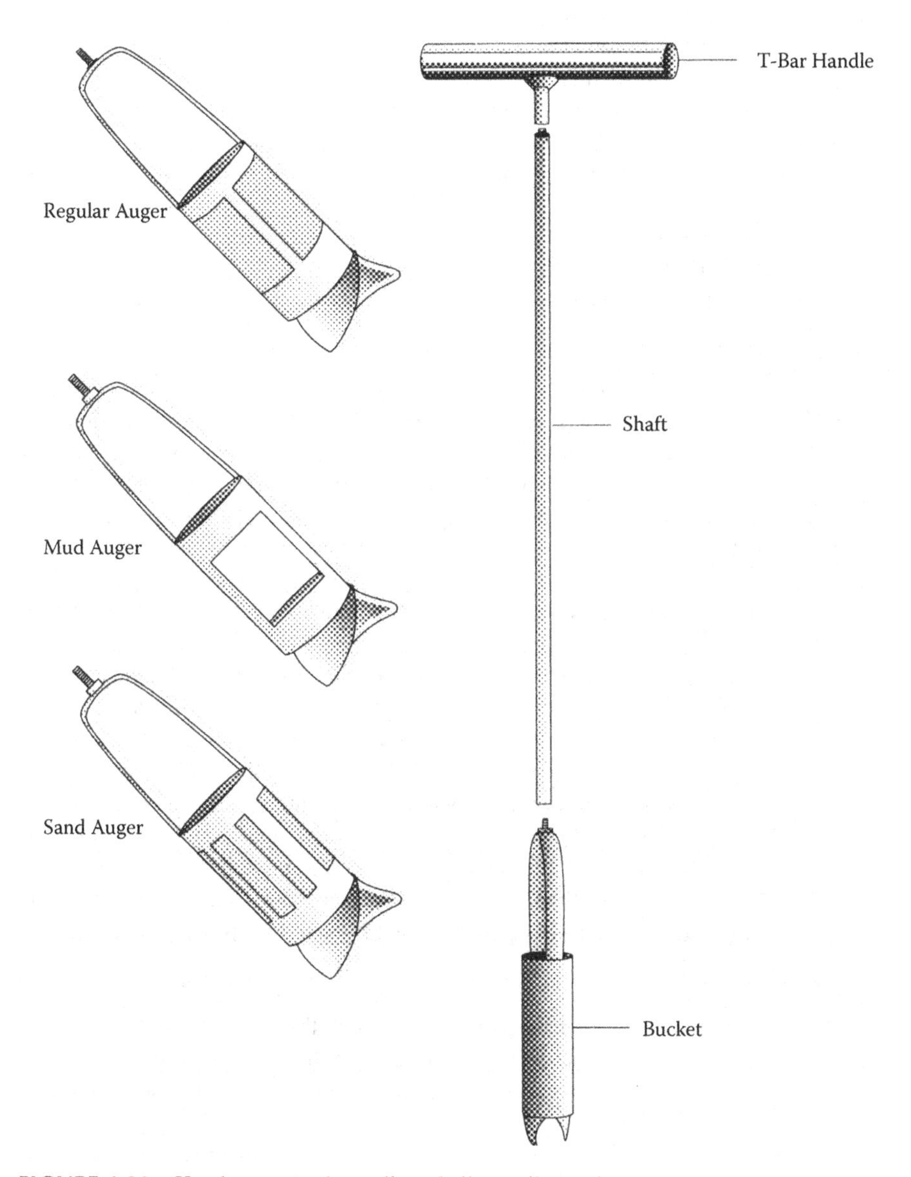

FIGURE 4.28 Hand auger used to collect shallow soil samples.

(From Byrnes, 2001. *Sampling and Surveying Radiological Environments.* Lewis Publishers, Boca Raton, FL.)

6. Cooler packed with Blue Ice® (Blue Ice® is not required for radiological analysis)
7. Trip blank (only required for volatile organic analyses)
8. Coolant blank (not required for radiological analysis)
9. Sample logbook
10. Chain-of-custody forms

11. Chain-of-custody seals
12. Permanent ink marker
13. Health and safety instruments
14. Chemical and/or radiological field screening instruments
15. Health and safety clothing
16. Waste container (e.g., 55-gal drum)
17. Sampling table
18. Plastic sheeting
19. Plastic waste bags

Sampling Procedure

1. In preparation for sampling, confirm that all necessary preparatory work has been completed, including obtaining property access agreements, meeting health and safety and equipment decontamination requirements, and checking the calibration of all health and safety and chemical and/or radiological field screening instruments.
2. Cut a 1-ft-diameter hole in the center of the plastic sheeting, and center the hole over the sampling point. The purpose of this sheeting is to help prevent the spread of contamination.
3. To collect a composite sample from one continuous depth interval (e.g., 0 to 1 ft), advance the hand auger by applying a downward pressure while rotating the hand auger clockwise. When the auger if full of soil, remove it from the borehole and transfer the soil into a stainless steel bowl using a stainless steel spoon. Composite the soil in a stainless steel bowl using the stainless steel spoon to break apart any large chunks of soil; then mix and stir the soil enough to homogenize the sample thoroughly. Transfer the soil into a sample jar using a stainless steel spoon.
4. To collect a vertical composite sample from multiple depths within the same borehole (e.g., 0 to 1 ft, 2 to 3 ft, 4 to 5 ft), advance the hand auger to the bottom of the first sample interval (e.g., 0 to 1 ft). Remove the auger from the borehole and transfer the soil into a stainless steel bowl using a stainless steel spoon. Return the hand auger to the borehole and advance the auger to the top of the next sample interval. Dispose of unused soil in a waste container. Use a decontaminated hand auger to advance the hand auger to the bottom of the second sampling interval (e.g., 2 to 3 ft). Remove the auger from the borehole and transfer the soil into the stainless steel bowl using a stainless steel spoon. Repeat this procedure until all the soil sampling intervals have been collected and have been transferred into the stainless steel bowl. Composite the soil in the stainless steel bowl by using the stainless steel spoon to break apart any large chunks of soil; then mix and stir the soil enough to homogenize the sample thoroughly. Transfer the soil into a sample jar using a stainless steel spoon.
5. To collect an areal composite sample, at the first location to be composited, advance the hand auger, to the bottom of the desired sample

depth. Remove the auger from the borehole and transfer the soil into a stainless steel bowl using a stainless steel spoon. Dispose of unused soil in a waste container. Move to the second location to be composited. Advance a decontaminated hand auger to the bottom of the desired sample depth. Remove the auger from the borehole and transfer the soil into the stainless steel bowl using a stainless steel spoon. Dispose of unused soil in a waste container. Repeat this procedure at each of the soil sampling locations to be composited. Composite the soil from all the locations in the stainless steel bowl by using the stainless steel spoon to break apart any large chunks of soil; then mix and stir the soil enough to homogenize the sample thoroughly. Transfer the soil into a sample jar using a stainless steel spoon.

6. After the sample jar is capped, attach a sample label and custody seal to the jar and immediately place it into a sample cooler. Samples for chemical analysis should be packed in Blue Ice®.
7. Scan the hole where the sample was collected using chemical and/or radiological field screening instruments, and record the results in a bound logbook.
8. See Chapter 5 for details on preparing sample jars and coolers for sample shipment.
9. Transfer any soil left over from the sampling into a waste container. Before leaving the site, all waste containers should be sealed, labeled, and handled appropriately (see Chapter 11).
10. Transfer any other sampling-related wastes (e.g., gloves and foil) into a plastic waste bag.
11. Have a professional surveyor survey the coordinates of the sampling point to preserve the exact sampling location.

4.2.4.1.3 Core Barrel (Slide Hammer) Method

For collecting shallow soil core samples for chemical and/or radiological analysis, the core barrel (slide hammer) method is recommended. This tool consists of a core barrel, an extension rod, and a slide hammer (Figure 4.29). Most core barrels have an inside diameter of 2 or 2.5 in., and are 1 or 2 ft in length, but can be specially ordered in other sizes. The top of the barrel is threaded so that it can be screwed into an extension rod. The barrel is constructed in such a way that it can also accept sample liners to facilitate the collection of grab samples.

The sample liners are used to facilitate the removal of soil from the barrel without disturbing the sample (Figure 4.30). Without sample liners, soil must be extracted from the barrel using a spoon or knife, and then transferred into a sample jar. This can be a problem if samples are to be analyzed for volatile organic compounds because a good portion of the volatiles can be lost into the ambient air in the process of transferring the sample into the sample jar. In contrast, when collecting a grab sample, sample liners can be quickly removed from the barrel and sealed with airtight Teflon caps. After labeling the liners, they can be shipped directly to the laboratory for analysis. If grab samples are being collected for lithology description only, clear plastic liners are available. When collecting vertical or areal composite samples, sample liners are

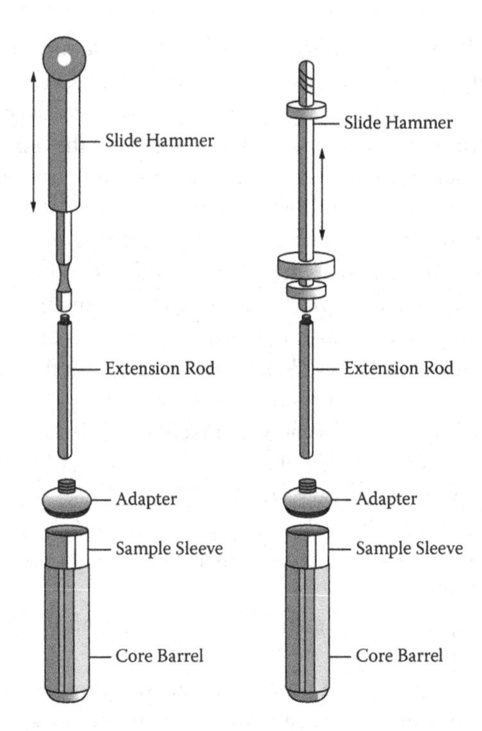

FIGURE 4.29 Core barrels (slide hammers) used to collect shallow soil samples.

(From Byrnes, 1994. *Field Sampling Methods for Remedial Investigations.* Lewis Publishers, Boca Raton, FL.)

not required. Soil from the intervals or locations to be composited is transferred from the core barrel into a stainless steel bowl and homogenized before filling a sample jar. (Note: a decontaminated sampler should be used at each location being composited)

Extension rods are available in various lengths to allow sampling at depths greater than the length of the core barrel. These rods are screwed into the core barrel at one end, and into the slide hammer at the other end. A slide hammer is used to beat the core barrel into the ground. The hammer is available in different shapes and weights to accommodate the needs of the sampler. For additional information on soil core samplers, see www.ams-samplers.com.

For most sampling programs, four people are sufficient for this sampling procedure. Two are needed for sample collection, lithology description, labeling, and documentation; a third is needed for health and safety; and a fourth is needed for miscellaneous tasks such as waste management and equipment decontamination.

The following equipment and procedure can be used to collect shallow soil samples for chemical and/or radiological analysis.

Equipment

1. Slide hammer and extension rods
2. Core barrel

FIGURE 4.30 AMS core barrel sample liners.

3. Sample liners
4. Teflon end-caps for sample liners
5. Stainless steel bowl
6. Stainless steel spoon
7. Stainless steel knife
8. Sample jars
9. Sample labels
10. Cooler packed with Blue Ice® (Blue Ice® is not required for radiological analysis.)
11. Trip blank (only required for volatile organic analyses)
12. Coolant blank (not required for radiological analysis)
13. Sample logbook
14. Chain-of-custody forms
15. Chain-of-custody seals
16. Permanent ink marker
17. Health and safety instruments
18. Chemical and/or radiological field screening instruments
19. Health and safety clothing
20. Waste container (e.g., 55-gal drum)
21. Sampling table
22. Plastic sheeting
23. Aluminum foil
24. Plastic waste bags

Sampling Procedure

1. In preparation for sampling, confirm that all necessary preparatory work has been completed, including obtaining property access agreements, meeting health and safety and equipment decontamination requirements, and checking the calibration of all health and safety and chemical and/or radiological field screening instruments.
2. Cut a 1-ft-diameter hole in the center of the plastic sheeting, and center the hole over the sampling point. The purpose of this sheeting is to help prevent the spread of contamination.
3. To collect a grab sample, unscrew the core barrel from the slide hammer and load it with sample liners of the desired length. Avoid touching the inside surface of the core barrel and sleeves, for this will contaminate the sampler. Screw the core barrel back onto the slide hammer and extension rod. Use the slide hammer to beat the core barrel to the desired depth, and record the blow count in a sample logbook. Remove the core barrel from the hole by rocking it from side to side several times before lifting or reverse-beating the core barrel from the hole. Unscrew the core barrel from the sampler, and slide the sample liners out onto a piece of aluminum foil. Using a stainless steel knife, separate the sample liners. Place Teflon caps over the ends of the sleeves to be sent to the laboratory. If sample liners are not being used, spoon soil from the core barrel directly into a sample jar.
4. After each sample liner or sample bottle is capped, attach a sample label and custody seal, and immediately place it into a sample cooler. Samples for chemical analysis should be packed in Blue Ice®. Proceed to Step 7.
5. To collect a composite sample, sample liners are not needed to line the core barrel of the sampler. Use the slide hammer to beat the core barrel to the desired depth, and record the blow count in a sample logbook. Remove the core barrel from the hole by rocking it from side to side several times before lifting or reverse-beating the core barrel from the hole. Unscrew the core barrel from the sampler. Use a stainless steel knife or spoon to transfer the soil from the core barrel into a stainless steel bowl. Repeat this procedure for all of the intervals to be composited. Use a stainless steel spoon to homogenize (composite) all of the sampling intervals together in the stainless steel bowl before filling a sample jar. (Note: a decontaminated sampler should be used at each location being composited).
6. After the jar is capped, attach a sample label and custody seal to the jar and immediately place it into a sample cooler. Samples for chemical analysis should be packed in Blue Ice®.
7. Scan the hole where the sample was collected using chemical and/or radiological field screening instruments, and record the results in a bound logbook.
8. See Chapter 5 for details on preparing samples and coolers for sample shipment.

9. Any soil left over from the sampling should be transferred into a waste container. Before leaving the site, all waste containers should be sealed, labeled, and handled appropriately (see Chapter 11).
10. Transfer any other sampling-related wastes (e.g., gloves and foil) into a plastic waste bag.
11. Have a professional surveyor survey the coordinates of the sampling point to preserve the exact sampling location.

4.2.4.1.4 Open-Tube Sampler Method

For collecting shallow soil samples for only lithology description, the open-tube sampler is recommended. This tool consists of an open-core barrel, extension rod, and T-bar handle. The AMS Soil Recovery Probe presented in Figure 4.31 is a type of open-tube sampler. Because the sampling tube is open on one side, the soil lithology can be described without removing the sample from the tube. This tool works most effectively in moist nongravelly soils. Clear plastic sleeve liners can be inserted into the open-tube sampler. When the sleeve liner is removed from the sampler, it can be capped to preserve the sample. For additional information on open-tube samplers, see www.ams-samplers.com.

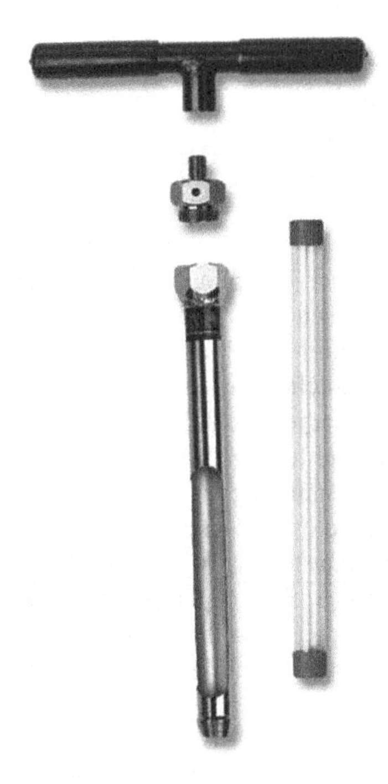

FIGURE 4.31 Soil recovery probe.

(From Byrnes, 2001. *Sampling and Surveying Radiological Environments*. Lewis Publishers, Boca Raton, FL.)

For a large sampling program, three people are sufficient for this procedure. One is needed for sample collection and description, a second is needed for health and safety, and a third is needed for miscellaneous tasks such as waste management and equipment decontamination.

The following equipment and procedure can be used to collect shallow soil samples for lithology description:

Equipment

1. Open-tube sampler and extension rods
2. Clear plastic sleeve liner with end-caps
3. Stainless steel knife
4. Sample labels
5. Sample logbook
6. Chain-of-custody seals
7. Permanent ink marker
8. Health and safety instruments
9. Chemical and/or radiological field screening instruments
10. Health and safety clothing
11. Waste container (e.g., 55-gal drum)
12. Sample table
13. Plastic sheeting
14. Aluminum foil
15. Plastic waste bags

Sampling Procedure

1. In preparation for sampling, confirm that all necessary preparatory work has been completed, including obtaining property access agreements, meeting health and safety and equipment decontamination requirements, and checking the calibration of all health and safety and chemical and/or radiological field screening instruments.
2. Cut a 1-ft-diameter hole in the center of the plastic sheeting, and center the hole over the sampling point. The purpose of this sheeting is to help prevent the spread of contamination.
3. Use the T-bar handle to rotate the sampler while pushing it into the ground. Continue advancing the sampler until it has reached the desired depth.
4. Remove the sampler from the ground by pulling upward on the T-bar. To avoid injury, be certain to keep your back straight and lift with your legs.
5. If a clear plastic sleeve liner is not being used, lay the sampler on the sampling table underlain by a piece of aluminum foil. Because the sampling tube is open-sided, the soil can be described without removing it from the tube. When describing the lithology, it is recommended that a stainless steel knife be used to slice open the sample, to reveal the sample texture.

6. If the sample is to be archived, it is recommended that a clear plastic sleeve liner be inserted into the sampling tube before sample collection. After the sample has been collected, remove the clear plastic sleeve liner from the sampler and cap the ends. After describing the lithology, add a sample label to the liner noting the name of the sampler, sampling time, date, location, and depth, and seal the end-caps with chain-of-custody tape before archiving.
7. To collect a deeper sample from the same hole, attach an extension rod to a clean sampling tube and repeat the preceding procedure.
8. Scan the hole where the sample was collected using chemical and/or radiological field screening instruments, and record the results in a bound logbook.
9. Transfer any soil left over from the sampling into a waste container. Before leaving the site, all waste containers should be sealed, labeled, and handled appropriately (see Chapter 11).
10. Transfer any other sampling-related wastes (e.g., gloves and foil) into a plastic waste bag.
11. Have a professional surveyor survey the coordinates of the sampling point to preserve the exact sampling location.

4.2.4.1.5 Split-Tube or Solid-Tube Method

The split-tube or solid-tube method is very similar to the core barrel (slide hammer) method (Section 4.2.4.1.3), except that a drill rig is typically used to beat the sampler into the ground. These samplers are composed of a split or solid sample tube, hardened shoe, soil catcher, and ball check (Figure 4.32). These samplers are available in two standard sizes, where the tubes are either 18 or 24 in. in length and have an outer diameter of 2, 3, or 4 in. Other sizes can be specially ordered.

Using sample liners to line the sample tube is not necessary; however, they are strongly recommended when collecting grab samples for volatile organic analysis. Without sample liners, soil must be extracted from the tube and transferred into a sample jar using a stainless steel spoon. In this procedure, volatile organics can be lost into the ambient air. In contrast, sample liners can be quickly removed from the barrel, sealed with airtight Teflon caps, labeled, custody-sealed, and then shipped to the laboratory for analysis. When samples are being analyzed for constituents other than volatile organic compounds, sample liners are not essential. When collecting vertical or areal composite samples, sample liners are not recommended. Soil from the intervals or locations to be composited is transferred from the split tube or solid tube into a stainless steel bowl and homogenized before filling a sample jar. (Note: a decontaminated sampler should be used at each location being composited).

For most sampling programs, three people are sufficient for this sampling procedure in addition to the drill rig operators. One is needed for sample collection, lithology description, labeling, and documentation; a second is needed for health and safety, and a third is needed for miscellaneous tasks such as waste management and equipment decontamination.

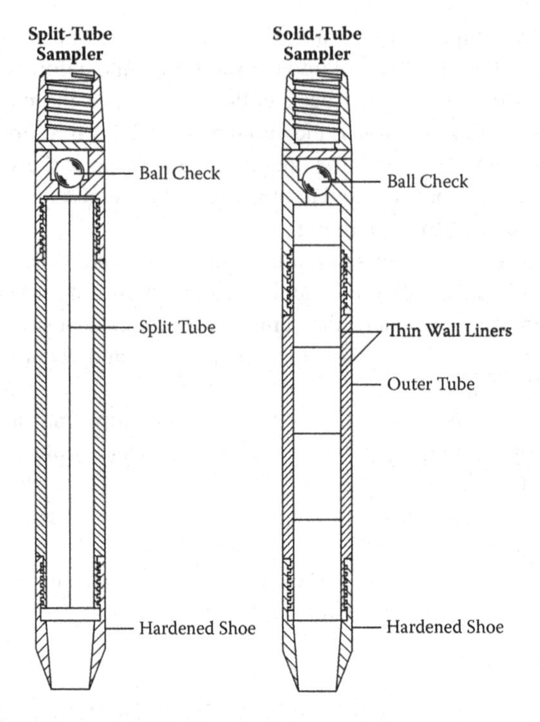

FIGURE 4.32 Split-tube and solid-tube sampler.

(From Byrnes, 1994. *Field Sampling Methods for Remedial Investigations.* Lewis Publishers, Boca Raton, FL.)

The following equipment and procedure can be used to collect shallow soil samples for chemical and/or radiological analysis:

Equipment

1. Split-tube or solid-tube sampler
2. Sample liners
3. Teflon end-caps for sample liners
4. Drill rig with slide hammer
5. Stainless steel bowl
6. Stainless steel spoon
7. Stainless steel knife
8. Soil sample jars
9. Sample labels
10. Cooler packed with Blue Ice® (Blue Ice® is not required for radiological analysis.)
11. Trip blank (only required for volatile organic analyses)
12. Coolant blank (not required for radiological analysis)
13. Sample logbook
14. Chain-of-custody forms

15. Chain-of-custody seals
16. Permanent ink marker
17. Health and safety instruments
18. Chemical or radiological field screening instruments
19. Health and safety clothing
20. Waste container (e.g., 55-gal drum)
21. Sampling table
22. Plastic sheeting
23. Plastic waste bags

Sampling Procedure

1. In preparation for sampling, confirm that all necessary preparatory work has been completed, including obtaining property access agreements, meeting health and safety and equipment decontamination requirements, and checking the calibration of all health and safety and chemical and/or radiological field screening instruments.
2. Cut a 1-ft-diameter hole in the center of the plastic sheeting, and center the hole over the sampling point. The purpose of this sheeting is to help prevent the spread of contamination.
3. Have the drillers back the drill rig up to the sampling location, carefully raise the mast, and then drill down to the top of the desired sampling interval.
4. Attach the sampler to a length of drilling rod, and position it above the interval to be sampled. Using the drill rig hammer, beat the sampler into the ground. Record the blow count in a sample logbook.
5. After removing the drill rod from the hole, detach the split-tube or solid-tube sampler.
6. To collect a grab sample using a split tube, break the tube open to reveal the sample liners. Using a stainless steel knife, separate the individual sleeves. Place Teflon caps over the ends of those to be sent to the laboratory for analysis. If sample liners are not being used, spoon soil from the split tube directly into a sample jar.
7. To collect a grab sample using a solid tube, slide the sample liners out of one end of the solid tube. Using a stainless steel knife, separate the individual sleeves. Place Teflon caps over the ends of those to be sent to the laboratory for analysis. If sample liners are not being used, spoon soil from the solid-tube directly into a sample jar.
8. To collect a composite sample from either a split or solid tube, there is no need to use sample liners. Rather, soil from each of the intervals to be composited should be transferred into a stainless steel bowl and homogenized before filling a sample jar. (Note: a decontaminated sampler should be used at each location being composited).
9. After the sleeve/jar is capped, attach a sample label and custody seal to the sleeve/jar, and immediately place it into a sample cooler. Samples for chemical analysis should be packed in Blue Ice®.

10. Scan the hole where the sample was collected using chemical and/or radiological field screening instruments, and record the results in a bound logbook.
11. See Chapter 5 for details on preparing samples and coolers for sample shipment.
12. Transfer any soil left over from the sampling into a waste container. Before leaving the site, all waste containers should be sealed, labeled, and handled appropriately (see Chapter 11).
13. Transfer any other sampling-related wastes (e.g., gloves and foil) into a plastic waste bag.
14. Have a professional surveyor survey the coordinates of the sampling point to preserve the exact sampling location.

4.2.4.1.6 Thin-Walled Tube (Shelby Tube) Method

What is unique about this method is that the sample tube is hydraulically pushed into the ground using a drill rig, as opposed to being driven into the ground with a hammer. The advantage of pushing the sampler into the ground is that the soil is not artificially compacted in the sampling process. Consequently, the thin-walled tube (Shelby tube) is the preferred method for collecting samples for geotechnical analysis. Some of the more common geotechnical tests run on soil samples include porosity, hydraulic conductivity, specific gravity, grain size distribution, Atterberg limits, compaction, consolidation, compression, and shear.

The thin-walled tube (Shelby tube) method utilizes a thin-walled sampling tube that has a standard outer diameter of 3 in., and length that allows the collection of a 30-in. sample (Figure 4.33). There are four holes at the top of the tube, which are used to connect the sampler to a sampling rod. Thin-walled tubes (Shelby tubes) are available in either low-carbon steel, or stainless steel. If only geotechnical analyses are to be performed on the sample, low-carbon steel is acceptable. However, if the soil sample is to be tested for chemical and/or radiological composition, the sampler should preferably be made of stainless steel.

For most sampling programs, three people in addition to the drill rig operators are sufficient for this sampling procedure. One is needed for sample collection, lithology description, labeling, and documentation; a second is needed for health and safety; and a third is needed for miscellaneous tasks such as waste management and equipment decontamination.

The following equipment and procedure can be used to collect shallow soil samples for geotechnical testing:

Equipment

1. Thin-walled tube (Shelby tube) sampler
2. Sample tube end-caps
3. Drill rig and drill rod
4. Sampling knife
5. Sample labels
6. Sample logbook

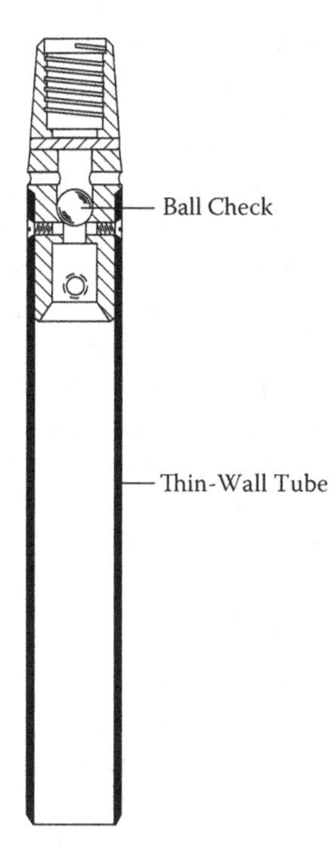

FIGURE 4.33 Thin-walled tube (Shelby tube) sampler.

(From Byrnes, 1994. *Field Sampling Methods for Remedial Investigations.* Lewis Publishers, Boca Raton, FL.)

7. Chain-of-custody forms
8. Chain-of-custody seals
9. Permanent ink marker
10. Health and safety instruments
11. Chemical and/or radiological field screening instruments
12. Health and safety clothing
13. Waste container (e.g., 55-gal drum)
14. Sampling table
15. Plastic sheeting
16. Paraffin wax
17. Plastic waste bags

Sampling Procedure

1. In preparation for sampling, confirm that all necessary preparatory work has been completed, including obtaining property access agreements, meeting health and safety and equipment decontamination requirements, and checking the calibration of all health and safety and chemical and/or radiological field screening instruments.

2. Cut a 1-ft-diameter hole in the center of the plastic sheeting, and center the hole over the sampling point. The purpose of this sheeting is to help prevent the spread of contamination.
3. Have the drillers back the drill rig up to the sampling location, carefully raise the mast, and then advance the hole down to the top of the desired sampling interval.
4. Attach the thin-walled tube (Shelby tube) to a length of drill rod, and lower it into the hole. Using the weight of the drill rig, hydraulically push the sample tube into the ground.
5. After removing the drill rod from the hole, detach the sample tube. Using a sampling knife, shave approximately 0.5 in. of soil from each end of the sample tube. Fill the space with melted paraffin wax, then place a cap over each end of the tube. The purpose of the wax is to prevent the shifting of soil in the tube during shipment to the geotechnical laboratory.
6. Attach a sample label to the tube and then place custody seals over each end cap.
7. Place sample tubes in a vertical position for transport to the geotechnical laboratory to preserve the soil compaction characteristics. If samples must be shipped, it is best to mark the sample box "FRAGILE." Also, denote on the outside of the box which end is "UP." If no chemical analyses are being performed, there is no need to chill the sample.
8. Scan the hole where the sample was collected using chemical or radiological field screening instruments, and record the results in a bound logbook.
9. Transfer any soil left over from the sampling into a waste container. Before leaving the site, all waste containers should be sealed, labeled, and handled appropriately (see Chapter 11).
10. Transfer any other sampling-related wastes (e.g., gloves and foil) into a plastic waste bag.
11. Have a professional surveyor survey the coordinates of the sampling point to preserve the exact sampling location.

4.2.4.1.7 En Core® Soil Sampler Method

When collecting surface soils, or shallow soils brought to the ground surface using methods identified in Section 4.2.4.1 for VOC analysis, the En Core® Sampler (Figure 4.34) should be considered since it minimizes the handling of the sample which in turn reduces the loss of VOCs. The En Core® Sampler works by collecting and storing soil samples in an airtight self-contained coring body that is made of an inert composite polymer. This coring body is available in two sizes (5 g and 25 g). An En Core® T-handle is used to push the coring body into the soil. This coring body is particularly effective in collecting soil samples that contain moist clay and silt. This coring body has challenges collecting dry sand since it will not form a cohesive plug.

There is a viewing hole in the T-handle where one can see if the plunger has been fully pushed back which means that the coring body is full of soil. If you are able to twist the plunger as the procedure requires below, then you also know the coring body

FIGURE 4.34 En Core® Soil sampler.

is full of soil. One will have to push the coring body into the soil again if it is not full. A sealing cap is placed over the coring body, twisted, and then two locking arms are snapped in place to complete the seal. The hold time for this method is currently 48 hours when stored at 4°C.

Advantages of this sampling method include:

- Minimizes the handling of soil samples which reduces the loss of VOCs
- A sealing cap placed over the coring body ensures the sampler is secure
- Coring body is made of an inert composite polymer

Disadvantages of this sampling method include:

- Has challenges collecting cores of dry sand since it will not form a cohesive plug

The following equipment and procedure can be used to collect soil samples for VOC analysis.

Equipment

1. En Core® sealed sample bag that contains coring body and sealing cap
2. En Core® T-Handle
3. Stainless steel Spatula
4. Project specific sampling and analysis plan (see Section 3.2.7)
5. Plastic sheeting ground cover (optional)
6. Sample labels

7. Cooler packed with Blue Ice®
8. Trip blank
9. Coolant blank
10. Field logbook
11. Chain-of custody forms
12. Chain-of custody seals
13. Permanent ink marker
14. Paper towels
15. Health and safety instruments and clothing (see Chapter 10)
16. Waste container (e.g., 55-gal drum)
17. Sampling table (optional)
18. Plastic waste bag

Procedure

1. In preparation for sampling, confirm that all necessary preparatory work has been completed, including obtaining property access agreements, meeting health and safety and equipment decontamination requirements, and checking the calibration of all health and safety instruments.
2. Verify all equipment has been properly decontaminated (see Chapter 9).
3. Remove En Core® coring body from sample bag.
4. Holding En Core® T-Handle, depress locking lever, then slide coring body (plunger end first) into open end of T-Handle (see Figure 4.34).
5. Align two locking pins on inside of T-Handle with two slots on coring body.
6. Twist coring body clockwise to lock pins into slots.
7. Verify coring body is locked into T-Handle.
8. Use health and safety instruments (e.g., photoionization detector, flame-ionization detector) to screen air quality as soil is brought to the ground surface. Based on results, adjust personal protective equipment if needed in accordance with project health and safety plan (see Chapter 10).
9. Use permanent ink marker to record in field logbook the date and time the sample was collected. Label En Core® sample bag with site name, sampling location, sample number, sample date/time, and analyses to be performed.
10. Gripping T-Handle in hand, push coring body several inches into soil being sampled.
11. Look through the first viewing hole on T-Handle if 5 g sample is being collected, or look through second viewing hole if 25 g sample is being collected.
12. If the O-ring found on plunger is not in center of viewing hole on T-Handle, push coring body into soil being sampled a second time. Repeat until O-ring is in center of viewing hole on T-Handle.
13. Use stainless steel spatula to scrap off excess soil from end of coring body.
14. Use paper towel to wipe off excess soil from sides of coring body.

15. Slide sealing cap (that contains two wings) on to coring body. Coring body has two flat sections which is where two grooved wings on sealing cap slide on. Once sealing cap is slide on, give sealing cap ¼ of a twist clockwise to lock in place.
16. Depress locking lever on T-Handle and remove coring body.
17. Slide plunger end of coring body into grooves in the T-Handle and twist core body ¼ of a turn counterclockwise to lock plunger and seal the coring body.
18. Attach custody seal to the coring body, then place into labeled En Core® sample bag and use zip-lock to close the bag.
19. Place En Core® sample bag into cooler packed in Blue Ice®, coolant blank, and trip blank.
20. See Chapter 5 for details on preparing sample containers and coolers for sample shipment
21. Containerize any waste in a waste container. Before leaving the site, all waste containers should be sealed, labeled, and handled appropriately (see Chapter 11).
22. Have a professional surveyor survey coordinates of sampling point to preserve the exact sampling location.

4.2.4.2 Deep Soil Sampling

Soil samples collected at depths greater than 5 ft are generally referred to as *deep*. These samples are most commonly collected by driving a split-tube or solid-tube sampler (Figure 4.32) or hydraulically pushing a thin-walled tube (Shelby tube) (Figure 4.33) into the ground with the assistance of a drill rig. These sampling methods are similar to those described for shallow soil sampling (Sections 4.2.4.1.5 and 4.2.4.1.6) except samples are collected from deeper intervals.

Table 4.2 summarizes the effectiveness of the three recommended sampling methods. A number "1" in the table indicates that a particular procedure is most effective in collecting samples for a particular laboratory analysis, sample type, or sampling depth. A number "2" indicates that the procedure is acceptable, but less preferred, whereas an empty cell indicates that the procedure is not recommended. For example, Table 4.2 indicates that the split-tube and solid-tube samplers are both effective in collecting soil samples for volatile organic analysis and all other chemical analyses, whereas the En Core® sampler is only recommended when collecting soil samples for volatile organic analysis and the thin-walled tube (Shelby tube) sampler is only recommended to be used to collect geotechnical samples.

The following sections provide background information about common drilling methods used to support deep soil sampling and further details and SOPs for deep soil sampling methods.

4.2.4.2.1 Common Drilling Methods Used to Support Deep Soil Sampling

The purpose of this section is to give a high-level overview of the advantages and disadvantages of several common drilling methods used to support environmental investigations. The drilling methods discussed in this section include:

TABLE 4.2
Rating Table for Deep Soil Sampling Methods

	Laboratory Analyses								Sample Type			Depth	
	Radionuclides	Volatiles	Semivolatiles	Metals	Pesticides	PCBs	TPH	Geo technical	Grab	Composite (Vertical)	Composite (Areal)	Deep (> 5.0 ft)	Lithology Description
Split tube/ solid tube	1/1	1/1	1/1	1/1	1/1	1/1	1/1		1/1	1/2	1/2	1/1	1/2
Thin-walled tube (Shelby tube)								1	1			1	
En Core® Sampler		1							1			1[a]	

Note: 1 = preferred method; 2 = acceptable method; empty cell = method not recommended.

[a] The En Core® sampler may be used to collect a soil core sample from any depth at which a soil core is brought to the ground surface using either the core barrel (slide hammer) or split tube/solid tube sampler.

- Direct Push
- Hollow-stem auger
- Reverse circulation air rotary
- Sonic
- Cable tool

Prior to selecting a preferred drilling method, it is important to perform a thorough literature study to become familiar with the local geology, depth to groundwater, depth to bedrock, type of bedrock, cohesiveness of the material to be drilled, presence of cobbles and boulders, presence of heaving sands, etc. If boreholes have been drilled nearby, study the borehole logs.

Since the direct push method is less expensive than other drilling methods and offers many additional sampling options, one should always look at this method closely. Direct push uses a hydraulic press and slide hammer mounted on the rear end of a truck or other vehicle and provides the advantage of generating very little investigation-derived waste, it gathers less public attention than a drill rig, and it provides very little disturbance to the surrounding environment. Because it only generates a very small amount of waste it is more protective of worker's health than other methods. Direct push has the ability to collect depth-discrete soil samples, groundwater samples (using an exposed screen, sealed screen, or groundwater profiler sampler), soil gas samples, geophysical sensing and geochemical sensing data, and can install very small diameter wells or piezometers. A Cone Penetrometer Test (CPT) probe can be used with the direct push method to collect pressure and friction data to produce a stratigraphic and soil properties log. A membrane interface probe (MIP) can be used to identify hydrocarbon and other VOCs (including light nonaqueous phase liquid [LNAPL] and dense nonaqueous phase liquid [DNAPL]) in the vadose and saturated zones using a combination of photoionization (for aromatic hydrocarbons), electron capture (for chlorinated contaminants), and flame ionization (for straight-chain hydrocarbons). The Laser-Induced Fluorescence (LIF) and Ultra-Violet Optical Screening Tool (UVOST) are direct push tools that provide screening-level qualitative and quantitative information without directly sampling LNAPL and heavier hydrocarbons including petroleum, oil, and lubricant contamination (California EPA 2012). One disadvantage of the direct push method is that the soil core samples collected are of very small diameter so there may not be enough sample volume to run a full suite of laboratory analyses. Direct push is not effective when cobbles or boulders are present and it can only install wells to a maximum depth of between 100 and 150 ft (see Section 4.2.7.1 for more details on the direct push method).

The hollow-stem auger is a common drilling method used to set groundwater wells in unconsolidated sediment as deep as 150 ft below the ground surface. This is a moderate speed drilling method that advances a "hollow stem", with a claw bit attached to the end, through the formation. The hollow-stem makes it easy to collect shallow and deep soil samples by lowering split-spoon/solid tube samplers and/or thin-walled tube (Shelby tube) samplers down the inside of the hollow-stem (see Sections 4.2.4.2.2 and 4.2.4.2.3). When constructing a groundwater well, the hollow-stem acts as a temporary casing which keeps the sides of the borehole from collapsing. Split-spoon/solid tube samplers and thin-walled tube (Shelby tube) samplers (see Sections 4.2.4.2.2 and 4.2.4.2.3) are most typically used to collect soil samples with this drilling method. Common downhole screening methods that can be used with this drilling method are presented in Section 4.2.3. One advantage of this drilling method is that fluid is rarely required to assist in the drilling process. Also, this drilling method is effective in collecting representative soil samples from the formation for all types of analyses (including volatile organics). The disadvantages of this method are that it is limited to drilling through unconsolidated sediment, it is not effective in drilling through cobbles, high hydrostatic pressures can cause problems with sand heaving up into the auger during soil sampling and well installation, and it has a maximum drilling depth of 150 ft.

The Reverse Circulation air rotary method is a common drilling method used to drill boreholes and set groundwater wells at depths >500 ft below ground surface. This is a high-speed drilling method that uses dual wall drill rods, with one outer drill rod and one inner tube. A hammer that contains the drill bit is attached to the end of the drill rod. During drilling, when air is blown down the annulus of the outer drill rod, the pressure shift generates a reverse circulation which brings the drill cuttings from the drill bit up the inner tube to the cyclone located at the ground surface. This keeps the drill cuttings free from cross-contamination. After the drill cuttings pass through the cyclone they fall into either a cutting container or a sample bag. Split-spoon/solid tube samplers and thin-walled tube (Shelby tube) samplers (see Sections 4.2.4.2.2 and 4.2.4.2.3) are most typically used to collect soil samples with this drilling method. Common downhole screening methods that can be used with this drilling method are presented in Section 4.2.3. The disadvantages of this drilling method are that it collects more public attention than the direct push method, and using the air rotary method could impact the concentration of volatile organic compounds in soil and groundwater samples collected during drilling.

Sonic drilling is a common method used to drill boreholes and set groundwater wells at depths up to 300 ft below the ground surface. With this moderate speed drilling method, high-frequency resonant energy is generated by a sonic head which is used to advance a core barrel through the formation. A larger diameter drill string is often advanced along with the core barrel to keep the borehole open. To achieve maximum drilling rates, the driller controls the sonic head's resonant energy to match the formation being drilled. This minimizes the friction of the soil immediately adjacent to the drill string. One of the advantages of this method is that it produces a relatively undisturbed sample of the formation with often close to 100% core recovery. This allows for a very thorough and accurate logging of the lithology of the borehole, and it allows the geologist to carefully select the optimum depth interval to sample for laboratory analysis. One disadvantage of this drilling method is that it generates heat as the sonic head and core barrel is advanced. If the heat gets too high, it can impact the chemistry of the samples being collected for laboratory analysis. As a general rule, the temperature of the core samples should be maintained below 140° F (60° C) to maintain sample integrity. THERMAX® temperature strips or an infrared thermometer temperature gun should be used to measure the temperature of core samples.

The cable tool drilling method is one of the earliest known drilling methods. With this slow speed drilling method, a borehole can be drilled >500 ft in depth by the repeated raising and dropping of a heavy drill string with a bit attached at the bottom. This up-and-down motion combined with a left-lay cable and rope socket causes the drill string and bit to rotate slightly on each vertical stroke. This motion loosens the underlying formation and breaks up rock into cuttings that are then removed from the borehole using a bailing device. When drilling through unconsolidated material, temporary drill casing must be driven as the borehole is advanced to keep the borehole from caving in. One of the advantages of this method is that it is suitable for drilling in most geologic conditions. Also, this drilling method is effective in collecting representative soil samples from the formation for all types of analyses (including volatile organics). This method is often used for drilling boreholes and wells at the Hanford Nuclear Site when radionuclides are expected to be present. This is because this method minimizes the mobilization of contaminants into the breathing zone. The primary disadvantage of this method is the slow speed of drilling.

4.2.4.2.2 Split-Tube or Solid-Tube Method

The split-tube or solid-tube method used to collect deep soil samples is identical to that described for shallow soil sampling (Section 4.2.4.1.5), with the exception that samples are collected from a depth greater than 5 ft.

4.2.4.2.3 Thin-Walled Tube (Shelby Tube) Method

The thin-walled tube (Shelby tube) method used to collect deep soil samples is identical to the procedure described for shallow soil sampling (Section 4.2.4.1.6), with the exception that samples are collected from a depth greater than 5 ft.

4.2.4.2.4 En Core® Soil Sampler

When deep soil samples are collected during drilling (e.g., split-tube sampler, solid-tube sampler) and need to be run for VOC analysis, the En Core® Sampler (Figure 4.34) should be considered to help minimize the loss of VOCs in the sample collection process. In this case the En Core® Sampler would be used to collect one or more grab soil samples from the deep soil sample for VOC analysis. The En Core® Sampler works by collecting and storing soil samples in an airtight self-contained coring body that is made of an inert composite polymer. The sampling method, equipment list, and procedure for using the En Core® Sampler to collect samples of deep soil are identical to that found in Section 4.2.4.1.7 for shallow soil sampling and is not repeated here.

4.2.5 Sediment Sampling

This section provides the reader with guidance on selecting sediment sampling methods for remedial investigation studies. The criteria used in selecting the most appropriate method include the analyses to be performed on the sample, the type of sample being collected (grab or composite), and the sampling depth. Standard operating procedures have been provided for each of the methods to facilitate implementation.

One objective of a remedial investigation should be to determine if chemical and/or radiological contamination is present in the sediment of nearby surface water units, such as streams, rivers, surface water drainages, ponds, lakes, retention basins, or tanks. It is particularly important to characterize the sediment in streams, rivers, and other surface water drainages because they provide avenues for rapid contaminant migration, and provide points where receptors are readily exposed to contamination. Ponds, lakes, retention basins, and tanks do not provide the same opportunity for rapid contaminant migration; however, they similarly provide exposure points for receptors.

The DQO process (Chapter 3, Section 3.2.5) should be used to define the sampling approach, required number of samples, analyses to be performed, and analytical performance requirements.

Sampling tools used to collect sediment samples for laboratory analysis should be constructed of materials that are compatible with the media and the constituents that are being tested for. In other words, the sampler should *not* be constructed of a material that could cause a loss or gain in contaminant concentrations measured due to sorption, desorption, degradation, or corrosion (EPA 2002). Because stainless steel, Teflon, and glass are inert substances, sampling equipment made of these materials should be preferentially selected over equipment made of other materials. USGS (2003) notes that fluorocarbon polymers (Teflon®, Kynar®, and Tefzel®), stainless steel 316-grade, and borosilicate glass (laboratory grade) are acceptable for both inorganic and organic analyses; polypropylene, polyethylene, PVC, silicone, and nylon are only acceptable for inorganic analyses; and stainless steel 304-grade, and other metals (brass, iron, copper, aluminum, and galvanized and carbon steels) are only acceptable for organic analyses. Having a rigorous quality control sampling program (e.g., collecting rinsate blanks) in place will help ensure that the sample integrity is not being impacted by the sampling equipment (see Chapter 6).

The following subsections present preferred sampling methods and procedures for collecting sediment samples from streams, rivers, surface water drainages, ponds, lakes, retention basins, and tanks.

4.2.5.1 Stream, River, and Surface Water Drainage Sampling

Although a number of sophisticated sampling devices are available to collect sediment samples, not all of these tools are effective in collecting samples through the shallow, fast-moving water typical of streams, rivers, and surface water drainages. The methods that have proved to be the most effective when sampling these environments include the scoop or dipper method, core barrel (slide hammer) method, WaterMark® Russian sediment borer method, box sampler (e.g., Ekman dredge) method, integrating sediment sampler method, and the Helley-Smith sampler method. The first four of the preceding methods collect samples of the sediment below the water–sediment interface. On the other hand, the latter two methods are designed to collect a sample of the sediment carried in suspension by a moving stream or river water.

Of these six methods, the scoop or dipper method is the easiest to implement because it just involves pushing the sampler into the sediment, and then either transferring the sediment directly into a sample jar or a stainless steel bowl and compositing with other sampling locations before filling the sample jar. This technique is only effective in collecting sediment samples 0.5 ft below the sediment –water interface, where the water depth is less than 2 ft.

The core barrel (slide hammer) method involves beating a sampling tube into the sediment. This technique removes a core of sediment for analytical testing. Because the sampling tube is typically lined with sampling sleeves, either individual sleeves can be sent to the laboratory as grab samples or the sediment can be removed from the sleeves and composited before filling a sample jar. This method is effective in collecting sediment samples as deep as 5 ft below the sediment– water interface.

In the WaterMark® Russian Sediment Borer method, the borer (in the closed position) is advanced to the preferred sampling depth by pushing downward on the sampler while the bar handle is rotated, or driving it downward with the optional slide hammer assembly. The bore is then rotated clockwise 180° so that the sharpened edge of the chamber cuts a semicylindrical sediment core that is contained by the pivotal cover plate. During retrieval, the cover plate's counterclockwise rotation extrudes an undisturbed sediment sample. This method is effective in collecting sediment samples as deep as 5 ft below the sediment–water interface.

The box sampler (e.g., Ekman dredge) method utilizes a spring-loaded sample box attached to a sampling pole to collect grab sediment samples. This method involves pushing the sampling box into the sediment, and releasing the spring-loaded sample jaws. After the sampler is retrieved, sediment from the sampling box is transferred into a sample jar. Similar to the scoop or dipper method, this method is only effective in collecting sediment samples 0.5 ft below the sediment–water interface.

The integrating sediment sampler method is designed to collect an integrated sample of sediment being carried in suspension by a moving stream or river. This sampler is composed of the sampler body, a glass sample bottle, a nozzle where the sample enters the sampler, and an air exhaust port where air escapes from the sample bottle as the sediment sample is collected. Once the sampler is retrieved, the filled sample bottle is removed, capped, and labeled in preparation for analysis. This method is only used to collect a sample of the sediment carried in suspension by a moving stream or river water.

The Helley-Smith sampler method utilizes a stainless steel sampler and nylon mesh bag to catch a sample of silt, sand, or gravel carried in suspension by surface water. This sampler is often placed just above the drainage, stream, or riverbed to collect a sample of the bed load material being transported. After the sampler is retrieved, sediment is transferred into a sample jar. This method is only used to collect a sample of the sediment carried in suspension by moving drainage, stream, or river water.

Depending on the depth of the water overlying the sediment, samplers may need a pair of waders, a raft, or a boat to access the sampling point. If waders are used, the sampler should face upstream while collecting the sample to ensure that the sampler's boots do not contaminate the sample. Similarly, samples should be collected from the upstream side of the raft or boat. As a general rule, the sample located farthest downstream should always be the first collected. Sampling should then proceed upstream. By collecting samples in this manner, any sediment disturbed by the samplers will not contaminate downstream sampling points.

Table 4.3 summarizes the effectiveness of each of the six recommended sampling methods. A number "1" in the table indicates that a particular procedure is most effective in collecting samples for a particular laboratory analysis, sample type, or

TABLE 4.3
Rating Table for Sediment Sampling Methods for Streams, Rivers, and Surface Water Drainages

	Laboratory Analyses								Sample Type				Depth			
	Radionuclides	Volatiles	Semivolatiles	Metals	Pesticides	PCBs	TPH	Geotechnical	Grab	Composite (Vertical)	Composite (Areal)	Integrated	Surface (0.0–0.5 ft)	Shallow (0.0–5.0 ft)	Lithology Description	Suspended sediment in water
Methods for collecting sediment below sediment–water interface																
Scoop or dipper	2	2	2	2	2	2	2		2		2		2		2	
Core barrel (slide hammer)	1	1	1	1	1	1	1		1	1	2		1	1	2	
WaterMark® Russian sediment borer	1	2	1	1	1	1	1		1	1	1		1	1	1	
Box sampler	1	2	1	1	1	1	1		1		1		1		1	
Methods for collecting sediment carried by surface water in suspension																
Integrating sediment sampler	1	2	1	1[a]	1	1	1					1				1
Helley-Smith sampler	2	2	2	2	2	2	2					2				2

Note: 1 = preferred method; 2 = acceptable method; empty cell = method not recommended.

[a] The optional brass nozzle should not be used when collecting samples to be analyzed for metals.

sampling depth. A number "2" indicates that the procedure is acceptable, but less preferred, whereas an empty cell indicates that the procedure is not recommended. For example, Table 4.3 indicates that the core barrel (slide hammer) method is most effective in collecting sediment samples for volatile organic analysis.

4.2.5.1.1 Scoop or Dipper Method

The scoop or dipper is the simplest sampling tool for collecting grab or areal composite sediment samples from streams, rivers, and surface water drainages, and is available in many shapes and sizes. Figure 4.26 presents several varieties of scoop samplers. A dipper is basically a deep cup attached to a long handle. This method involves lowering a scoop or dipper by hand through the surface water and pushing the sampler deep into the underlying sediment. When collecting a grab sample, sediment is transferred from the sampler directly into a sample jar. If a composite sample is collected, the sediment to be composited is transferred into a stainless steel bowl and homogenized with a stainless steel spoon before filling a sample jar. This technique is generally effective in collecting samples of the top 0.5 ft of sediment, at locations where the water depth is less than 2 ft.

This method is most effective when collecting samples from water bodies with relatively slow flow velocities, because a significant amount of the finer-grained sediment tends to be lost when sampling higher-energy environments. The dipper typically works more effectively than the scoop in preventing the loss of fine-grained sediment when retrieving the sample; however, because the dipper has no cutting edge, it is only effective in sampling soft sediment. The scoop or dipper method is most commonly used for preliminary sediment sampling activities. If contaminants are identified during the preliminary sampling, the core barrel (slide hammer) (Section 4.2.5.1.2), WaterMark® Russian Sediment Borer (Section 4.2.5.1.3), or box sampler (Section 4.2.5.1.4) methods should be considered to define the distribution of contaminants more accurately.

For most sampling programs, four people are sufficient for this sampling procedure. Two are needed for sample collection, labeling, and documentation; a third is needed for health and safety; and a fourth is needed for waste management and equipment decontamination.

The following equipment and procedure can be used to collect sediment samples for chemical or radiological analysis:

Equipment

1. Scoop or dipper
2. Stainless steel bowl
3. Stainless steel spoon
4. Sample jars
5. Sample labels
6. Cooler packed with Blue Ice® (Blue Ice® is not required for radiological analysis.)
7. Trip blank (only required for volatile organic analyses)
8. Coolant blank (not required for radiological analysis)

9. Sample logbook
10. Chain-of-custody forms
11. Chain-of-custody seals
12. Permanent ink marker
13. Health and safety instruments
14. Chemical and/or radiological field screening instruments
15. Health and safety clothing (including life jackets)
16. Waste container (e.g., 55-gal drum)
17. Sampling table
18. Plastic waste bags

Sampling Procedure

1. In preparation for sampling, confirm that all necessary preparatory work has been completed, including obtaining property access agreements, meeting health and safety and equipment decontamination requirements, and checking the calibration of all health and safety and chemical and/or radiological field screening instruments.
2. Approach the sampling point from downstream, being careful of not to disturb the underlying sediment.
3. Push the scoop or dipper firmly downward into the sediment, and then lift upward. Quickly raise the sampler out of the water in an effort to reduce the amount of fine-grained sediment lost to the water current. If a grab sample is being collected, transfer the sediment from the scoop or dipper directly into a sample jar. If a composite sample is being collected, transfer the sediment from each composite interval or location into a stainless steel bowl and homogenize with a stainless steel spoon before filling a sample jar. (Note: a decontaminated sampler should be used at each location being composited)
4. Scan the sediment in the jar using chemical and/or radiological screening instruments. Record the results in a bound logbook.
5. After the jar is capped, attach a sample label and custody seal to the jar, and immediately place it into a sample cooler. Samples for chemical analysis should be packed in Blue Ice®.
6. See Chapter 5 for details on preparing sample jars and coolers for sample shipment.
7. Transfer any sediment left over from the sampling into a waste container. Before leaving the site, all waste containers should be sealed, labeled, and handled appropriately (see Chapter 11).
8. Have a professional surveyor survey the coordinates of the sampling point to preserve the exact sampling location.

4.2.5.1.2 Core Barrel (Slide Hammer) Method

The core barrel (slide hammer) is an effective tool for collecting core samples of sediment from streams, rivers, and surface water drainages. This tool consists of a core barrel, extension rod, and slide hammer (see Figure 4.29). Stock core barrels have an inside diameter of 2 or 2.5 in., and are 1–3 ft in length. However, the core barrel can be specially ordered to meet project-specific volume requirements. The top of the barrel is threaded so that it can be screwed into an extension rod to allow sampling through deeper water.

The barrel is constructed to accept sample liners (see Figure 4.30), which are commonly used to facilitate the removal of sediment from the barrel without disturbing the sample. The use of liners is not necessary; however, they are recommended when grab samples are to be analyzed for volatile organic compounds. Without sample liners, sediment must be extracted from the barrel and transferred into a sample jar. In this process, volatile organics can be lost into the ambient air. In contrast, sample liners can be quickly removed from the barrel and sealed with airtight Teflon caps. After labeling and custody-sealing the liners, they can be shipped directly to the laboratory for analysis. When collecting a composite sample, one should transfer sediment from the intervals or locations to be composited into a stainless steel bowl and homogenize it before filling a sample jar.

Extension rods can be ordered in various lengths to allow sampling through various depths of water. The rods are screwed into the core barrel at one end and into the slide hammer at the other end. The slide hammer is used to beat the sampler into the sediment. The hammer is available in various weights to accommodate the needs of the sampler. If samples are to be collected from locations where water depths exceed several feet, a raft or boat will be required to assist the sampling procedure.

For most sampling programs, four people are sufficient for this sampling procedure. Two are needed for sample collection, labeling, and documentation; a third is needed for health and safety; and a fourth is needed for waste management and equipment decontamination. If a raft or boat is used to assist the sampling procedure, at least one additional person will be needed.

The following equipment and procedure can be used to collect sediment samples for chemical or radiological analysis:

Equipment

1. Core barrel (slide hammer) and extension rods
2. Sample liners
3. Teflon end-caps for sample liners
4. Stainless steel bowl
5. Stainless steel spoon
6. Stainless steel knife
7. Sample jars
8. Sample labels
9. Cooler packed with Blue Ice® (Blue Ice® is not required for radiological analysis.)
10. Trip blank (only required for volatile organic analyses)
11. Coolant blank (not required for radiological analysis)
12. Sample logbook
13. Chain-of-custody forms

14. Chain-of-custody seals
15. Permanent ink marker
16. Health and safety instruments
17. Chemical and/or radiological field screening instruments
18. Health and safety clothing (including life jackets)
19. Waders, raft, or boat
20. Sampling table
21. Waste container (e.g., 55-gal drum)
22. Plastic waste bags

Sampling Procedure

1. In preparation for sampling, confirm that all necessary preparatory work has been completed, including obtaining property access agreements, meeting health and safety and equipment decontamination requirements, and checking the calibration of all health and safety and chemical and/or radiological field screening instruments.
2. Approach the sampling point from downstream, being careful of not to disturb the underlying sediment.
3. Lower the sampler through the water, beat the core barrel to the desired depth, and record the blow counts in a sample logbook.
4. Remove the core barrel from the hole by either rocking it from side to side several times before lifting or reverse-beating the sampler from the hole.
5. To collect a grab sample, unscrew the core barrel from the sampler and slide the sample liners out onto the sampling table. Using a stainless steel knife, separate the sample liners. Scan the sediment exposed at the ends of the sample liners using chemical and/or radiological screening instruments. Record the results in a bound logbook. Place Teflon caps over the ends of the sleeves to be sent to the laboratory. If sampling sleeves are not being used, spoon sediment from the core barrel directly into a sample jar.
6. To collect a composite sample, sample liners are not needed; rather, sediment from each of the intervals to be composited should be transferred into a stainless steel bowl. The sediment in the bowl should be scanned using chemical and/or radiological screening instruments and then homogenized before filling a sample jar. (Note: a decontaminated sampler should be used at each location being composited.)
7. After the sleeve/jar is capped, attach a sample label and custody seal to it and immediately place it into a sample cooler. Samples for chemical analysis should be packed in Blue Ice®.
8. See Chapter 5 for details on preparing sample liners/jars and coolers for sample shipment.
9. Transfer any sediment left over from the sampling into a waste container. Before leaving the site, all waste containers should be sealed, labeled, and handled appropriately (see Chapter 11).

10. Transfer any other sampling-related wastes (e.g., gloves and foil) into a plastic waste bag.
11. Have a professional surveyor survey the coordinates of the sampling point to preserve the exact sampling location.

4.2.5.1.3 WaterMark® Russian Sediment Borer

The WaterMark® Russian Sediment Borer (Figure 4.35) is a very effective tool for collecting grab and composite (vertical and areal) samples of sediments from streams, rivers, and surface water drainages. This sampler is composed of a borer head (that opens and closes), extension rods, and a turning handle (or optional slide hammer assembly). With this method, the borer (in the closed position) is advanced to preferred sampling depth by pushing downward on the sampler while rotating the handle, or driving it downward with the optional slide hammer assembly. The bore is then rotated clockwise 180° so that the sharpened edge of the chamber cuts a semi-cylindrical sediment core that is contained by the pivotal cover plate. During retrieval, the cover plate's counterclockwise rotation extrudes the undisturbed sample. When collecting a grab sample, the sediment is transferred directly from the sampler into a sample jar. When collecting a composite sample, the sediment from each of the intervals to be composited is transferred into a stainless steel bowl and is homogenized before filling a sample jar. The standard borer measures 5 cm in diameter and 50 cm in length. This method is effective in collecting sediment samples as deep as 5 ft below the sediment–water interface.

For most sampling programs, four people are sufficient for this sampling procedure. Two are needed for sample collection, labeling, and documentation; a third is needed for health and safety; and a fourth is needed for waste management and equipment decontamination. If a raft or boat is used to assist the sampling procedure, at least one additional person will be needed.

The following equipment and procedure can be used to collect sediment samples for chemical and/or radiological analysis.

Equipment

1. WaterMark® Russian Sediment Borer
2. Stainless steel bowl
3. Stainless steel spoon
4. Sample jars
5. Sample labels

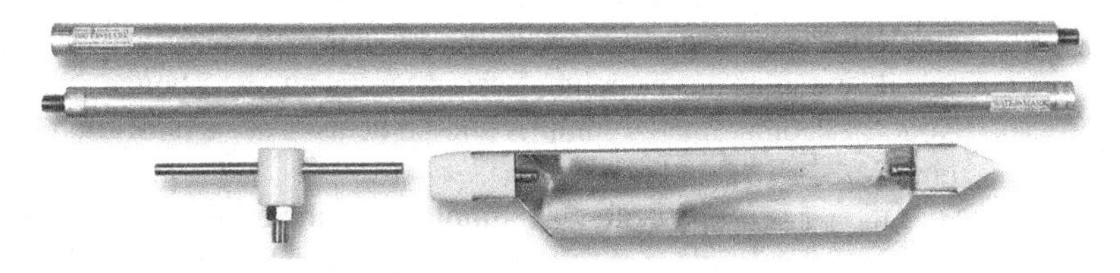

FIGURE 4.35 WaterMark® Russian sediment borer.

6. Cooler packed with Blue Ice® (Blue Ice® is not required for radiological analysis.)
7. Trip blank (only required for volatile organic analyses)
8. Coolant blank (not required for radiological analysis)
9. Sample logbook
10. Chain-of-custody forms
11. Chain-of-custody seals
12. Permanent ink marker
13. Health and safety instruments
14. Chemical or radiological field screening instruments
15. Health and safety clothing (including life jackets)
16. Waders, raft, or boat
17. Sampling table
18. Waste container (e.g., 55-gal drum)
19. Plastic waste bags

Sampling Procedure

1. In preparation for sampling, confirm that all necessary preparatory work has been completed, including obtaining property access agreements, meeting health and safety and equipment decontamination requirements, and checking the calibration of all health and safety and chemical and/or radiological field screening instruments.
2. Approach the sampling point from downstream, being careful of not to disturb the underlying sediment.
3. With the borer in the closed position, advance the head of the sampler through the sediment to the preferred sampling depth by either pushing downward while rotating the handle attached to the extension rod, or by driving the head of the sampler downward with the optional slide hammer assembly.
4. Rotate the bore clockwise 180° so that the sharpened edge of the chamber cuts a semicylindrical sediment core that is contained by the pivotal cover plate.
5. During retrieval, the cover plate's counterclockwise rotation extrudes the undisturbed sample.
6. If a grab sample is being collected, transfer the sediment from the sampler directly into a sample jar. If a composite sample is being collected, transfer the sediment from the locations to be composited into a stainless steel bowl and homogenize with a stainless steel spoon before filling a sample jar. Scan the sediment in the jar using chemical and/or radiological screening instruments. Record the results in a bound logbook. (Note: a decontaminated sampler should be used at each location being composited)
7. After the jar is capped, attach a sample label and custody seal to the jar and immediately place it into a sample cooler. Samples for chemical analysis should be packed in Blue Ice®.

8. See Chapter 5 for details on preparing sample jars and coolers for sample shipment.
9. Transfer any sediment left over from the sampling into a waste container. Before leaving the site, all waste containers should be sealed, labeled, and handled appropriately (see Chapter 11).
10. Transfer any other sampling-related wastes (e.g., gloves and foil) into a plastic waste bag.
11. Have a professional surveyor survey the coordinates of the sampling point to preserve the exact sampling location.

4.2.5.1.4 Box Sampler Method

The box sampler (e.g., Ekman dredge) is a very effective tool for collecting grab and areal composite samples of sediments from streams, rivers, and surface water drainages. This sampler is composed of a sample box, spring-loaded sample jaws, and a pole containing a spring release mechanism (Figure 4.36). After the sample box is pushed firmly into the sediment, the jaws are released to seal off the bottom of the box. The closed box is then retrieved from the water. The advantage of this sampling method is that fine-grained sediment is not stripped from the sample as it is removed from the water. When collecting a grab sample, the sediment is transferred directly

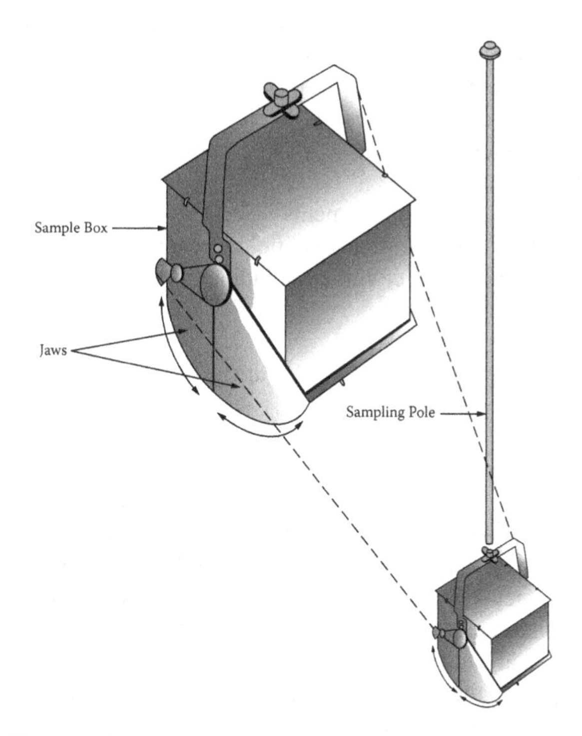

FIGURE 4.36 Box sampler.

(From Byrnes, 1994. *Field Sampling Methods for Remedial Investigations.* Lewis Publishers, Boca Raton, FL.)

from the sample box into a sample jar. When collecting a composite sample, the sediment from each of the intervals to be composited is transferred into a stainless steel bowl and is homogenized before filling a sample jar.

For most sampling programs, four people are sufficient for this sampling procedure. Two are needed for sample collection, labeling, and documentation; a third is needed for health and safety; and a fourth is needed for waste management and equipment decontamination. If a raft or boat is used to assist the sampling procedure, at least one additional person will be needed.

The following equipment and procedure can be used to collect sediment samples for chemical and/or radiological analysis:

Equipment

1. Box sampler (e.g., Ekman dredge)
2. Stainless steel bowl
3. Stainless steel spoon
4. Sample jars
5. Sample labels
6. Cooler packed with Blue Ice® (Blue Ice® is not required for radiological analysis.)
7. Trip blank (only required for volatile organic analyses)
8. Coolant blank (not required for radiological analysis)
9. Sample logbook
10. Chain-of-custody forms
11. Chain-of-custody seals
12. Permanent ink marker
13. Health and safety instruments
14. Chemical and/or radiological field screening instruments
15. Health and safety clothing (including life jackets)
16. Waders, raft, or boat
17. Sampling table
18. Waste container (e.g., 55-gal drum)
19. Plastic waste bags

Sampling Procedure

1. In preparation for sampling, confirm that all necessary preparatory work has been completed, including obtaining property access agreements, meeting health and safety and equipment decontamination requirements, and checking the calibration of all health and safety and chemical and/or radiological field screening instruments.
2. Approach the sampling point from downstream, being careful of not to disturb the underlying sediment.
3. Hold the sampling pole so that the open sampler jaws are positioned several inches above the surface of the sediment; then firmly thrust the sampler downward. Depress the button at the top of the sampling pole to release the spring-loaded jaws. Retrieve the box sampler.

4. If a grab sample is being collected, transfer the sediment from the box sampler directly into a sample jar. If a composite sample is being collected, transfer the sediment from the locations to be composited into a stainless steel bowl and homogenize with a stainless steel spoon before filling a sample jar. Scan the sediment in the jar using chemical and/or radiological screening instruments. Record the results in a bound logbook. (Note: a decontaminated sampler should be used at each location being composited.)
5. After the jar is capped, attach a sample label and custody seal to the jar and immediately place it into a sample cooler. Samples for chemical analysis should be packed in Blue Ice®.
6. See Chapter 5 for details on preparing sample jars and coolers for sample shipment.
7. Transfer any sediment left over from the sampling into a waste container. Before leaving the site, all waste containers should be sealed, labeled, and handled appropriately (see Chapter 11).
8. Transfer any other sampling-related wastes (e.g., gloves and foil) into a plastic waste bag.
9. Have a professional surveyor survey the coordinates of the sampling point to preserve the exact sampling location.

4.2.5.1.5 Integrating Sediment Sampler Method

The integrating sediment sampler (Figure 4.37) is designed to collect an integrated sample of sediment being carried in suspension by a moving stream or river. Most often this sampler is used to collect a sample of the sediment in a stream or river just above the sediment–water interface. Once the sampler is retrieved, the filled sample bottle is removed, capped, and labeled in preparation for analysis.

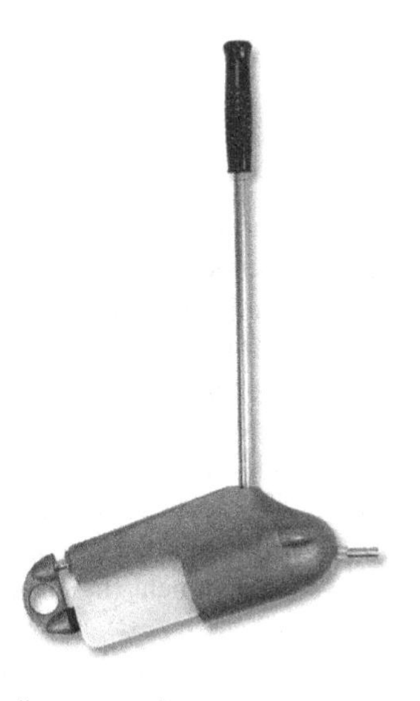

FIGURE 4.37 Integrating sediment sampler.

This sampler is composed of the sampler body, a glass sample bottle, a nozzle where the sample enters the sampler, and an air exhaust port where air escapes from the sample bottle as it fills with water and sediment.

For most sampling programs, four people are sufficient for this sampling procedure. Two are needed for sample collection, labeling, and documentation, a third is needed for health and safety, and a fourth is needed for waste management and equipment decontamination. If a raft or boat is used to assist the sampling procedure, at least one additional person will be needed.

The following equipment and procedure can be used to collect integrated sediment samples for chemical or radiological analysis:

Equipment

1. Integrating sediment sampler
2. Sample jars
3. Sample labels
4. Cooler packed with Blue Ice® (Blue Ice® is not required for radiological analysis.)
5. Trip blank (only required for volatile organic analyses)
6. Coolant blank (not required for radiological analysis)
7. Sample logbook
8. Chain-of-custody forms
9. Chain-of-custody seals
10. Permanent ink marker
11. Health and safety instruments
12. Chemical and/or radiological field screening instruments
13. Health and safety clothing (including life jackets)
14. Waders, raft, or boat
15. Sampling table
16. Waste container (e.g., 55-gal drum)
17. Plastic waste bags

Sampling Procedure

1. In preparation for sampling, confirm that all necessary preparatory work has been completed, including obtaining property access agreements, meeting health and safety and equipment decontamination requirements, and checking the calibration of all health and safety and chemical and/or radiological field screening instruments.
2. Approach the sampling point from downstream, being careful of not to disturb the underlying sediment.
3. Lower the integrating sediment sampler through the water to the desired sampling depth. Hold the sampler entrance opening so that it faces upstream and allows the sediment (in suspension) and water to fill the sample bottle.
4. When the sample bottle is full, remove the sampler from the water.

5. Remove the sample bottle from sampler and attach a bottle cap, sample label, and custody seal. Immediately place the bottle into a sample cooler. Samples for chemical analysis should be packed in Blue Ice®.
6. See Chapter 5 for details on preparing sample bottles and coolers for sample shipment.
7. Transfer any waste material left over from the sampling into a waste container. Before leaving the site, all waste containers should be sealed, labeled, and handled appropriately (see Chapter 11).
8. Transfer any other sampling-related wastes (e.g., gloves and foil) into a plastic waste bag.
9. Have a professional surveyor survey the coordinates of the sampling point to preserve the exact sampling location.

4.2.5.1.6 Helley-Smith Method

The Helley-Smith sampler (Figure 4.38) is a sediment sampling tool that is designed to collect a sample of the silt, sand, or gravel carried in suspension by a surface water drainage, stream, or river. This sampler is often placed just above the drainage, stream, or riverbed to collect a sample of the bed load material that is being transported. The handheld Helley-Smith sampler has a 3 in. × 3 in. entrance opening for the sediment to enter a 250-micron nylon mesh bag that captures the sediment. The disadvantage of this method is that the very fine portion of the sediment sample is lost through the mesh bag.

For most sampling programs, four people are sufficient for this sampling procedure. Two are needed for sample collection, labeling, and documentation; a third is needed for health and safety; and a fourth is needed for waste management and

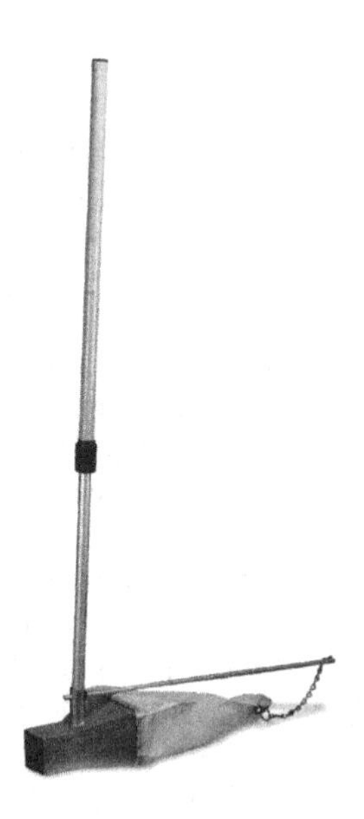

FIGURE 4.38 Helley-Smith sampler.

equipment decontamination. If a raft or boat is used to assist the sampling procedure, at least one additional person will be needed.

Equipment

1. Helley-Smith Sampler
2. Stainless steel spoon
3. Sample jars
4. Sample labels
5. Cooler packed with Blue Ice® (Blue Ice® is not required for radiological analysis.)
6. Trip blank (only required for volatile organic analyses)
7. Coolant blank (not required for radiological analysis)
8. Sample logbook
9. Chain-of-custody forms
10. Chain-of-custody seals
11. Permanent ink marker
12. Health and safety instruments
13. Chemical or radiological field screening instruments
14. Health and safety clothing (including life jackets)
15. Waders, raft, or boat
16. Sampling table
17. Waste container (e.g., 55-gal drum)
18. Plastic waste bags

Sampling Procedure

1. In preparation for sampling, confirm that all necessary preparatory work has been completed, including obtaining property access agreements, meeting health and safety and equipment decontamination requirements, and checking the calibration of all health and safety and chemical and/or radiological field screening instruments.
2. Approach the sampling point from downstream, being careful of not to disturb the underlying sediment.
3. Lower the Helley-Smith Sampler through the water to the desired sampling depth. Hold the sampler entrance opening so that it faces upstream and allows the sediment (in suspension) to pass through the mesh bag.
4. When the mesh bag is full of sediment, remove the sampler from the water.
5. Remove the mesh bag from the rest of the sampler, and use a stainless steel spoon to transfer the sediment from the mesh bag into a sample jar.
6. Attach the cap on the sample jar along with a sample label, and custody seal. Immediately place the jar into a sample cooler. Samples for chemical analysis should be packed in Blue Ice®.
7. See Chapter 5 for details on preparing sample jars and coolers for sample shipment.

8. Transfer any waste material left over from the sampling into a waste container. Before leaving the site, all waste containers should be sealed, labeled, and handled appropriately (see Chapter 11).
9. Transfer any other sampling-related wastes (e.g., gloves and foil) into a plastic waste bag.
10. Have a professional surveyor survey the coordinates of the sampling point to preserve the exact sampling location.

4.2.5.2 Pond, Lake, Retention Basin, and Tank Sampling

Sampling methods that have proved to be the most effective when sampling sediments from ponds, lakes, retention basins, and tanks include the scoop or dipper, core barrel (slide hammer), WaterMark® Russian sediment borer, box sampler, and dredge sampler. The first four methods are all effective when collecting sediment samples from around the edges of ponds, lakes, and retention basins, where the water is relatively shallow. On the other hand, the box sampler and dredge sampler, used in combination with a wire line, are effective methods for collecting sediment samples through deep water.

Of these five methods, the scoop or dipper method is the easiest to implement because it simply involves lowering a scoop or dipper by hand through the surface water and pushing the sampler deep into the underlying sediment. As the sampler is retrieved, the sediment is transferred into a sample jar. This technique is generally effective in collecting samples of the top 0.5 ft of sediment at locations where the water depth is less than 2 ft.

The core barrel (slide hammer) method involves beating a sampling tube into the sediment. This technique removes a core of sediment for analytical testing. Because the sampling tube is typically lined with sampling sleeves, either individual sleeves can be sent to the laboratory as grab samples or the sediment can be removed from the sleeves and composited before filling a sample jar. Of the five sampling methods, only the core barrel (slide hammer) and the WaterMark® Russian Sediment Borer are effective in collecting sediment samples deeper than the top 0.5 ft. When used in combination with extension rods, the core barrel (slide hammer) can be used to collect a 2- to 3-ft sediment core through 10 to 15 ft of water. This method is effective in preventing the loss of fine-grained sediment as the sample is retrieved.

In the WaterMark® Russian Sediment Borer method, the borer (in the closed position) is advanced to preferred sampling depth by pushing downward on the sampler while the bar handle is rotated, or driving it downward with the optional slide hammer assembly. The bore is then rotated clockwise 180° so that the sharpened edge of the chamber cuts a semicylindrical sediment core that is contained by the pivotal cover plate. During retrieval, the cover plate's counterclockwise rotation extrudes an undisturbed sediment sample. This method is effective in collecting sediment samples as deep as 5 ft below the sediment–water interface. This method is effective in preventing the loss of fine-grained sediment as the sample is retrieved.

The box sampler (e.g., Ekman dredge) method utilizes a spring-loaded sample box attached to either a sampling pole or wire line to collect grab sediment samples. This method involves pushing the box sampler (e.g., Ekman dredge) into the sediment and releasing the spring-loaded sampler jaws, or letting the sampler free-fall (when attached to a wire line) through deep water to embed itself in the underlying sediment before triggering the spring-loaded sampler jaws. After the sampler is retrieved, sediment from the sampling box is transferred into a sample jar. This method is only effective in collecting samples from the top 0.5 ft of sediment. When using a sampling pole, one can collect a sediment sample through 5 ft of water. When using a wire line, there is no limit to the depth of water through which a sediment sample can be collected. This method is effective in preventing the loss of fine-grained sediment as the sample is retrieved.

The dredge sampler method is composed of two jaws connected by a lever, and is dropped through the water using a wire line. As the sampler is dropped through the water, the jaws open and embed themselves in the underlying sediment. As the wire line is raised, the jaws close to capture a sample. Sediment from the dredge is then transferred into a sample jar using a stainless steel spoon. Similar to the scoop or dipper and box sampler, this method is only effective in collecting samples from the top 0.5 ft of sediment. This method is effective in preventing the loss of fine-grained sediment as the sample is retrieved.

Depending on the depth of the water overlying the sediment, samplers will need a pair of waders, a raft, or a boat to access the sampling point. Table 4.4 summarizes the effectiveness of each of the five recommended sampling methods. A number "1" in the table indicates that a particular procedure is most effective in collecting samples for a particular laboratory analysis, sample type, or sampling depth. A number "2" indicates that the procedure is acceptable but less preferred, whereas an empty cell indicates that the procedure is not recommended. For example, Table 4.4 indicates that the core barrel (slide hammer) is the most effective method for collecting sediment samples for volatile organic analysis. Although the scoop or dipper, WaterMark® Russian Sediment Borer, box sampler, and dredge sampler are acceptable sampling methods for volatile organic analysis, they are less preferred than the core barrel (slide hammer) method.

4.2.5.2.1 *Scoop or Dipper Method*

The scoop or dipper method used to collect sediment samples from ponds, lakes, and retention basins is identical to the procedure described for streams, rivers, and surface water drainage sampling (see Section 4.2.5.1.1), with the following modification:

- Modify Step 2 to read "Approach the sampling point being careful of not to disturb the underlying sediment."

4.2.5.2.2 *Core Barrel (Slide Hammer) Method*

The core barrel (slide hammer) method used to collect sediment samples from ponds, lakes, and retention basins is identical to the procedure described for streams, rivers, and surface water drainage sampling (see Section 4.2.5.1.2), with the following modification:

- Modify Step 2 to read "Approach the sampling point being careful of not to disturb the underlying sediment."

TABLE 4.4
Rating Table for Sediment Sampling Methods for Ponds, Lakes, and Retention Basins

	Laboratory Analyses							Sample Type				Depth		
	Radio-nuclides	Volatiles	Semi-volatiles	Metals	Pesticides	PCBs	TPH	Geotech-nical	Grab	Compo-site (Vertical)	Compo-site (Areal)	Surface (0.0–0.5 ft)	Shallow (0.0–5.0 ft)	Lithology description
Methods for collecting sediment below sediment–water interface														
Scoop or dipper	2	2	2	2	2	2	2		2		2	2		2
Core barrel (slide hammer)	1	1	1	1	1	1	1		1	1	2	1	1	2
WaterMark® Russian sediment borer	1	2	1	1	1	1	1		1	1	1	1	1	1
Box sampler	1	2	1	1	1	1	1		1		1	1		1
Dredge sampler	1	2	1	1	1	1	1		1		1	1		1

Note: 1 = preferred method; 2 = acceptable method; empty cell = method not recommended.

4.2.5.2.3 WaterMark® Russian Sediment Borer Method

The WaterMark® Russian sediment borer method used to collect sediment samples from ponds, lakes, and retention basins is identical to the procedure described for streams, rivers, and surface water drainage sampling (see Section 4.2.5.1.3), with the following modification:

- Modify Step 2 to read "Approach the sampling point being careful of not to disturb the underlying sediment."

4.2.5.2.4 Box Sampler Method

The box sampler method used to collect sediment samples from ponds, lakes, and retention basins is identical to the procedure described for streams, rivers, and surface water drainage sampling (see Section 4.2.5.1.4), with the following modifications:

- A wire line may be used to lower the sampler through the water, as opposed to the sampling pole.
- Add "wire line" to the equipment list.
- Modify Step 3 to read as follows:

> When using a sampling pole, hold the sampling pole so the open sampler jaws are positioned several inches above the surface of the sediment, and then firmly thrust the sampler downward. Depress the button at the top of the sampling pole to release the spring-loaded jaws. When using a cable as opposed to a sampling pole, allow the sampler to free-fall through the water to assure that the sampler jaws become deeply embedded into the sediment. Then, slide a trip weight down the cable to trip the spring-loaded jaws.

4.2.5.2.5 Dredge Sampler Method

The dredge sampler is a very common and effective tool in collecting grab or areal composite samples of sediment from ponds, lakes, and retention basins. This sampler is composed of two jaws connected by a lever (Figure 4.39). To use this sampling method, the dredge is attached to a cable and allowed to free-fall through the water to ensure that the sampler jaws deeply embed themselves into the underlying sediment. As the cable is retrieved, the jaws to the sampler are forced closed. When collecting a grab sample, the sediment is transferred directly from the dredge into a sample jar using a stainless steel spoon. When collecting a composite sample, the sediment from each of the intervals to be composited is transferred into a stainless steel bowl and is homogenized using a stainless steel spoon before filling a sample jar. As with the core barrel (slide hammer), WaterMark® Russian Sediment Borer, and box sampler, this technique is effective in preventing the loss of fine-grained sediment as the sample is retrieved.

For most sampling programs, five people are sufficient for this sampling procedure. Two are needed for sample collection, labeling, and documentation; a third is needed for health and safety; a fourth is needed for waste management and equipment decontamination; and a fifth is needed to maneuver the raft or boat.

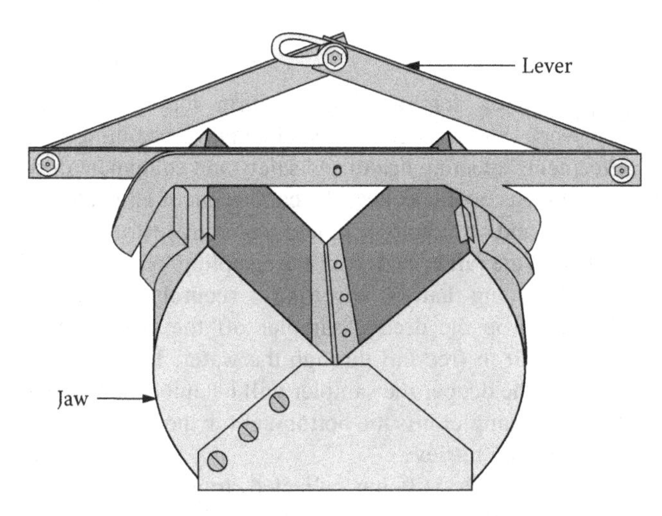

FIGURE 4.39 Dredge sampler.

(From Byrnes, 1994. *Field Sampling Methods for Remedial Investigations*. Lewis Publishers, Boca Raton, FL.)

The following equipment and procedure can be used to collect sediment samples for chemical and/or radiological analysis:

Equipment

1. Stainless steel dredge sampler
2. Wire line
3. Stainless steel spoon
4. Sample jars
5. Sample label
6. Cooler packed with Blue Ice® (Blue Ice® is not required for radiological analysis.)
7. Trip blank (only required for volatile organic analyses)
8. Coolant blank (not required for radiological analysis)
9. Sample logbook
10. Chain-of-custody forms
11. Chain-of-custody seals
12. Permanent ink marker
13. Health and safety instruments
14. Chemical and/or radiological field screening instruments
15. Health and safety clothing (including life jackets)
16. Boat large enough for five people and sampling equipment
17. Sampling table
18. Waste container (e.g., 55-gal drum)
19. Plastic waste bags

Sampling Procedure

1. In preparation for sampling, confirm that all necessary preparatory work has been completed, including obtaining property access agreements, meeting health and safety and equipment decontamination requirements, and checking the calibration of all health and safety and chemical and/or radiological field screening instruments.
2. Maneuver the raft or boat over the sampling location.
3. After verifying that the wire line is securely fastened to the dredge sampler, drop the dredge sampler off the side of the raft or boat and allow it to free-fall through the water. The faster the sampler is dropped, the deeper the sampler will be embedded into the sediment. When the sampler hits the bottom, allow the line to go slack for a few seconds, then retrieve.
4. If a grab sample is being collected, transfer the sediment from the dredge directly into a sample jar using a stainless steel spoon. If a composite sample is being collected, transfer the sediment from the locations to be composited into a stainless steel bowl and homogenize with a stainless steel spoon before filling a sample jar. (Note: a decontaminated dredge sampler should be used at each location to be composited).
5. Scan the sediment in the jar using chemical and/or radiological field screening instruments. Record the results in a bound logbook.
6. After the jar is capped, attach a sample label and custody seal to the jar and immediately place it into a sample cooler. Samples for chemical analysis should be packed in Blue Ice®.
7. See Chapter 5 for details on preparing sample jars and coolers for sample shipment.
8. Transfer any sediment left over from the sampling into a waste container. Before leaving the site, all waste containers should be sealed, labeled, and handled appropriately (see Chapter 11).
9. Transfer any other sampling-related wastes (e.g., gloves and foil) into a plastic waste bag.
10. Have a professional surveyor survey the coordinates of the sampling point(s) to preserve the exact sampling location.

4.2.6 Surface Water and Liquid Waste Sampling

This section provides the reader with guidance on selecting surface water and liquid waste sampling methods for site characterization. The criteria used in selecting the most appropriate method include the analyses to be performed on the sample, the type of sample being collected (grab, composite, or integrated), and the sampling depth. Standard operating procedures have been provided for each of the methods to facilitate implementation.

One objective of an initial site characterization study should be to determine if contamination is present in nearby surface water or liquid waste units, such as streams, rivers, surface and storm sewer drainages, ponds, lakes, retention basins, and tanks. It is particularly important to characterize the surface water in streams, rivers, and other surface water drainages because they provide avenues for rapid contaminant migration, and provide points where receptors are readily exposed to contamination. Ponds, lakes, retention basins, and tanks do not provide the same opportunity for rapid contaminant migration; however, they similarly provide exposure points for receptors and are potential sources for groundwater contamination.

The seven-step DQO process (Chapter 3, Section 3.2.5) should be used to define the optimum sampling design, required number of samples, analyses to be performed, and analytical performance requirements. When performing an initial site characterization study (Stage 1), one grab sample is commonly collected from each surface water and storm sewer drainage, sump, and retention pond at the site. These samples are often collected from sampling locations that are downstream from potential sources of contamination. One upstream water sample should also be collected from surface water and storm sewer drainages to assist in defining background surface water chemistry (Figure 4.40).

The objective in collecting a sample from each surface water and storm sewer drainage is to determine if contaminants are currently migrating off-site. If contaminants are identified from this Stage 1 round of sampling, one should consider collecting additional upstream samples to assist in locating the contamination source, and downstream samples to further define the extent of contamination (Stage 2). The downstream sampling should include collecting samples from nearby streams, ponds, or lakes that receive water from the contaminated drainage. These samples are commonly collected just downstream from where each surface water drainage enters a stream (Figure 4.40), and near the inflow and outflow points in a pond or lake (Figure 4.41). If the contaminants of concern do not include volatile organics, initial characterization of ponds and lakes could be performed by collecting an areal composite sample (Figure 4.41). The specific number and location of samples to be collected to complete the surface water characterization effort will be defined in Step 7 of the DQO process (see Chapter 3, Section 3.2.5.7).

Most retention ponds are not watertight, and slowly release contaminated water into the underlying soil and groundwater. For this reason, it is important to characterize the water held in these ponds. Because retention ponds are typically small in size, it is rarely necessary to collect more than one surface water sample from each.

The depth at which surface water samples are collected should be based on the suspected concentration and density of the contaminants of concern. If contaminants are suspected to be in high enough concentrations to form an NAPL, a surface water sample should be collected from the bottom of the water column if the contaminant has a specific gravity greater than water (DNAPL) or at the top of the water column if the contaminant has a specific gravity less than water (LNAPL). Otherwise, samples are most commonly collected from the top of the water column. If historical information leads one to believe that contamination could be layered, due to varying specific gravities of the contaminants, one should consider collecting either a vertical composite sample, or several grab samples from different depth intervals (Figure 4.42).

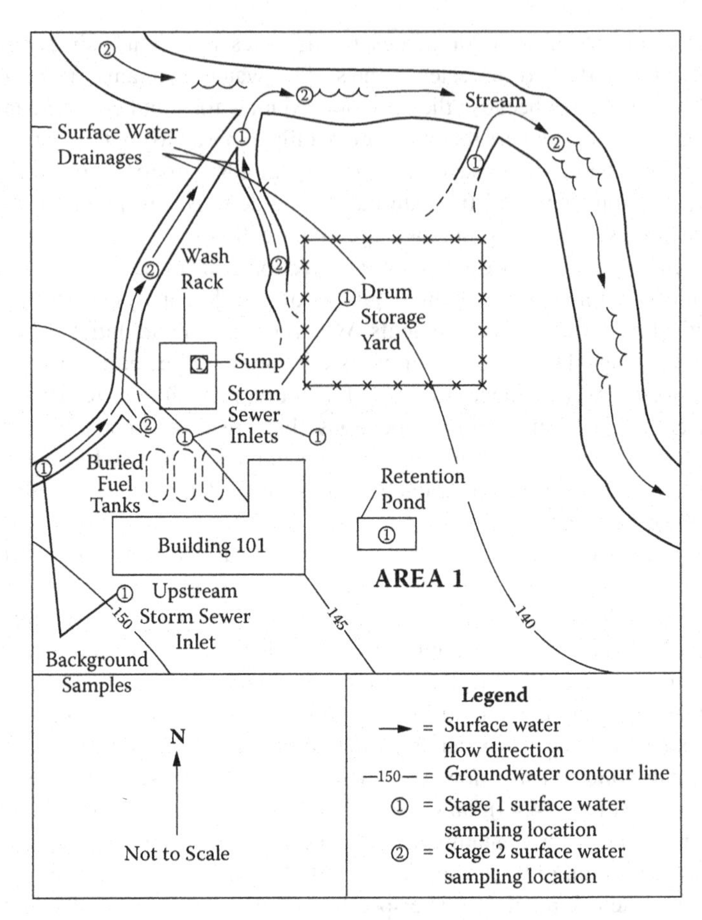

FIGURE 4.40 Example of a Stage 1 and Stage 2 surface water sampling effort.

(From Byrnes, 1994. *Field Sampling Methods for Remedial Investigations.* Lewis Publishers, Boca Raton, FL.)

Sampling tools used to collect surface water samples for laboratory analysis should be constructed of materials that are compatible with the media and the constituents that are being tested for. In other words, the sampler should *not* be constructed of a material that could cause a loss or gain in contaminant concentrations measured due to sorption, desorption, degradation, or corrosion (EPA 2002). Because stainless steel, Teflon, and glass are inert substances, sampling equipment made of these materials should be preferentially selected over equipment made of other materials. USGS (2003) notes that fluorocarbon polymers (Teflon®, Kynar®, and Tefzel®), stainless steel 316-grade, and borosilicate glass (laboratory grade) are acceptable for both inorganic and organic analyses; polypropylene, polyethylene, PVC, silicone, and nylon are only acceptable for inorganic analyses; and stainless steel 304-grade, and other metals (brass, iron, copper, aluminum, and galvanized and carbon steels) are only

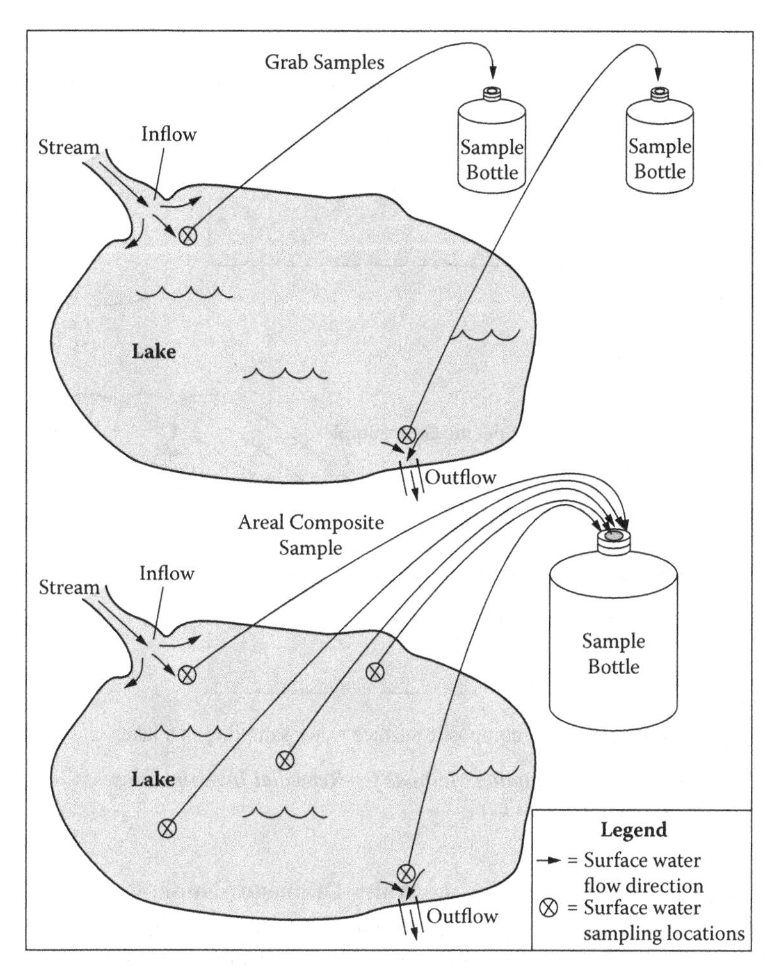

FIGURE 4.41 Possible initial surface water sampling strategy for lakes and ponds.

(From Byrnes, 1994. *Field Sampling Methods for Remedial Investigations.* Lewis Publishers, Boca Raton, FL.)

acceptable for organic analyses. Having a rigorous quality control sampling program (e.g., collecting rinsate blanks) in place will help ensure that the sample integrity is not being impacted by the sampling equipment (see Chapter 6).

When preparing to collect a surface water sample, one should always collect a pre-sample of the surface water in a clean glass jar, and it should be analyzed for the field parameters specified in the sampling and analysis plan (see Section 3.2.7) (e.g., pH, temperature, conductivity, turbidity, dissolved oxygen).

The following subsections present preferred sampling methods and procedures for collecting surface water or liquid waste samples from streams, rivers, and surface water drainages, and ponds, lakes, retention basins, and tanks.

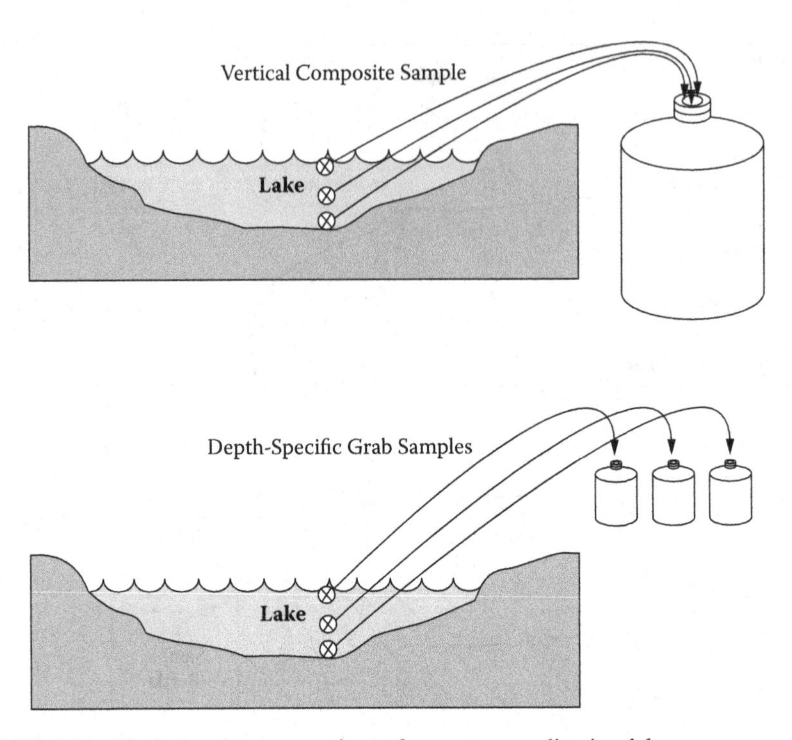

FIGURE 4.42 Vertical grab or composite surface water sampling in a lake.

(From Byrnes, 1994. *Field Sampling Methods for Remedial Investigations.* Lewis Publishers, Boca Raton, FL.)

4.2.6.1 Stream, River, and Surface Water Drainage Sampling

Although a number of sophisticated sampling devices are available to collect surface water samples, not all of these tools are effective in collecting samples in the shallow, fast-moving water that typifies streams, rivers, and surface water drainages. Because the movement of water in these environments tends to homogenize the water column naturally, in most cases only grab samples or integrated samples (see Section 4.2.1.1) are collected from the water surface. However, vertical or areal composite samples can also be collected.

The methods that have proved to be the most effective when sampling these environments include the bottle submersion method, dipper method, and subsurface grab method. The bottle submersion, dipper, and subsurface grab sampler are all effective methods for collecting grab, areal composite, and integrated samples. The subsurface grab sampler is also an effective method for collecting vertical composite samples.

Of these methods, the bottle submersion method is the easiest to implement, because it just involves submerging a sample bottle beneath the water surface. With this method, a telescoping extension rod is often used to hold the sample bottle as it is lowered into the water. The dipper method is also very easy to implement because it just involves lowering a stainless steel dipper below the water surface, and then transferring the water into sample bottles. The bottle submersion and dipper methods are only effective in collecting samples from the water surface.

The subsurface grab sampler is composed of a sample bottle, sampler head that attaches to the sample bottle, sampling pole, handle, and seal ring. The seal ring is connected by a wire to the sampler head and is used to open and close a valve that allows water to enter the sample bottle. After the sampler is retrieved, the sample bottle is removed from the sampler, capped, labeled, and shipped to the laboratory for analysis. This method is effective in collecting samples as deep as 5 ft below the water surface.

Table 4.5 summarizes the effectiveness of each of the three recommended sampling methods. A number "1" in the table indicates that a particular procedure is most effective in collecting samples for a particular laboratory analysis, sample type, or sampling depth. A number "2" indicates that the procedure is acceptable but less preferred, whereas an empty cell indicates that the procedure is not recommended. For example, Table 4.5 indicates that bottle submersion is the preferred procedure for collecting samples for volatile organic analysis. Although the dipper and subsurface grab sampler are acceptable methods for volatile organic analysis, they are less desirable because the water from the sampler must be poured into sample bottles, which tends to aerate the sample.

Whichever sampling method is selected, sample bottles for volatile organic analysis should always be the first to be filled. Bottles for other analyses should be filled consistent with their relative importance to the sampling program. As a general rule, the sample located farthest downstream should always be the first to be collected. Sampling should then proceed upstream. By collecting samples in this manner, any sediment disturbed by the samplers will not contaminate downstream sampling points. Depending on the depth of the water, samplers may need a pair of waders, a raft, or a boat to access the sampling point. If waders are used, the sampler should face upstream while collecting the sample to ensure that the waders do not contaminate the sample. Similarly, samples should be collected from the upstream side of a raft or boat.

4.2.6.1.1 Bottle Submersion Method

The bottle submersion method is the simplest and one of the most commonly used surface water sampling methods for collecting samples from streams, rivers, and surface water drainages. This method utilizes a water sample bottle and an optional telescoping extension rod with an adjustable beaker clamp (Figure 4.43). This method is effective in collecting grab, composite (areal), or integrated samples from the top few inches of the water column.

The following procedure is written to include the use of a telescoping extension rod. If an extension rod is not available or not needed, the bottle can be lowered into the water by hand. Because this procedure is both simple and effective, one should consider using this method or the dipper method (Section 4.2.6.1.2) for initial surface water sampling activities. If contaminants are identified from this preliminary sampling, the subsurface grab sampler method (Section 4.2.6.1.3) should be considered to define the vertical distribution of contaminants. For a summary of the effectiveness and limitations of this sampling method, see Table 4.5.

For most sampling programs, four people are sufficient for this sampling procedure. Two are needed for field testing, sample collection, labeling, and documentation; a third is needed for health and safety; and a fourth is needed for miscellaneous tasks such as waste management and equipment decontamination.

TABLE 4.5
Rating Table for Surface Water Sampling Methods for Streams, Rivers, and Drainages

	Laboratory Analyses							Sample Type				Depth		
	Radionuclides	Volatiles	Semi-volatiles	Metals	Pesticides	PCBs	TPH	Grab	Composite (Vertical)	Composite (Areal)	Integrated	Surface (0.0–0.5 ft)	Shallow (0.0–5.0 ft)	Deep (> 5.0 ft)
Bottle submersion	1	1	1	1	1	1	1	1		1	1	1		
Dipper	1	2	1	1	1	1	1	1		1	1	1		
Subsurface grab sampler	1	2	1	1	1	1	1	1	1	2	2	2	1	

Note: 1 = preferred method; 2 = acceptable method; empty cell = method not recommended.

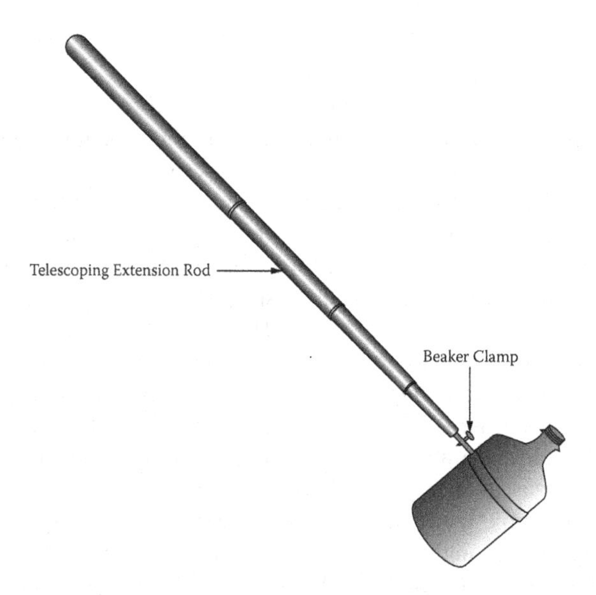

FIGURE 4.43 Telescoping extension rod used to assist bottle submersion method.

(From Byrnes, 1994. *Field Sampling Methods for Remedial Investigations.* Lewis Publishers, Boca Raton, FL.)

The following equipment and procedure can be used to collect surface water samples for chemical and/or radiological analysis:

Equipment

1. Telescoping extension rod with adjustable beaker clamp
2. Sample bottles
3. Sample preservatives
4. pH, temperature, conductivity, turbidity, dissolved oxygen meters
5. Sample labels
6. Cooler packed with Blue Ice® (Blue Ice® is not required for radiological analysis.)
7. Trip blank (only required for volatile organic analyses)
8. Coolant blank (not required for radiological analysis)
9. Sample logbook
10. Chain-of-custody forms
11. Chain-of-custody seals
12. Permanent ink marker
13. Health and safety instruments
14. Chemical and/or radiological field screening instruments
15. Health and safety clothing (including life jackets)
16. Waste container (e.g., 55-gal drum)
17. Sampling table
18. Plastic waste bags

Sampling Procedure

1. In preparation for sampling, confirm that all necessary preparatory work has been completed, including obtaining property access agreements, meeting health and safety and equipment decontamination requirements, and checking the calibration of all health and safety and chemical or radiological field screening instruments.
2. Before sample collection, fill a clean glass jar with sample water, and measure the field parameters specified in the sampling and analysis plan (see Section 3.2.7)(e.g., pH, temperature, conductivity, turbidity, dissolved oxygen). Record this information in a sample logbook.
3. Secure a clean sample bottle to the end of the telescoping extension rod using a beaker clamp. Remove the bottle cap just before sampling.
4. To collect a grab sample, while standing on the bank of the stream, river, or surface water drainage, extend the rod out over the water and lower the sample bottle just below the water surface, being careful of not to disturb the underlying sediment. Allow the sample bottle to fill completely with water. When the bottle is full, retrieve the sampler. (Note: If the analyses to be performed require the sample to be preserved, preservative should be added to the bottle before it is filled.) Remove the sample bottle from the telescoping extension rod. Scan the sample using chemical and/or radiological screening instruments, and record the results in a bound logbook. Screw on the bottle cap.
5. To collect an areal composite sample, while standing on the bank of the stream, river, or surface water drainage, extend the rod out over the water and lower the sample bottle just below the water surface, being careful of not to disturb the underlying sediment. Allow the sample bottle to fill partially with water at this location. Cap the sample bottle. Move to the next location to be composited. Uncap the sample bottle. Extend the rod out over the water, and lower the sample bottle just below the water surface. Allow the sample bottle to fill partially with water at this location. Cap the sample bottle. Repeat this procedure until the sample bottle is completely full. (Note: If the analyses to be performed require the sample to be preserved, preservative should be added to the bottle before it is filled.) Remove the sample bottle from the telescoping extension rod. Scan the sample using chemical and/or radiological screening instruments, and record the results in a bound logbook. Screw on the bottle cap.
6. To collect an integrated sample, while standing on the bank of the stream, river, or surface water drainage, extend the rod out over the water and lower the sample bottle just below the water surface, being careful of not to disturb the underlying sediment. Allow the sample bottle to fill partially with water. Cap the sample bottle. Wait a specified length of time. At the same location, uncap the sample bottle. Extend the rod out over the water, and lower the sample bottle just below the water surface. Allow the sample bottle to fill partially with water from the same location. Cap the sample bottle. Repeat this procedure until the sample bottle is

completely full. (Note: If the analyses to be performed require the sample to be preserved, preservative should be added to the bottle before it is filled.) Remove the sample bottle from the telescoping extension rod. Scan the sample using chemical and/or radiological screening instruments, and record the results in a bound logbook. Screw on the bottle cap.

7. After the bottle is capped, attach a sample label and custody seal to the bottle and immediately place it into a sample cooler. Samples for chemical analysis should be packed in Blue Ice®.
8. See Chapter 5 for details on preparing sample bottles and coolers for sample shipment.
9. Transfer any water left over from the sampling into a waste container. Before leaving the site, all waste containers should be sealed, labeled, and handled appropriately (see Chapter 11).
10. Transfer any other sampling-related wastes (e.g., gloves and foil) into a plastic waste bag.
11. Have a professional surveyor survey the coordinates of the sampling points to preserve the exact sampling location.

4.2.6.1.2 Dipper Method

The dipper method is a simple but effective method for collecting samples of surface water from streams, rivers, and surface water drainages for both chemical and radiological analysis. With this method, a dipper (Figure 4.44) is used to collect a water sample, which is then transferred into a sample bottle. This method is effective in collecting grab, composite (areal), or integrated samples from the top few inches of the water column.

Because this procedure is both simple and effective, it is recommended that initial surface water sampling be performed using either this method or the bottle submersion method (Section 4.2.6.1.1). If contaminants are identified from this preliminary sampling, the subsurface grab sampler method (Section 4.2.6.1.3) should be

FIGURE 4.44 Surface water sampling dipper.

(From Byrnes, 1994. *Field Sampling Methods for Remedial Investigations*. Lewis Publishers, Boca Raton, FL.)

considered to define the vertical distribution of contaminants. For a summary of the effectiveness and limitations of this sampling method, see Table 4.5.

For most sampling programs, four people are sufficient for this sampling procedure. Two are needed for field testing, sample collection, labeling, and documentation; a third is needed for health and safety; and a fourth is needed for waste management and equipment decontamination.

The following equipment and procedure can be used to collect surface water samples for chemical and/or radiological analysis:

Equipment

1. Dipper
2. Sample bottles
3. Sample preservatives
4. pH, temperature, conductivity, turbidity, dissolved oxygen meters
5. Sample labels
6. Cooler packed with Blue Ice® (Blue Ice® is not required for radiological analysis.)
7. Trip blank (only required for volatile organic analyses)
8. Coolant blank (not required for radiological analysis)
9. Sample logbook
10. Chain-of-custody forms
11. Chain-of-custody seals
12. Permanent ink marker
13. Health and safety instruments
14. Chemical and/or radiological field screening instruments
15. Health and safety clothing (including life jackets)
16. Waste container (e.g., 55-gal drum)
17. Sampling table
18. Plastic waste bags

Sampling Procedure

1. In preparation for sampling, confirm that all necessary preparatory work has been completed, including obtaining property access agreements, meeting health and safety and equipment decontamination requirements, and checking the calibration of all health and safety and chemical or radiological field screening instruments.
2. Before sample collection, fill a clean glass jar with sample water, and measure the field parameters specified in the sampling and analysis plan (see Section 3.2.7)(e.g., pH, temperature, conductivity, turbidity, dissolved oxygen). Record this information in a sample logbook.
3. To collect a grab sample, while standing on the bank of the stream, river, or surface water drainage, extend the dipper out over the water and lower it just below the water surface, being careful of not to disturb

the underlying sediment. When the dipper is full, retrieve the dipper and carefully transfer the water into a sample bottle. If a larger volume of water is required to fill the sample bottle, repeat this procedure until the bottle is full, being careful to collect water from the same location. (Note: If the analyses to be performed require the sample to be preserved, preservative should be added to the bottle before it is filled.) Scan the sample using chemical and/or radiological screening instruments, and record the results in a bound logbook. Screw on the bottle cap.

4. To collect an areal composite sample, while standing on the bank of the stream, river, or surface water drainage, extend the dipper out over the water and lower it just below the water surface, being careful of not to disturb the underlying sediment. When the dipper is full, retrieve the dipper and carefully transfer the desired volume of water into a sample bottle. Cap the sample bottle. Move to the next location to be composited. Uncap the sample bottle. Extend a second decontaminated dipper out over the water, and lower it just below the water surface, being careful of not to disturb the underlying sediment. When the dipper is full, retrieve it and carefully transfer the desired volume of water into a sample bottle. Cap the sample bottle. Repeat this procedure until the sample bottle is completely full. (Note: If the analyses to be performed require the sample to be preserved, preservative should be added to the bottle before it is filled.) Scan the sample using chemical and/or radiological screening instruments, and record the results in a bound logbook. Screw on the bottle cap.
5. To collect an integrated sample, while standing on the bank of the stream, river, or surface water drainage, extend the dipper out over the water and lower it just below the water surface, being careful of not to disturb the underlying sediment. When the dipper is full, retrieve the dipper and carefully transfer the desired volume of water into a sample bottle. Cap the sample bottle. Wait a specified length of time. At the same location, uncap the sample bottle. Extend a second decontaminated dipper out over the water and lower it just below the water surface, being careful of not to disturb the underlying sediment. When the dipper is full, retrieve the dipper and carefully transfer the desired volume of water into a sample bottle. Cap the sample bottle. Repeat this procedure until the sample bottle is completely full. (Note: If the analyses to be performed require the sample to be preserved, preservative should be added to the bottle before it is filled.) Scan the sample using chemical and/or radiological screening instruments, and record the results in a bound logbook. Screw on the bottle cap.
6. After the bottle is capped, attach a sample label and custody seal to the bottle and immediately place it into a sample cooler. Samples for chemical analysis should be packed in Blue Ice®.
7. See Chapter 5 for details on preparing sample bottles and coolers for sample shipment.

8. Transfer any water left over from the sampling into a waste container. Before leaving the site, all waste containers should be sealed, labeled, and handled appropriately (see Chapter 11).
9. Transfer any other sampling-related wastes (e.g., gloves and foil) into a plastic waste bag.
10. Have a professional surveyor survey the coordinates of the sampling points to preserve the exact sampling location.

4.2.6.1.3 Subsurface Grab Sampler

The subsurface grab sampler is an effective tool for collecting samples of surface water from streams, rivers, and surface water drainages for both chemical and radiological analysis. This sampler is composed of a vertical sampling pole, handle, bottle seal ring, and sampling head that attaches to a glass sample bottle (Figure 4.45).

This method is particularly effective for characterizing streams and rivers that are suspected of having stratified zones of contamination, because this sampler can collect grab samples from various depth intervals within a water column. Although this method is most effective in collecting grab samples, it can also be used to collect composite (vertical and areal), and integrated samples as deep as 5 ft below the water surface. A composite sample should not be collected when the analyses to be performed on the sample include volatile organics, because volatilization is facilitated by the compositing process. For a summary of the effectiveness and limitations of this sampling method, see Table 4.5. For more details on this sampler, see www.forestry-suppliers.com.

For most sampling programs, four people are sufficient for this sampling procedure. Two are needed for field testing, sample collection, labeling, and documentation; a third is needed for health and safety; and a fourth is needed for waste management and equipment decontamination. If a raft or boat is used to assist the sampling procedure, at least one additional person will be needed.

The following equipment and procedure can be used to collect surface water samples for chemical and/or radiological analysis:

Equipment

1. Subsurface grab sampler
2. Sample bottles
3. Sample preservatives
4. pH, temperature, conductivity, turbidity, and dissolved oxygen meters
5. Sample labels
6. Cooler packed with Blue Ice® (Blue Ice® is not required for radiological analysis.)
7. Trip blank (only required for volatile organic analyses)
8. Coolant blank (not required for radiological analysis)
9. Sample logbook
10. Chain-of-custody forms
11. Chain-of-custody seals
12. Permanent ink marker
13. Health and safety instruments

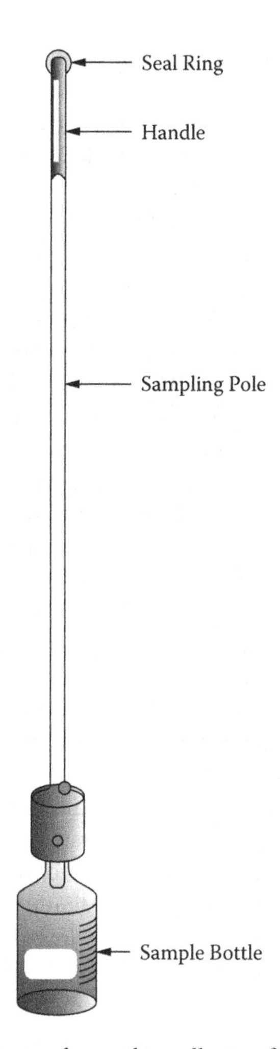

FIGURE 4.45 Subsurface grab sampler used to collect surface water samples.

(From Byrnes, 1994. *Field Sampling Methods for Remedial Investigations.* Lewis Publishers, Boca Raton, FL.)

14. Chemical and/or radiological field screening instruments
15. Health and safety clothing (including life jackets)
16. Waders, raft, or boat
17. Waste container (e.g., 55-gal drum)
18. Sampling table
19. Plastic waste bags

Sampling Procedure

1. In preparation for sampling, confirm that all necessary preparatory work has been completed, including obtaining property access

agreements, meeting health and safety and equipment decontamination requirements, and checking the calibration of all health and safety and chemical and/or radiological field screening instruments.

2. Before sample collection, fill a clean glass jar with sample water and measure the pH, temperature, conductivity, turbidity, and dissolved oxygen of the water. Record this information in a sample logbook.
3. To collect a grab sample, lower the sampler to the desired sampling interval and lift up on the bottle seal ring to allow the sample bottle to fill with water. (Note: If the analyses to be performed require the sample to be preserved, add preservative to the sample bottle.) When the bottle is full, depress the seal ring and retrieve the sampler. Remove the sample bottle from the sampling head. Scan the sample using chemical or radiological screening instruments, and record the results in a bound logbook. Screw on the bottle cap.
4. To collect a vertical or areal composite sample, lower the sampler to the first desired sampling interval and lift up on the bottle seal ring to allow the sample bottle to fill partially with water. Close the bottle seal ring, and move the sampler to the next depth interval or next sampling location. Lift up on the bottle seal ring to allow the sample bottle to fill partially with water at this interval/location. Repeat this procedure until the sample bottle is full. Remove the sample bottle from the sampling head. Scan the sample using chemical and/or radiological screening instruments, and record the results in a bound logbook. (Note: If the analyses to be performed require the sample to be preserved, add preservative to the sample bottle). Screw on the bottle cap.
5. To collect an integrated sample, lower the sampler to the desired sampling interval and lift up on the bottle seal ring to allow the sample bottle to fill partially with water. Close the bottle seal ring. Wait a specified length of time (minutes/hours/days). At the same location, lower the sampler to the same sampling interval and lift up on the bottle seal ring to allow the sample bottle to fill partially with water. Repeat this procedure until the sample bottle is full. Remove the sample bottle from the sampling head. Scan the sample using chemical and/or radiological screening instruments, and record the results in a bound logbook. (Note: If the analyses to be performed require the sample to be preserved, add preservative to the sample bottle). Screw on the bottle cap.
6. After the bottle is capped, attach a sample label and custody seal to the sample bottle and immediately place it into a sample cooler. Samples for chemical analysis should be packed in Blue Ice®.
7. See Chapter 5 for details on preparing sample bottles and coolers for sample shipment.
8. Transfer any water left over from the sampling into a waste container. Before leaving the site, all waste containers should be sealed, labeled, and handled appropriately (see Chapter 11).

9. Transfer any other sampling-related wastes (e.g., gloves and foil) into a plastic waste bag.
10. Have a professional surveyor survey the coordinates of the sampling points to preserve the exact sampling location.

4.2.6.2 Pond, Lake, Retention Basin, and Tank Sampling

As part of an initial site characterization study, one surface water/liquid sample is commonly collected from each retention basin and tank, and as many as two samples (e.g., one near the inlet and outlet) from each pond or lake (Figure 4.41). If contamination is identified in any of these samples, additional sampling may be required. The number of additional samples needed to complete the characterization is dependent on the size of the water body and the specific decision statements that need to be resolved (see Chapter 3, Section 3.2.5.2.4).

The depth at which surface water samples are collected should be based on the suspected concentration and specific gravity of the contaminants of concern. If historical information leads the investigator to believe that contamination could be layered, due to varying specific gravities of the contaminants, collecting either a vertical composite sample or several grab samples from different depth intervals should be considered (see Figure 4.42).

Although a number of sophisticated sampling devices are available to collect surface water and liquid waste samples, not all of these tools are effective in collecting samples from ponds, lakes, retention basins, and tanks. The methods that have proved to be the most effective when sampling these environments include the bottle submersion method, dipper method, subsurface grab sampler method, bailer method, Kemmerer bottle method, WaterMark® vertical polycarbonate water bottle method, and bomb sampler method. The first three of these methods are identical to those used to collect samples from streams, rivers, and surface water drainages (see Section 4.2.6.1), with a few minor modifications.

All seven of these methods are effective in collecting grab, areal composite, and integrated samples. Vertical composite samples can be collected using either the subsurface grab sampler, Kemmerer bottle, WaterMark® vertical polycarbonate water bottle, or bomb sampler. Of these methods, the bottle submersion method is the easiest to implement because it simply involves submerging a sample bottle just beneath the water surface. A telescoping extension rod is often used with this method to hold the sample bottle as it is lowered into the water. The dipper method is also very easy to implement because it just involves lowering a dipper below the water surface, and then transferring the water into sample bottles. The bottle submersion and dipper methods are only effective in collecting samples from the top few inches of the water surface.

The subsurface grab sampler is composed of a vertical sampling pole, handle, bottle seal ring, and sampling head that attaches to a glass sample bottle. Once the sampler has been lowered through the water to the desired sampling interval, the bottle seal ring is lifted to allow the sample bottle to fill with water. When the bottle is full, the seal ring is depressed to seal the bottle. After the sampler has been retrieved, the sample bottle is removed from the sampling head and capped. This method is effective in collecting samples as deep as 5 ft below the water surface.

The bailer is composed of a sampling tube, pouring spout, and a bottom check valve. The bailer is lowered by hand just below the water surface. When the bailer is full, it is retrieved and the water is transferred into a sample bottle. This method collects a sample representative of the top several feet of the water column.

The Kemmerer bottle is composed of a vertical sampling tube, center rod, trip head, and bottom plug. The Kemmerer bottle is lowered through the water by means of a sampling line to the desired sampling depth. The sampling tube is then opened with the assistance of a trip weight. When the sampler is full, it is retrieved. The water is then transferred from the sampler into a sample bottle. The depth at which a sample can be collected with this method is only restricted by the length of the sampling line.

The WaterMark® vertical polycarbonate water bottle is composed of polycarbonate (polyethylene and silicone) sample bottle, nylon cord, line reel, and a trip weight. The sample bottle is lowered to the desired sampling depth using a nylon sample line. The sampling bottle is then opened with the assistance of a trip weight. When the sampler is full, it is retrieved. The water is then transferred from the sampler into a sample bottle. The depth at which a sample can be collected with this method is only restricted by the length of the sampling line.

The bomb sampler is composed of a sampling tube, center rod, and support ring. The sampler is lowered to the desired sampling depth using a support line. The center rod is then lifted using a second line, which allows the sampler to fill with water. The water is then retrieved and transferred into a sample bottle. The depth at which a sample can be collected with this method is only restricted by the length of the sampling line.

Table 4.6 summarizes the effectiveness of each of the seven sampling methods. A number "1" in the table indicates that a particular procedure is most effective in collecting samples for a particular laboratory analysis, sample type, or sampling depth. A number "2" indicates that the procedure is acceptable but less preferred, whereas an empty cell indicates that the procedure is not recommended. For example, Table 4.6 indicates that bottle submersion is the preferred method for collecting samples for volatile organic analysis. Although the dipper, subsurface grab sampler, bailer, Kemmerer bottle, WaterMark® vertical polycarbonate water bottle, and bomb sampler are acceptable methods for volatile organic analysis, they are less desirable because the water from the sampler must be transferred into sample bottles, which tends to aerate the sample.

Whichever sampling method is selected, sample bottles for volatile organic analysis should always be the first to be filled. Bottles for the remaining parameters should be filled consistent with their relative importance to the sampling program.

4.2.6.2.1 Bottle Submersion Method

The bottle submersion method used to collect grab, composite (areal), or integrated water samples from ponds, lakes, and retention basins is identical to the procedure described for stream, river, and water drainage sampling (Section 4.2.6.1.1), with the following modifications:

- If contaminants are identified from the preliminary sampling, the subsurface grab sampler, Kemmerer bottle, WaterMark® vertical polycarbonate water

TABLE 4.6
Rating Table for Surface Water Sampling Methods for Ponds, Lakes, Retention Basins, and Tanks

	Laboratory Analyses							Sample Type				Depth		
	Radio-nuclides	Volatiles	Semi-volatiles	Metals	Pesti-cides	PCBs	TPH	Grab	Composite (Vertical)	Composite (Areal)	Integrated	Surface (0.0–0.5 ft)	Shallow (0.0–5.0 ft)	Deep (> 5.0 ft)
Bottle submersion	1	1	1	1	1	1	1	1		1	1	1		
Dipper	1	2	1	1	1	1	1	1		1	1	1		
Subsurface grab sampler	1	2	1	1	1	1	1	1	1	2	2	2	1	
Bailer	1	2	1	1	1	1	1	1		1	1	1	2[a]	
Kemmerer bottle	1	2	1	1	1	1	1	1	1	2	2		2	1
WaterMark® vertical polycarbonate water bottle method	1	2	1	1	1	1	1	1	1	2	2		2	1
Bomb sampler	1	2	1	1	1	1	1	1	1	2	2		2	1

Note: 1 = preferred method; 2 = acceptable method; empty cell = method not recommended.

[a] Able to collect a sample equal to the length of the bailer.

bottle, or bomb sampler should be considered to delineate clearly the vertical distribution of contaminants.

- If a raft or boat is used to assist the sample collection, a fifth person will be needed on the sampling team. This person's responsibilities are to maneuver and steady the boat.
- Add "waders, raft, or boat" to the equipment list.

4.2.6.2.2 Dipper Method

The dipper method used to collect grab, composite (areal), or integrated surface water samples from ponds, lakes, retention basins, and tanks is identical to the procedure described for stream, river, and water drainage sampling (Section 4.2.6.1.2), with the following modifications:

- If contaminants are identified from the preliminary sampling, the subsurface grab sampler, Kemmerer bottle, WaterMark® vertical polycarbonate water bottle, or bomb sampler should be considered to delineate clearly the vertical distribution of contaminants.
- If a raft or boat is used to assist the sample collection, a fifth person will be needed on the sampling team. This person's responsibilities are to maneuver and steady the boat.
- Add "waders, raft, or boat" to the equipment list.

4.2.6.2.3 Subsurface Grab Sampler

The subsurface grab sampler method used to collect grab, composite (vertical and areal), or integrated water samples from ponds, lakes, retention basins, and tanks is identical to the procedure described for stream, river, and water drainage sampling (Section 4.2.6.1.3).

4.2.6.2.4 Bailer Method

A bailer is most commonly used to collect groundwater samples; however, it can also be used to collect water/liquid samples from ponds, lakes, retention basins, and tanks. A standard bailer is composed of a bailer body, which is available in various lengths and diameters, a pouring spout, and a bottom check valve that contains a check ball (Figure 4.46). This method can be used to collect grab, composite (areal), or integrated water samples.

As the bailer is lowered into the water, water flows into the bailer through the bottom check valve. When the sampler is retrieved, a check ball prevents water from escaping through the bottom of the sampler. The sample is then poured from the bailer through a pouring spout into sample bottles. Depending on the depth of the water at the sampling point, samplers may need a pair of waders, raft, or boat to access the sampling point.

Some common modifications to the bailer include the use of extension couples to increase the length of the bailer, and a controlled flow bottom assembly. The controlled flow bottom assembly allows the bailer to be emptied through the bottom of the sampler, which reduces the opportunity for volatilization to occur. For more

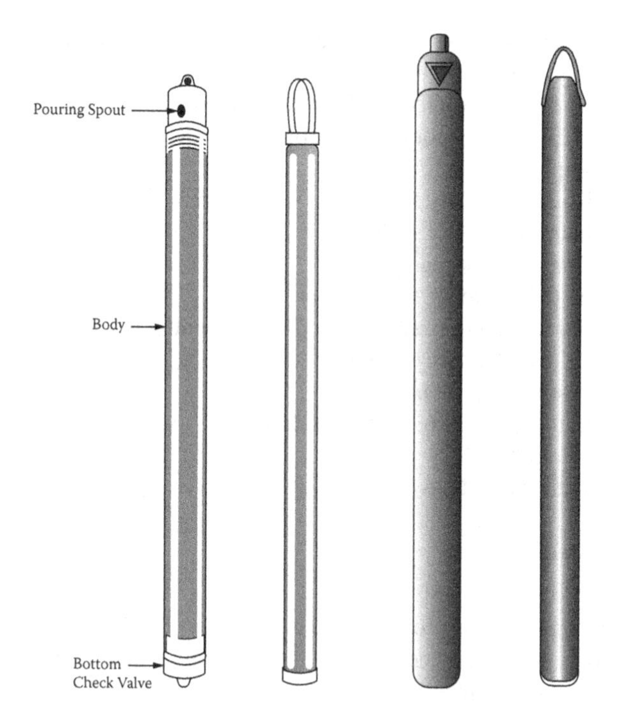

FIGURE 4.46 Various types of bailers available to collect surface water and groundwater samples.

(From Byrnes, 1994. *Field Sampling Methods for Remedial Investigations.* Lewis Publishers, Boca Raton, FL.)

information about bailers, visit www.ams-samplers.com or www.forestry-suppliers.com.

For most sampling programs, four people are sufficient for this sampling procedure. Two are needed for field testing, sample collection, labeling, and documentation; a third is needed for health and safety; and a fourth is needed for miscellaneous tasks such as managing wastewater drums and equipment decontamination. If a raft or boat is used to assist the sampling procedure, at least one additional person will be needed.

The following equipment and procedure can be used to collect water samples for chemical and/or radiological analysis:

Equipment

1. Bailer
2. Sample bottles
3. Sample preservatives
4. pH, temperature, conductivity, turbidity, and dissolved oxygen meters
5. Sample labels

6. Cooler packed with Blue Ice® (Blue Ice® is not required for radiological analysis.)
7. Trip blank (only required for volatile organic analyses)
8. Coolant blank (not required for radiological analysis)
9. Sample logbook
10. Chain-of-custody forms
11. Chain-of-custody seals
12. Permanent ink marker
13. Health and safety instruments
14. Chemical and/or radiological field screening instruments
15. Health and safety clothing (including life jackets)
16. Waders, raft, or boat
17. Waste container (e.g., 55-gal drum)
18. Sampling table
19. Plastic waste bags

Sampling Procedure

1. In preparation for sampling, confirm that all necessary preparatory work has been completed, including obtaining property access agreements, meeting health and safety and equipment decontamination requirements, and checking the calibration of all health and safety and chemical or radiological field screening instruments.
2. Before sample collection, fill a clean glass jar with sample water and measure the pH, temperature, conductivity, turbidity, and dissolved oxygen of the water. Record this information in a sample logbook.
3. To collect a grab sample, hold the bailer above the pouring spout, and slowly lower it just deep enough in the water to fill the bailer. Retrieve the bailer, and carefully transfer the water into a sample bottle. Scan the sample using chemical and/or radiological screening instruments, and record the results in a bound logbook. Screw on the bottle cap.
4. To collect an areal composite sample, hold the bailer above the pouring spout at the first sampling location. Slowly lower the bailer just deep enough in the water to fill the bailer. Retrieve the bailer, and carefully transfer a portion of the water into the sample bottle. Cap the sample bottle. Dispose of any extra water in the waste container. Move to the next sampling location. While holding the bailer above the pouring spout, slowly lower a second decontaminated bailer just deep enough in the water to fill the bailer at the second sampling location. Retrieve the bailer and carefully transfer a portion of the water into the sample bottle. Cap the sample bottle. Dispose of any extra water in the waste container. Repeat this procedure until the sample bottle is full. (Note: If the analyses to be performed require the sample to be preserved, preservative should be added to the bottle before it is filled.) Scan the sample using chemical and/or radiological screening instruments, and record the results in a bound logbook. Screw on the bottle cap.

5. To collect an integrated sample, hold the bailer above the pouring spout at the selected sampling location. Slowly lower the bailer just deep enough in the water to fill the bailer. Retrieve the bailer, and carefully transfer a portion of the water into the sample bottle. Cap the sample bottle. Dispose of any extra water in the waste container. After a specified length of time (minutes/hours/days), hold a second decontaminated bailer just above the pouring spout at the same sampling location. Slowly lower the bailer just deep enough in the water to fill the bailer. Retrieve the bailer, and carefully transfer a portion of the water into the sample bottle. Cap the sample bottle. Repeat this procedure until the sample bottle is full. (Note: If the analyses to be performed require the sample to be preserved, preservative should be added to the bottle before it is filled.) Scan the sample using chemical and/or radiological screening instruments, and record the results in a bound logbook. Screw on the bottle cap.
6. After the bottle is capped, attach a sample label and custody seal to the sample bottle and immediately place it into a sample cooler. Samples for chemical analysis should be packed in Blue Ice®.
7. See Chapter 5 for details on preparing sample bottles and coolers for sample shipment.
8. Transfer any water left over from the sampling into a waste container. Before leaving the site, all waste containers should be sealed, labeled, and handled appropriately (see Chapter 11).
9. Transfer any other sampling-related wastes (e.g., gloves and foil) into a plastic waste bag.
10. Have a professional surveyor survey the coordinates of the sampling points to preserve the exact sampling location.

4.2.6.2.5 Kemmerer Bottle Method

The Kemmerer bottle sampler method is effective for collecting at-depth grab, vertical composite, areal composite, and integrated samples of water from ponds, lakes, retention basins, and tanks. The sampler is composed of a vertical sampling tube, center rod, head plug, and bottom plug (Figure 4.47). A line attached to the top of the sampler is used to lower the sampler to the desired sampling depth. The head plug and bottom plug are then tripped open by sliding a trip weight down the sampling line. When the sampling tube is full of water, the sampler is retrieved and the water is transferred into a sample bottle.

The sampling line should be monofilament, such as common fishing line, and should be discarded between sampling points. The line should be cut to a length long enough to reach the desired sampling depth, and it must be strong enough to lift the weight of the Kemmerer bottle when it is full of water. Before using the sampling line, it should first be decontaminated in the same manner as other sampling equipment (see Chapter 9).

The effective sampling depth of this sampler is only limited by the length of the sampling line. Because this method is often used to collect deep water samples, a raft or boat may be required to assist the sampling procedure.

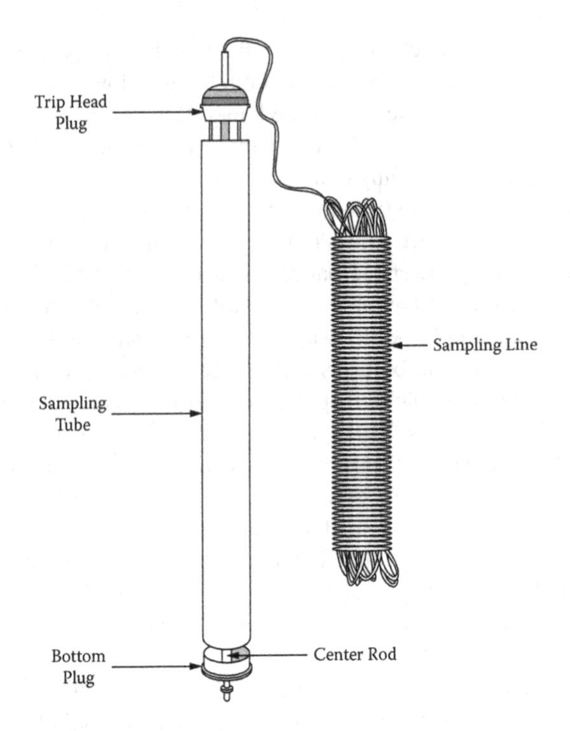

FIGURE 4.47 Kemmerer bottle sampler.

(From Byrnes, 1994. *Field Sampling Methods for Remedial Investigations.* Lewis Publishers, Boca Raton, FL.)

When samples are to be collected from depths of 5 ft or less, the subsurface grab sampler (Section 4.2.6.2.3) is recommended over the Kemmerer bottle, because it is easier to use.

This Kemmerer bottle method is particularly effective for characterizing ponds, lakes, retention basins, and tanks, which may contain vertically stratified contaminant layers. To characterize stratified conditions, the subsurface grab sampler, Kemmerer bottle, WaterMark® vertical polycarbonate water bottle, or bomb sampler (Sections 4.2.6.2.3, 4.2.6.2.5, 4.2.6.2.6, and 4.2.6.2.7) can be used to collect water samples from discrete intervals throughout the water column.

For most sampling programs, five people are sufficient for this sampling procedure. Two are needed for field testing, sample collection, labeling, and documentation; a third is needed for health and safety; a fourth is needed for waste management and equipment decontamination; and a fifth may be needed to operate the boat or maneuver the raft (if required).

The following equipment and procedure can be used to collect surface water samples for chemical and/or radiological analysis:

Equipment

1. Kemmerer bottle
2. Sampling line (monofilament)

3. Sample bottles
4. Sample preservatives
5. pH, temperature, conductivity, turbidity, and dissolved oxygen meters
6. Sample labels
7. Cooler packed with Blue Ice® (Blue Ice® is not required for radiological analysis.)
8. Trip blank (only required for volatile organic analyses)
9. Coolant blank (not required for radiological analysis)
10. Sample logbook
11. Chain-of-custody forms
12. Chain-of-custody seals
13. Permanent ink marker
14. Health and safety instruments
15. Chemical and/or radiological field screening instruments
16. Health and safety clothing (including life jackets)
17. Waders, raft, or boat
18. Waste container (e.g., 55-gal drum)
19. Sampling table
20. Plastic waste bags

Sampling Procedure

1. In preparation for sampling, confirm that all necessary preparatory work has been completed, including obtaining property access agreements, meeting health and safety and equipment decontamination requirements, and checking the calibration of all health and safety and chemical or radiological field screening instruments.
2. To collect a grab sample, when properly positioned over the sampling point, slowly lower the Kemmerer bottle to the desired sampling depth. Slide the trip weight down the sampling line to trip open the sample tube. When the sampler is full, retrieve it and transfer the water from the Kemmerer bottle directly into a sample bottle. (Note: If the analyses to be performed require the sample to be preserved, preservative should be added to the bottle before it is filled.) Scan the sample using chemical and/or radiological screening instruments, and record the results in a bound logbook. Screw on the bottle cap. Pour any remaining water from the sampler into a clean glass jar and measure the pH, temperature, conductivity, turbidity, and dissolved oxygen. If there is not sufficient water to run these field analyses, lower the sampler to the desired sampling depth a second time to collect the water volume needed to run these field analyses. Dispose of any extra water in the waste container.
3. To collect a vertical composite sample, at the first sampling location, slowly lower the Kemmerer bottle to the first sampling depth. Slide the trip weight down the sampling line to trip open the sample tube. When the sampler is full, retrieve it and transfer a portion of the water from the Kemmerer bottle directly into a sample bottle. Cap the sample bottle and transfer to a cooler packed with Blue Ice® (Blue Ice® is not required for radiological

analysis). Pour any remaining water from the sampler into a clean glass jar and measure the pH, temperature, conductivity, turbidity, and dissolved oxygen. Dispose of any extra water in the waste container. Lower a second decontaminated Kemmerer bottle to the next sampling depth. Slide the trip weight down the sampling line to trip open the sample tube. When the sampler is full, retrieve it and transfer a portion of the water from the Kemmerer bottle directly into a sample bottle. Cap the sample bottle and transfer to a cooler packed with Blue Ice® (Blue Ice® is not required for radiological analysis). Pour any remaining water from the sampler into a clean glass jar and measure the pH, temperature, conductivity, turbidity, and dissolved oxygen. Dispose of any extra water in the waste container. Repeat this procedure until the sample bottle is full. (Note: If the analyses to be performed require the sample to be preserved, preservative should be added to the bottle before it is filled.) Scan the sample using chemical and/or radiological screening instruments, and record the results in a bound logbook. Screw on the bottle cap.

4. To collect an areal composite sample, at the first sampling location, slowly lower the Kemmerer bottle to the desired sampling depth. Slide the trip weight down the sampling line to trip open the sample tube. When the sampler is full, retrieve it and transfer a portion of the water from the Kemmerer bottle directly into a sample bottle. Cap the sample bottle and transfer to a cooler packed with Blue Ice® (Blue Ice® is not required for radiological analysis). Pour any remaining water from the sampler into a clean glass jar and measure the pH, temperature, conductivity, turbidity, and dissolved oxygen. Dispose of any extra water in the waste container. Move to the next sampling location. Slowly lower a second decontaminated Kemmerer bottle to the desired sampling depth. Slide the trip weight down the sampling line to trip open the sample tube. When the sampler is full, retrieve it and transfer a portion of the water from the Kemmerer bottle directly into a sample bottle. Cap the sample bottle and transfer to a cooler packed with Blue Ice® (Blue Ice® is not required for radiological analysis). Pour any remaining water from the sampler into a clean glass jar and measure the pH, temperature, conductivity, turbidity, and dissolved oxygen. Repeat this procedure until the sample bottle is full. (Note: If the analyses to be performed require the sample to be preserved, preservative should be added to the bottle before it is filled.) Scan the sample using chemical and/or radiological screening instruments, and record the results in a bound logbook. Screw on the bottle cap.
5. To collect an integrated sample, at the selected sampling location, slowly lower the Kemmerer bottle to the desired sampling depth. Slide the trip weight down the sampling line to trip open the sample tube. When the sampler is full, retrieve it and transfer a portion of the water from the Kemmerer bottle directly into a sample bottle. Cap the sample bottle and transfer to a cooler packed with Blue Ice® (Blue Ice® is not required for radiological analysis). Pour any remaining water from the

sampler into a clean glass jar and measure the pH, temperature, conductivity, turbidity, and dissolved oxygen. Dispose of any extra water in the waste container. After a specified length of time (minutes/hours/days), slowly lower a second decontaminated Kemmerer bottle to the same location and same sampling depth as the first sample. Slide the trip weight down the sampling line to trip open the sample tube. When the sampler is full, retrieve it and transfer a portion of the water from the Kemmerer bottle directly into a sample bottle. Cap the sample bottle and transfer to a cooler packed with Blue Ice® (Blue Ice® is not required for radiological analysis). Pour any remaining water from the sampler into a clean glass jar and measure the pH, temperature, conductivity, turbidity, and dissolved oxygen. Dispose of any extra water in the waste container. Repeat this procedure until the sample bottle is full. (Note: If the analyses to be performed require the sample to be preserved, preservative should be added to the bottle before it is filled.) Scan the sample using chemical and/or radiological screening instruments, and record the results in a bound logbook. Screw on the bottle cap.

6. After the bottle is capped, attach a sample label and custody seal to the sample bottle and immediately place it into a sample cooler. Samples for chemical analysis should be packed in Blue Ice®.
7. See Chapter 5 for details on preparing sample bottles and coolers for sample shipment.
8. Transfer any water left over from the sampling into a waste container. Before leaving the site, all waste containers should be sealed, labeled, and handled appropriately (see Chapter 11).
9. Transfer any other sampling-related wastes (e.g., gloves and foil) into a plastic waste bag.
10. Have a professional surveyor survey the coordinates of the sampling point to preserve the exact sampling location.

4.2.6.2.6 WaterMark® Vertical Polycarbonate Water Bottle Method

The WaterMark® vertical polycarbonate water bottle method is effective for collecting at-depth grab, vertical composite, areal composite, and integrated samples of water from ponds, lakes, retention basins, and tanks. This sampler is composed of a polycarbonate sample bottle, nylon cord, line reel, and trip weight (Figure 4.48). The nylon cord attached to the bottle is used to lower the bottle to the desired sampling depth. The bottle is tripped open by sliding a trip weight down the nylon cord. This allows water to enter the bottle. When the bottle is full of water, it is retrieved and the water is transferred into a sample bottle.

The effective sampling depth of this sampler is only limited by the length of the nylon cord. Because this method is often used to collect deep water samples, a raft or boat may be required to assist the sampling procedure. When samples are to be collected from depths of 5 ft or less, the subsurface grab sampler (Section 4.2.6.2.3) is recommended over the WaterMark® vertical polycarbonate water bottle method, because it is easier to use.

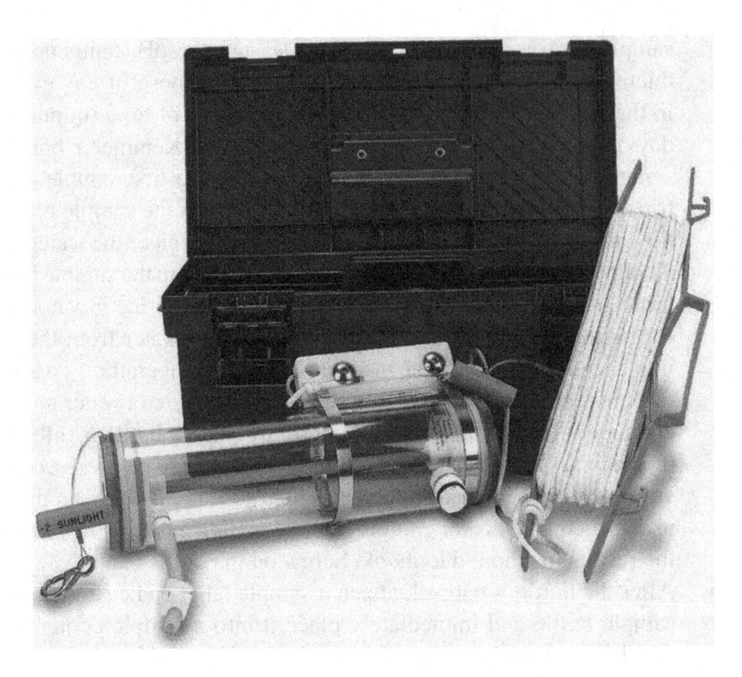

FIGURE 4.48 WaterMark® vertical polycarbonate water bottle.

This method is particularly effective for characterizing ponds, lakes, retention basins, and tanks, which may contain vertically stratified contaminant layers. To characterize stratified conditions, the subsurface grab sampler, Kemmerer bottle, WaterMark® vertical polycarbonate water bottle, or bomb sampler (Sections 4.2.6.2.3, 4.2.6.2.5, 4.2.6.2.6, and 4.2.6.2.7) can be used to collect water samples from discrete intervals throughout the water column.

For most sampling programs, five people are sufficient for this sampling procedure. Two are needed for field testing, sample collection, labeling, and documentation; a third is needed for health and safety; a fourth is needed for waste management and equipment decontamination; and a fifth may be needed to operate the boat or maneuver the raft (if required).

The following equipment and procedure can be used to collect surface water samples for chemical and/or radiological analysis:

Equipment

1. WaterMark® vertical polycarbonate water bottle
2. Nylon cord
3. Sample bottles
4. Sample preservatives
5. pH, temperature, conductivity, turbidity, and dissolved oxygen meters
6. Sample labels
7. Cooler packed with Blue Ice® (Blue Ice® is not required for radiological analysis.)

8. Trip blank (only required for volatile organic analyses)
9. Coolant blank (not required for radiological analysis)
10. Sample logbook
11. Chain-of-custody forms
12. Chain-of-custody seals
13. Permanent ink marker
14. Health and safety instruments
15. Chemical and/or radiological field screening instruments
16. Health and safety clothing (including life jackets)
17. Waders, raft, or boat
18. Waste container (e.g., 55-gal drum)
19. Sampling table
20. Plastic waste bags

Sampling Procedure

1. In preparation for sampling, confirm that all necessary preparatory work has been completed, including obtaining property access agreements, meeting health and safety and equipment decontamination requirements, and checking the calibration of all health and safety and chemical or radiological field screening instruments.
2. To collect a grab sample, when properly positioned over the sampling point, slowly lower the WaterMark® vertical polycarbonate water bottle to the desired sampling depth. Slide the trip weight down the nylon cord to trip open the bottle. When the bottle is full, retrieve it and transfer the water from the sampler directly into a sample bottle. (Note: If the analyses to be performed require the sample to be preserved, preservative should be added to the bottle before it is filled.) Scan the sample using chemical and/or radiological screening instruments, and record the results in a bound logbook. Screw on the bottle cap. Pour any remaining water from the sampler into a clean glass jar and measure the pH, temperature, conductivity, turbidity, and dissolved oxygen. If there is not sufficient water to run these field analyses, lower the sampler to the desired sampling depth a second time to collect the water volume needed to run these field analyses.
3. To collect a vertical composite sample, when properly positioned over the sampling point, slowly lower the WaterMark® vertical polycarbonate water bottle to the first sampling depth. Slide the trip weight down the nylon cord to trip open the bottle. When the bottle is full, retrieve it and transfer a portion of the water from the sampler directly into a sample bottle. Cap the sample bottle. Pour any remaining water from the sampler into a clean glass jar and measure the pH, temperature, conductivity, turbidity, and dissolved oxygen. Dispose of any extra water in the waste container. Slowly lower a second decontaminated vertical polycarbonate water bottle to the next sampling depth. Slide the trip weight down the nylon cord to trip open the bottle. When the bottle is full,

retrieve it and transfer a portion of the water from the bottle directly into a sample bottle. Cap the sample bottle. Pour any remaining water from the sampler into a clean glass jar and measure the pH, temperature, conductivity, turbidity, and dissolved oxygen. Repeat this procedure until the sample bottle is full. (Note: If the analyses to be performed require the sample to be preserved, preservative should be added to the bottle before it is filled.) Scan the sample using chemical and/or radiological screening instruments, and record the results in a bound logbook. Screw on the bottle cap.

4. To collect an areal composite sample, when properly positioned over the sampling point, slowly lower the WaterMark® vertical polycarbonate water bottle to the desired sampling depth. Slide the trip weight down the nylon cord to trip open the bottle. When the bottle is full, retrieve it and transfer a portion of the water from the sampler directly into a sample bottle. Cap the sample bottle. Pour any remaining water from the sampler into a clean glass jar and measure the pH, temperature, conductivity, turbidity, and dissolved oxygen. Dispose of any extra water in the waste container. Move to the next sampling location. Slowly lower a second decontaminated WaterMark® vertical polycarbonate water bottle to the desired sampling depth. Slide the trip weight down the nylon cord to trip open the bottle. When the bottle is full, retrieve it and transfer a portion of the water from the bottle directly into a sample bottle. Cap the sample bottle. Pour any remaining water from the sampler into a clean glass jar and measure the pH, temperature, conductivity, turbidity, and dissolved oxygen. Dispose of any extra water in the waste container. Repeat this procedure until the sample bottle is full. (Note: If the analyses to be performed require the sample to be preserved, preservative should be added to the bottle before it is filled.) Scan the sample using chemical and/or radiological screening instruments, and record the results in a bound logbook. Screw on the bottle cap.
5. To collect an integrated sample, at the selected sampling location, slowly lower the WaterMark® vertical polycarbonate water bottle to the desired sampling depth. Slide the trip weight down the nylon cord to trip open the bottle. When the bottle is full, retrieve it and transfer a portion of the water from the bottle directly into a sample bottle. Cap the sample bottle. Pour any remaining water from the sampler into a clean glass jar and measure the pH, temperature, conductivity, turbidity, and dissolved oxygen. Dispose of any extra water in the waste container. After a specified length of time (minutes/hours/days), slowly lower a second decontaminated WaterMark® vertical polycarbonate water bottle to the same location and same sampling depth. Slide the trip weight down the nylon cord to trip open the bottle. When the bottle is full, retrieve it and transfer a portion of the water from the bottle directly into a sample bottle. Cap the sample bottle. Pour any remaining water from the sampler into a clean glass jar and measure the pH, temperature, conductivity, turbidity, and dissolved oxygen. Dispose of any extra water in the waste

container. Repeat this procedure until the sample bottle is full. (Note: If the analyses to be performed require the sample to be preserved, preservative should be added to the bottle before it is filled.) Scan the sample using chemical and/or radiological screening instruments, and record the results in a bound logbook. Screw on the bottle cap.

6. After the bottle is capped, attach a sample label and custody seal to the sample bottle and immediately place it into a sample cooler. Samples for chemical analysis should be packed in Blue Ice®.
7. See Chapter 5 for details on preparing sample bottles and coolers for sample shipment.
8. Transfer any liquid left over from the sampling into a waste container. Before leaving the site, all waste containers should be sealed, labeled, and handled appropriately (see Chapter 11).
9. Transfer any other sampling-related wastes (e.g., gloves and foil) into a plastic waste bag.
10. Have a professional surveyor survey the coordinates of the sampling point to preserve the exact sampling location.

4.2.6.2.7 Bomb Sampler Method

The bomb sampler method is effective for collecting at-depth grab, composite (vertical and areal), and integrated samples of water/liquid from ponds, lakes, retention basins, and tanks. This sampler is composed of a sampling tube, center rod, and support ring (Figure 4.49). A line attached to the support ring is used to lower the

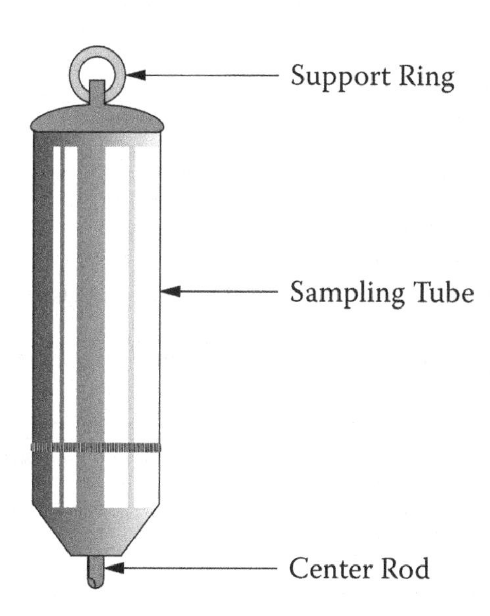

FIGURE 4.49 Bomb sampler.

(From Byrnes, 1994. *Field Sampling Methods for Remedial Investigations*. Lewis Publishers, Boca Raton, FL.)

sampler to the desired sampling depth. A second line is attached to the top of the spring-loaded center rod, and is used to open and close the sampling tube.

The support and sampling line should be monofilament, such as common fishing line, and should be discarded between sampling points. The line should be cut to a length long enough to reach the desired sampling depth, and must be strong enough to lift the weight of the bomb sampler when it is full of water. Before using the sampling line, it should first be decontaminated in the same manner as other sampling equipment (Chapter 9).

After lowering the sampler to the desired sampling depth, the sampling line is lifted to allow the sampling tube to fill with water. When the sampling line is released, the center rod drops to reseal the sampling tube. The sampler is then retrieved, and water is transferred into a sample bottle by placing the bottle beneath the center rod, and lifting up on the sampling line.

The effective sampling depth of this sampler is only limited by the length of the sampling line. Because this method is used primarily to collect deep water samples, a raft or boat may be required to assist the sampling procedure. When samples are to be collected from depths of 5 ft or less, the subsurface grab sampler (Section 4.2.6.2.3) is recommended over the bomb sampler, because it is easier to use.

For most sampling programs, five people are sufficient for this sampling procedure. Two are needed for field testing, sample collection, labeling, and documentation; a third is needed for health and safety; a fourth is needed for waste management and equipment decontamination; and a fifth may be needed to operate the boat or maneuver the raft.

The following equipment and procedure can be used to collect water samples for chemical or radiological analysis:

Equipment

1. Bomb sampler
2. Sampling line
3. Sample bottles
4. Sample preservatives
5. pH, temperature, conductivity, turbidity, and dissolved oxygen meters
6. Sample labels
7. Cooler packed with Blue Ice® (Blue Ice® is not required for radiological analysis.)
8. Trip blank (only required for volatile organic analyses)
9. Coolant blank (not required for radiological analysis)
10. Sample logbook
11. Chain-of-custody forms
12. Chain-of-custody seals
13. Permanent ink marker
14. Health and safety instruments
15. Chemical and/or radiological field screening instruments
16. Health and safety clothing (including life jackets)

17. Waders, raft, or boat
18. Waste container (e.g., 55-gal drum)
19. Sampling table
20. Plastic waste bags

Sampling Procedure

1. In preparation for sampling, confirm that all necessary preparatory work has been completed, including obtaining property access agreements, meeting health and safety and equipment decontamination requirements, and checking the calibration of all health and safety and chemical and/or radiological field screening instruments.
2. To collect a grab sample, when properly positioned over the sampling point, slowly lower the bomb sampler to the desired sampling depth. Lift up on the sampling line, and allow the sampling tube to fill with water. When the sampler is full, release the sampling line to reseal. Retrieve the sampler, and transfer the water into a sample bottle. (Note: If the analyses to be performed require the sample to be preserved, preservative should be added to the bottle before it is filled.) Scan the sample using chemical and/or radiological screening instruments, and record the results in a bound logbook. Screw on the bottle cap. Pour any remaining water from the sampler into a clean glass jar and measure the pH, temperature, conductivity, turbidity, and dissolved oxygen. If there is not sufficient water to run these field analyses, lower the sampler to the desired sampling depth a second time to collect the water volume needed to run these field analyses.
3. To collect a vertical composite sample, when properly positioned over the sampling point, slowly lower the bomb sampler to the first sampling depth. Lift up on the sampling line and allow the sampling tube to fill with water. When the sampler is full, release the sampling line to reseal. Retrieve the sampler, and transfer a portion of the water into a sample bottle. Cap the sample bottle. Pour any remaining water from the sampler into a clean glass jar and measure the pH, temperature, conductivity, turbidity, and dissolved oxygen. Dispose of any extra water in the waste container. Slowly lower a second decontaminated bomb sampler to the next sampling depth. Lift up on the sampling line, and allow the sampling tube to fill with water. When the sampler is full, release the sampling line to reseal. Retrieve the sampler, and transfer a portion of the water into a sample bottle. Cap the sample bottle. Pour any remaining water from the sampler into a clean glass jar and measure the pH, temperature, conductivity, turbidity, and dissolved oxygen. Repeat this procedure until the sample bottle is full. (Note: If the analyses to be performed require the sample to be preserved, preservative should be added to the bottle before it is filled.) Scan the sample using chemical and/or radiological screening instruments, and record the results in a bound logbook. Screw on the bottle cap.

4. To collect an areal composite sample, when properly positioned over the sampling point, slowly lower the bomb sampler to the desired sampling depth. Lift up on the sampling line, and allow the sampling tube to fill with water. When the sampler is full, release the sampling line to reseal. Retrieve the sampler, and transfer a portion of the water into a sample bottle. Cap the sample bottle. Pour any remaining water from the sampler into a clean glass jar and measure the pH, temperature, conductivity, turbidity, and dissolved oxygen. Dispose of any extra water in the waste container. Move to the next sampling location. Slowly lower a second decontaminated bomb sampler to the desired sampling depth. Lift up on the sampling line, and allow the sampling tube to fill with water. When the sampler is full, release the sampling line to reseal. Retrieve the sampler, and transfer a portion of the water into a sample bottle. Cap the sample bottle. Pour any remaining water from the sampler into a clean glass jar and measure the pH, temperature, conductivity, turbidity, and dissolved oxygen. Repeat this procedure until the sample bottle is full. (Note: If the analyses to be performed require the sample to be preserved, preservative should be added to the bottle before it is filled.) Scan the sample using chemical and/or radiological screening instruments, and record the results in a bound logbook. Screw on the bottle cap.
5. To collect an integrated sample, at the selected sampling location, slowly lower the bomb sampler to the desired sampling depth. Lift up on the sampling line, and allow the sampling tube to fill with water. When the sampler is full, release the sampling line to reseal. Retrieve the sampler, and transfer a portion of the water into a sample bottle. Cap the sample bottle. Pour any remaining water from the sampler into a clean glass jar and measure the pH, temperature, conductivity, turbidity, and dissolved oxygen. Dispose of any extra water in the waste container. After a specified length of time (minutes/hours/days), slowly lower a second decontaminated bomb sampler to the desired sampling depth. Lift up on the sampling line, and allow the sampling tube to fill with water. When the sampler is full, release the sampling line to reseal. Retrieve the sampler, and transfer a portion of the water into a sample bottle. Cap the sample bottle. Pour any remaining water from the sampler into a clean glass jar and measure the pH, temperature, conductivity, turbidity, and dissolved oxygen. Dispose of any extra water in the waste container. Repeat this procedure until the sample bottle is full. (Note: If the analyses to be performed require the sample to be preserved, preservative should be added to the bottle before it is filled.) Scan the sample using chemical and/or radiological screening instruments, and record the results in a bound logbook. Screw on the bottle cap.
6. After the bottle is capped, attach a sample label and custody seal to the sample bottle and immediately place it into a sample cooler. Samples for chemical analysis should be packed in Blue Ice®.
7. See Chapter 5 for details on preparing sample bottles and coolers for sample shipment.

8. Transfer any water left over from the sampling into a waste container. Before leaving the site, all waste containers should be sealed, labeled, and handled appropriately (see Chapter 11).
9. Transfer any other sampling-related wastes (e.g., gloves and foil) into a plastic waste bag.
10. Have a professional surveyor survey the coordinates of the sampling point to preserve the exact sampling location.

4.2.7 Groundwater Sampling/Testing

This section provides the reader with guidance on selecting the most appropriate groundwater sampling method for the site under investigation. The criteria used to select the most appropriate method include the analyses to be performed on the sample, the type of sample to be collected (grab, composite, or integrated), and the sampling depth. Standard operating procedures have been provided for each of the recommended sampling methods to facilitate implementation. The following groundwater characterization strategies are provided as a supplement to general guidance provided in Chapter 3, Section 3.2.5.7.

Whenever contamination is identified in soil, there is always the possibility of contaminants migrating to groundwater. This migration is possible through the transport mechanism of water percolating through the soil, whereas the rate of migration is controlled by soil physical properties (such as hydraulic conductivity and clay content) as well as soil geochemical properties (e.g., the distribution coefficient [*Kd*]). Once contaminants reach the groundwater, they commonly disperse into the saturated formation. Depending on their physical or chemical properties, contaminants can concentrate near the upper, middle, or lower portion of the aquifer, or may evenly distribute themselves throughout the aquifer. Chemicals such as benzene, toluene, and xylene have specific gravities less than water and therefore tend to concentrate near the upper portion of the aquifer. In high enough concentrations, these types of chemicals can occur in the form of LNAPLs, which is a pure product that floats on the top of the aquifer. Chemicals such as trichloroethylene (TCE), tetrachloroethylene (PCE), and carbon tetrachloride have specific gravities greater than water and therefore tend to concentrate in the lower portion of the aquifer. In high enough concentrations, these types of chemicals can occur in the form of DNAPLs, which is a pure product that will sink to the bottom of the aquifer. Radionuclides and metals are less predictable regarding their distribution within an aquifer.

When assessing the groundwater conditions at a site, serious consideration should be given to using the direct push method (Section 4.2.7.1) in combination with a quick turnaround mobile laboratory to perform a preliminary groundwater assessment before installing permanent monitoring wells (Section 4.2.7.2). The advantages of the direct push method are that groundwater samples can be collected quickly and inexpensively when compared with collecting the same data through monitoring well installation and sampling. Because a mobile laboratory can typically provide analytical results within hours after sampling, the direct push method facilitates the "observational approach," where the results from samples analyzed in the field are used to guide the characterization effort. The direct push method can also be used to collect groundwater samples from multiple depths throughout the aquifer, which is important for defining the three-dimensional distribution of contamination within the aquifer. Once preliminary groundwater characterization is complete, groundwater wells can be precisely positioned for long-term monitoring, aquifer testing, and remediation. Although groundwater samples can also be collected from multiple depths throughout the aquifer during well drilling (for defining the three-dimensional distribution of contamination), it is more complicated, more expensive, and creates larger volumes of purge water (than the direct push method) that will need to be disposed of.

If, for some reason, the direct push method cannot be used to perform preliminary groundwater characterization, one groundwater monitoring well should be installed downgradient from each suspected source of contamination, and one well upgradient (Figure 4.50). These wells should be built to screen the first water-bearing unit (typically an unconfined aquifer) because this is where contaminants should be most

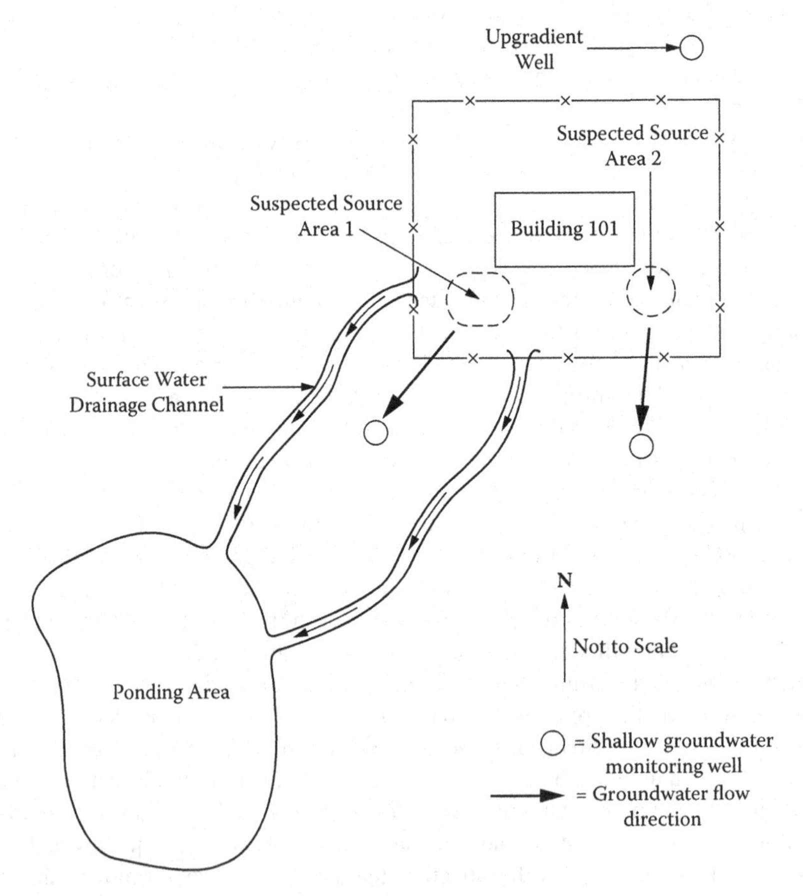

FIGURE 4.50 Common initial groundwater well configuration when the general groundwater flow direction is known.

(From Byrnes, 1994. *Field Sampling Methods for Remedial Investigations.* Lewis Publishers, Boca Raton, FL.)

concentrated. If the aquifer is more than 20 ft in thickness, depth-discrete groundwater samples should be collected at consistent depth intervals (e.g., every 10 to 20 ft) throughout the aquifer during well drilling to help define the depth where contamination is most concentrated. These depth-discrete groundwater samples should be analyzed by a quick-turnaround mobile laboratory, and the analytical results should be used to help select where the well screen should optimally be positioned within the aquifer. (Note: Each sampling intervals will first need to be purged [Section 4.2.7.4] before sample collection.) After all of the monitoring wells have been installed, they should be properly developed (Section 4.2.7.3), purged (Section 4.2.7.4), and sampled (Section 4.2.7.5) for standard laboratory analyses. When the analytical results are returned from the standard laboratory, any contaminants identified in the downgradient wells that are not also identified in the upgradient well are contaminants derived from the site.

Although unconfined aquifer groundwater flow contours often mimic topographic contours, this is not always the case. When groundwater flow direction at a site is unknown, a minimum of three monitoring wells (or piezometers) should be positioned to form an equilateral triangle around the site, one of which should be positioned topographically downslope from the largest suspected source of contamination (Figure 4.51). This arrangement allows for an accurate determination of groundwater

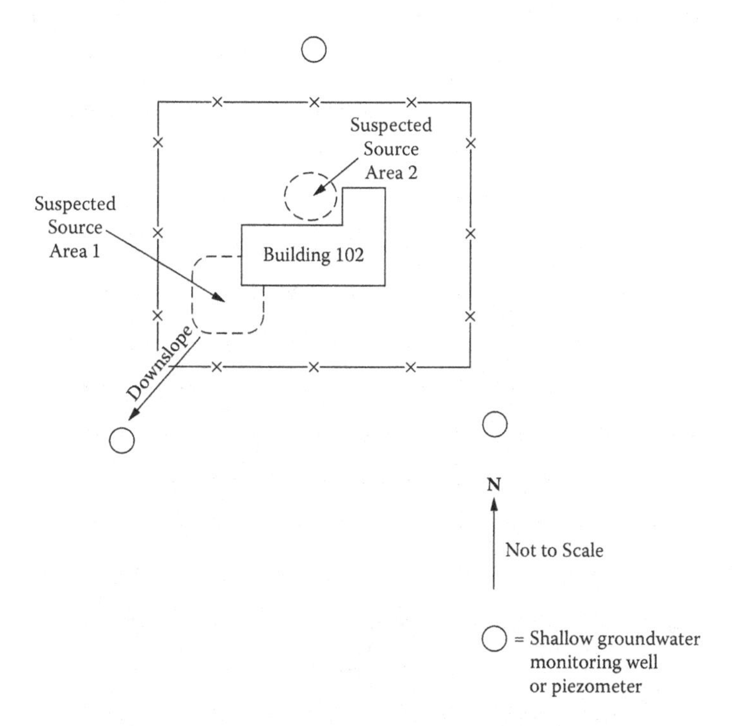

FIGURE 4.51 Common initial groundwater well configuration when the general groundwater flow direction is unknown.

(From Byrnes, 1994. *Field Sampling Methods for Remedial Investigations.* Lewis Publishers, Boca Raton, FL.)

flow direction and gradient. After groundwater flow direction has been determined, one well should be installed downgradient from each suspected contaminant source and one well should be installed upgradient of the site. The wells should then be properly developed (Section 4.2.7.3), purged (Section 4.2.7.4), and sampled (Section 4.2.7.5). These samples should be analyzed for all of the contaminants of concern.

If groundwater contamination is identified in the upper aquifer, it is important to define the source, and lateral and vertical extent of contaminant migration. This includes the sampling of water in the next lower aquifer. This is most effectively accomplished by using the direct push method to characterize the upper and lower aquifer thoroughly, followed by the installation of long-term monitoring wells. The next subsection provides examples of how this type of investigation is performed.

4.2.7.1 Direct Push Method

The direct push method utilizes a hydraulic press and slide hammer mounted on the rear end of a truck, to advance a sampling probe to a depth where a groundwater sample can be collected (Figures 4.52 and 4.53). Three sizes of trucks are available to perform this procedure. The size of the truck required is dependent on the depth of groundwater at the site. A lightweight van or truck is generally able to provide a reaction weight of 1,000 to 2,000 lb, which can advance a sampling probe 10 to 20 ft below the ground surface. The next larger truck provides a reaction weight of 3,000 to 5,000 lb, and is designed to sample as deep as 50 ft below the ground surface. The heaviest vehicle is the cone penetrometer, which has a reaction weight of 10 to 30 tons, and has been successful at penetrating through 200 ft (or more) of unconsolidated sediment.

FIGURE 4.52 PowerProbe™ direct push sampler.

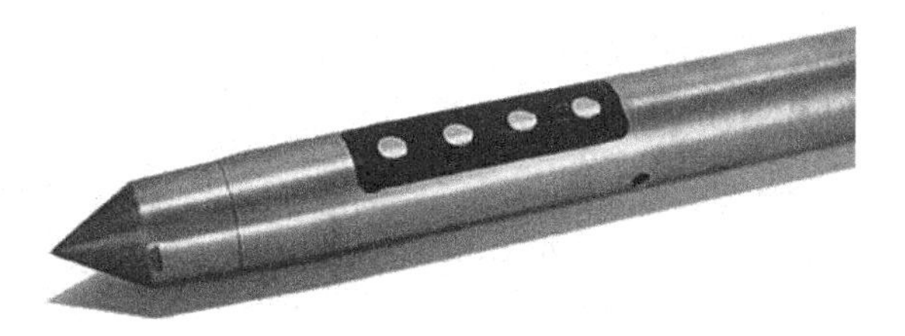

FIGURE 4.53 Example of direct push sample probe.

The primary advantages of using the direct push method to assist a groundwater characterization study include the following:

- There is very little investigation-derived waste generated with the procedure.
- Groundwater samples can be collected quickly and from multiple depths within the aquifer.
- The procedure is much less expensive than collecting the same data by installing and sampling monitoring wells.
- The equipment gathers less public attention than a drill rig.
- The procedure produces little disturbance to the surrounding environment.

Until recently, groundwater characterization has been performed by drilling and installing numerous groundwater monitoring wells in and around areas suspected of being contaminated. Because drilling procedures generate larger volumes of waste soil, which must be drummed, stored, and ultimately disposed of, the less drilling that is required, the better. Although the need to install groundwater wells has not gone away, the direct push method can assist a groundwater investigation by selecting the optimum location for fewer wells.

The direct push method is only effective in pushing through unconsolidated sediments. Other limitations of the direct push method include the sampling depth, volume of sample that can be retrieved, and difficulties in penetrating through soils that contain gravel. Groundwater samples can be collected in one of five different ways when using the direct push method:

1. Lowering a < 0.5-in.-diameter bailer down the inside of the probe multiple times to collect the volume of sample needed. This technique works well when only small volumes of water are needed.
2. Lowering a weighted sample vial under vacuum down the inside of the probe. A needle inside the probe punctures the septum and allows water to flow into the vial. This method works well for collecting small volumes of water.
3. A third method utilizes chambers in the probe that can be filled at depth, then brought to the surface. In most cases, the capacity of the chambers does not exceed 500 mL. To obtain larger volumes of water with this technique, the sampler must be advanced and retrieved repeatedly.
4. Lowering a sample tube down the inside of the probe and using a suction-lift pump to extract as much water as needed. This is the least preferred of all the

available methods because it is not effective in collecting samples deeper than 25 ft, and is reported by the EPA to cause the volatilization of the sample and, possibly, to affect the pH.

5. A fifth method can be used to collect samples from formations with very low permeability. This technique involves running a screened tube down the inside of the sampling hole, removing the steel rods, and packing sand around the screened section. This minimonitoring well can be left in place as long as is required for water to fill the hole. Tests have shown that analytical results from water samples collected from the temporary wells compare favorably with data from conventional wells, and the temporary wells are a fraction of the cost. In addition to installing minimonitoring wells, the direct push method can also be used to install temporary piezometers for water level monitoring.

If groundwater contaminants at a site may include both LNAPLs and DNAPLs, it is recommended that the groundwater samples be collected from both the top and bottom of the aquifer, because this is where these contaminants will concentrate. To characterize the distribution of contaminants more completely, samples should be collected at regular intervals throughout the depth of an aquifer. This type of depth interval sampling is commonly performed at locations close to the suspected contaminant sources, and the results are used to determine the most appropriate sampling intervals for more distant sampling points.

If contamination is identified in the upper aquifer, the direct push method can be used to collect samples from the next deeper aquifer, assuming it is within the depth penetration range of the truck. If possible, groundwater samples should also be collected from multiple depths throughout the lower aquifer.

To use the direct push method most cost-effectively, groundwater investigations should be performed in combination with a mobile laboratory and using the "observational approach." The observational approach utilizes the analytical data from each sampling point to decide where to position additional sampling points. This method is only possible when using a mobile laboratory that can provide analytical results shortly after sampling. This approach avoids the problem of collecting unnecessary or insufficient data.

An example of a successful groundwater investigation using the observational approach is illustrated in Figure 4.54. In this example, historical information led investigators to believe that buried tanks located south and west of Building 101 were potential sources of groundwater contamination at the site. The first step in this investigation involved collecting an initial row of groundwater samples in a "V" pattern, just inside the site property boundary, with the "V" pointing in the downgradient direction. Collecting initial groundwater samples from these locations will ensure that any groundwater contamination leaving the site will be detected by one or more of these sampling points. Based on the results from this first phase of sampling, the observational approach is used to track the extent of the contaminant plume. This approach involves sampling outward from a contaminated sampling point in a grid pattern until the edge of the contaminant plume is defined.

Once the plume has been defined, monitoring wells should be installed for long-term monitoring purposes (see Figure 4.54). These wells are commonly positioned

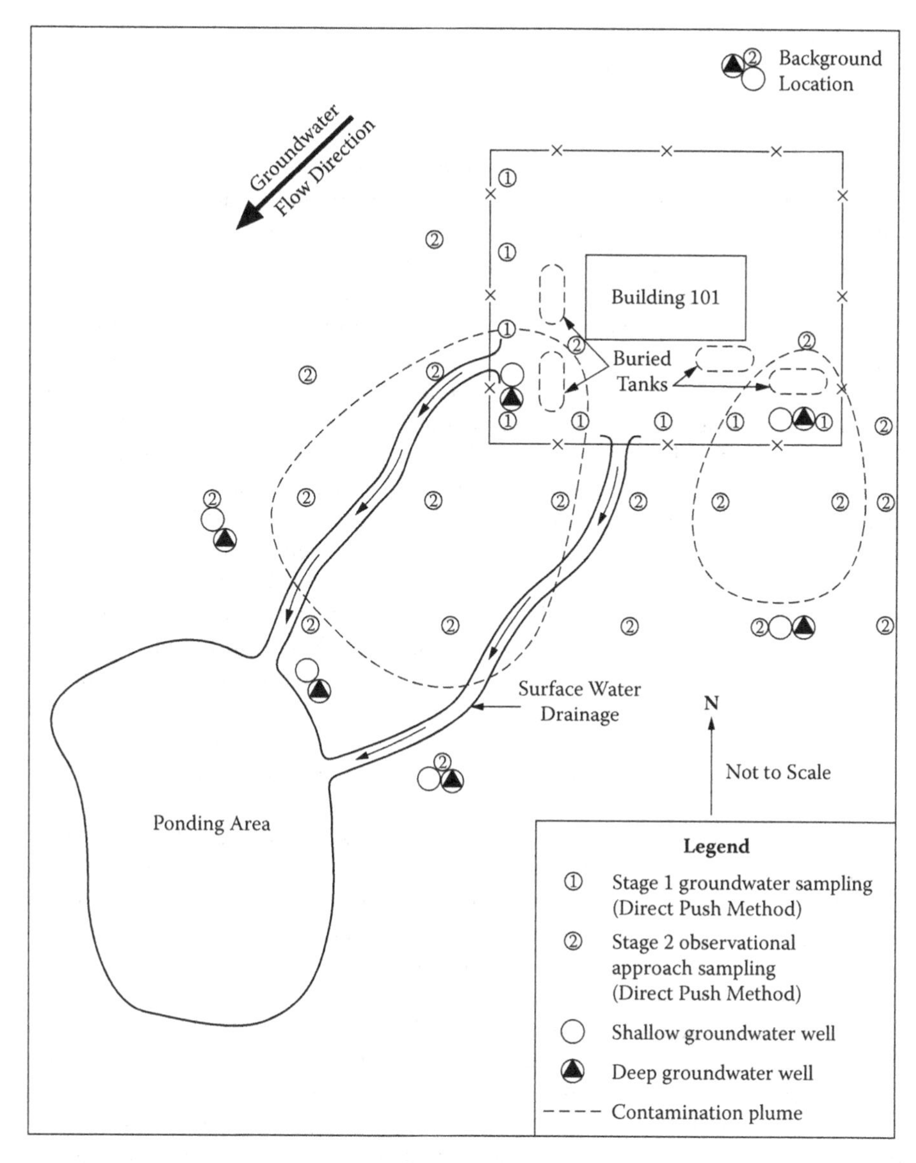

FIGURE 4.54 Example of how the direct push method is used to characterize groundwater contamination using the observational approach.

(From Byrnes, 1994. *Field Sampling Methods for Remedial Investigations.* Lewis Publishers, Boca Raton, FL.)

near the source and downgradient from the leading edge of the contaminant plumes to track the long-term migration of the contamination. Shallow and deep well pairs are recommended to track both the vertical and horizontal migration of contaminants. The number of wells required is based on the size of the contaminant plumes identified. If groundwater remediation is later determined to be necessary, extraction wells will be installed near the source of contamination.

4.2.7.2 Monitoring Wells

The primary objective of installing monitoring wells is to provide an access point where groundwater samples can be repeatedly collected, and groundwater elevations can be measured. When installing monitoring wells, it is important to minimize the disturbance to the surrounding formation, and to construct a well from materials that will not interfere with the chemistry of the groundwater.

The primary components of a groundwater monitoring well are the well screen, sump, riser pipe, well cap, protective steel casing, and lock (Figure 4.55). The well screen is by far the most critical component of a well. A well screen must have slots that are large enough to allow groundwater and contaminants to flow freely into a well, yet small enough to prevent formation soils from entering the well. The most common lengths of screen used to construct monitoring wells are 2, 5, 10, 15, and 20 ft. Well screens larger than 20 ft in length are not recommended for monitoring wells because they tend to dilute the chemistry of the well. In some regions of the country, EPA discourages the installation of monitoring well screens larger than 10 ft because of their concern over dilution. Another potential problem with using long screen lengths is that two aquifers can unintentionally be screened in the same well if one is not careful. Such an error would provide a conduit for cross-contamination between aquifers. Longer monitoring well screens (greater than 20 ft) are in some rare instances approved by regulatory agencies when unique conditions are present, such as if the monitoring wells are being installed at a site where static groundwater elevations are dropping significantly over time.

If the aquifer is more than 20 ft in thickness, depth-discrete groundwater samples should be collected at consistent depth intervals (e.g., every 10 to 20 ft) throughout the aquifer during well drilling (or before well drilling using the direct push method [see Section 4.2.7.1]) to help define the depth where contamination is most concentrated. See Section 4.2.7.2.1 for details on collecting depth-discrete groundwater samples during well drilling. These depth-discrete groundwater samples should be analyzed by a quick-turnaround mobile laboratory, and the analytical results should be used to help select where the well screen should optimally be positioned within the aquifer. (Note: Each sampling interval will first need to be purged [Section 4.2.7.4] before sample collection.) Note that, in most cases, the well screen will be set to screen the depth interval showing the highest contaminant concentrations based on the quick-turnaround mobile laboratory analytical results. If the aquifer is less than 20 ft in thickness, there is no need to collect depth-discrete groundwater samples to assist the positioning of the well screen. Rather, the well screen should in most cases be positioned to screen the full thickness of the aquifer.

If, based on the results from the analysis of depth-discrete groundwater samples, it is determined that contamination in the overlying soil has not yet reached the groundwater, one should position well screens at the top of the aquifer. For example, if a 10-ft well screen is selected, one should set approximately 8 ft of the screen below the mean static water level, and 2 ft above the mean static water level. By constructing the well in this manner, one will be able to detect contamination when

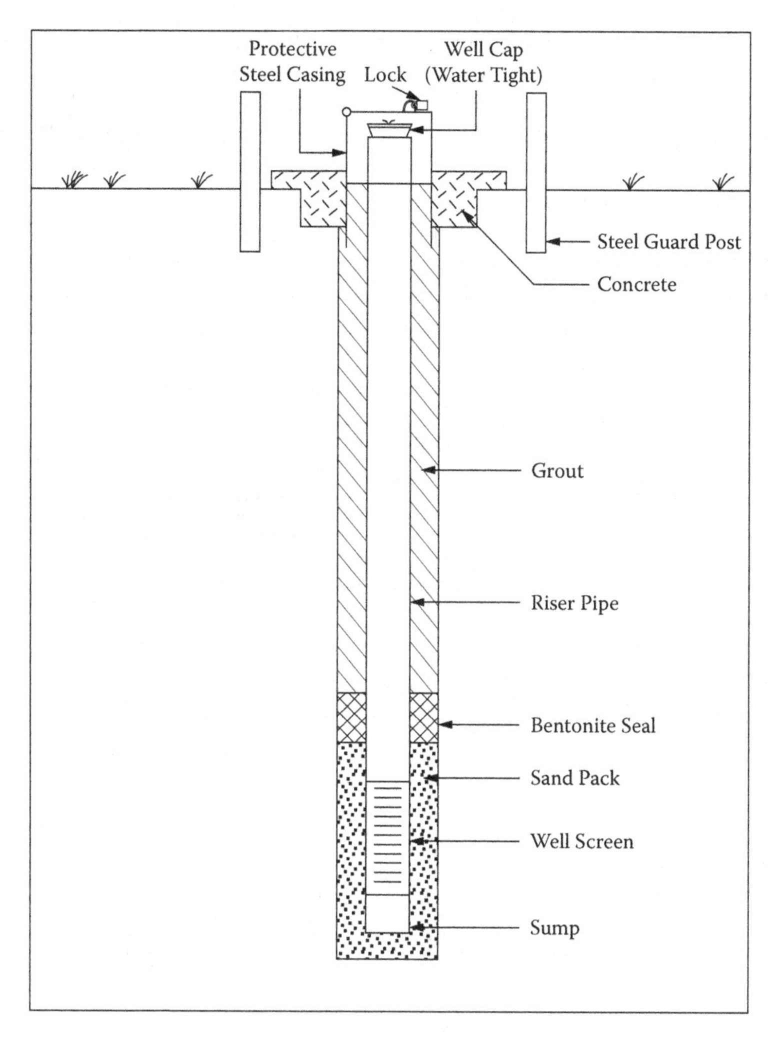

FIGURE 4.55 Primary components of a groundwater monitoring well.

(From Byrnes, 1994. *Field Sampling Methods for Remedial Investigations.* Lewis Publishers, Boca Raton, FL.)

it first contacts the groundwater. Also, if any floating hydrocarbons (LNAPL) are present in the formation, it will be visible in the well, even with seasonal fluctuations in groundwater levels. Similarly, if a 15- or 20-ft well screen is selected, 2 ft of the screen should be set above the mean static water level, with 13 and 18 ft of the screen below the static water level, respectively. Screen lengths of 2 and 5 ft are used to sample specific intervals within the aquifer and are not designed to screen the static water level.

If for some reason depth-discrete groundwater sampling cannot be performed to support the positioning of the well screen, one should set the well screen for the first set of monitoring wells at the top of the aquifer. If historical information suggests that

LNAPLs (e.g., benzene, toluene, and xylene) are likely present, the well should be constructed to screen the top of the aquifer, as described in the earlier paragraph, with 2 ft of screen above the water table and the remainder of the screen length below the water table. On the other hand, if historical information suggests that DNAPLs (e.g., TCE, PCE, and vinyl chloride) may be present, the bottom of the well screen should be set at the bottom of the aquifer (Figure 4.56).

In a relatively thick aquifer, where discrete layering of contamination is known or suspected to occur, several monitoring wells may need to be installed in clusters, with each monitoring well screening a different portion of the aquifer. Because the wells in a cluster are located close together, grouting of one well may impact the water

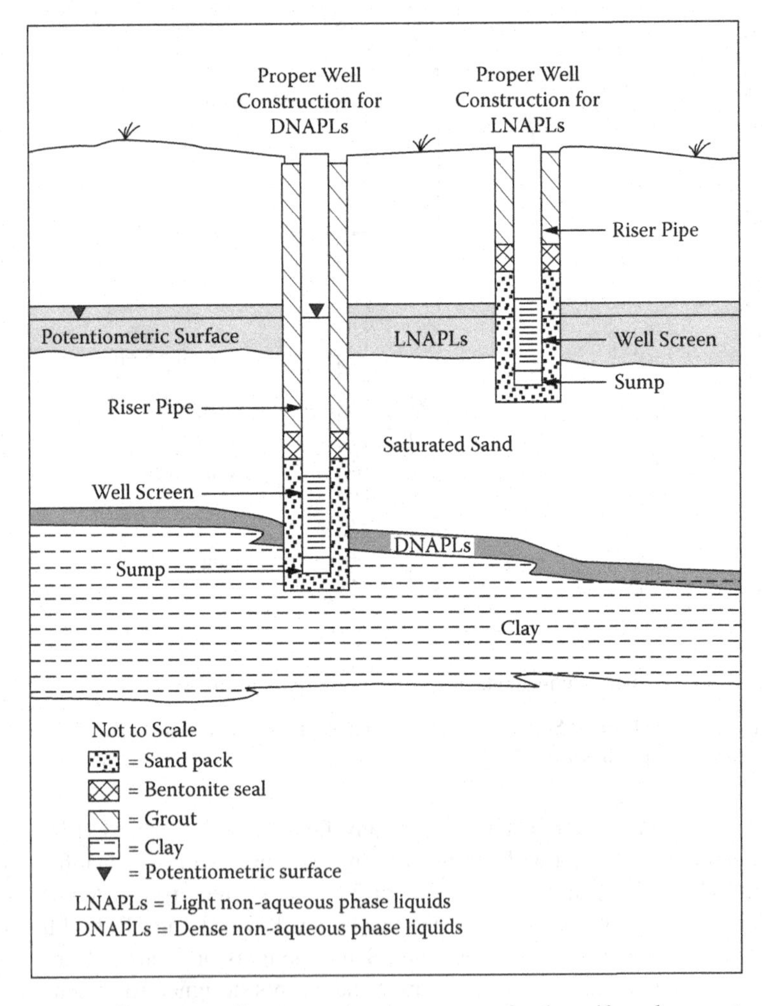

FIGURE 4.56 Common well construction in an unconfined aquifer when contaminants include DNAPLs and LNAPLs.

(From Byrnes, 1994. *Field Sampling Methods for Remedial Investigations*. Lewis Publishers, Boca Raton, FL.)

quality (e.g., pH) of the other wells in the cluster. For this reason the shallowest well in the cluster should be located upgradient of the deeper wells in the cluster. The exact positioning of each of the well screens within the aquifer should preferably be selected on the basis of the analytical results from depth-discrete groundwater sampling.

When installing monitoring wells within a confined aquifer, well screens should be preferably positioned to screen the high concentration portion of the confined aquifer on the basis of the results from depth-discrete groundwater sampling. However, if for some reason this is not possible, the top of the well screen should be set at the top of the confined aquifer if either LNAPLs are suspected to be present or if no information is known about the presence or absence of contamination in the confined aquifer (Figure 4.57). On the other hand, if DNAPLs are suspected to be present in the confined aquifer, the bottom of the well screen should be set at the bottom of

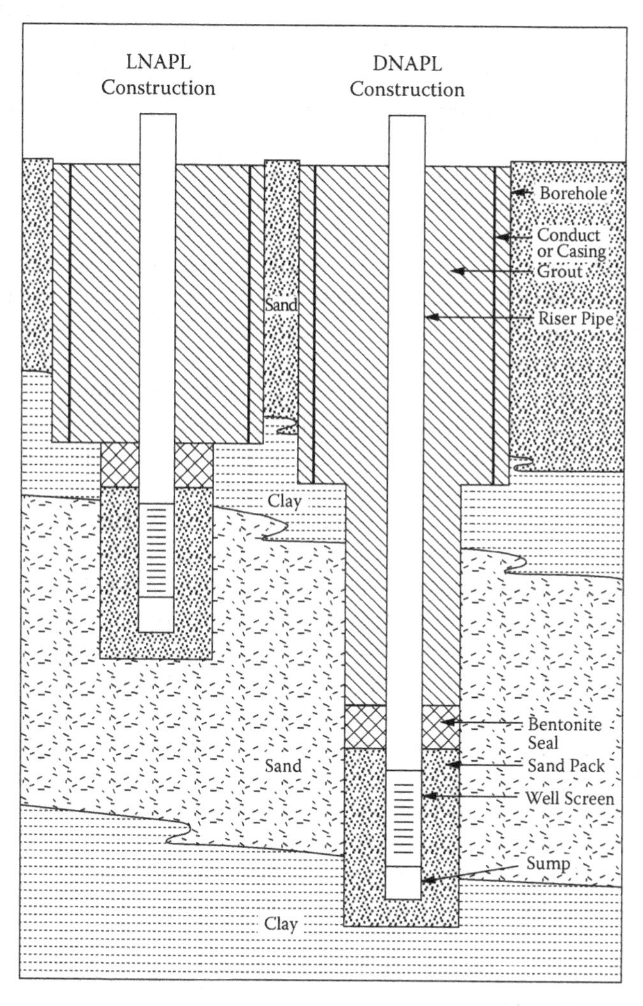

FIGURE 4.57 Common well construction in a confined aquifer when contaminants include DNAPLs and LNAPLs.

the confined aquifer. The installation of a monitoring well within a confined aquifer involves a confining layer being penetrated. Before penetrating a confining layer (in most cases) a conductor casing needs to be installed and sealed to prevent cross-contamination between aquifers. A smaller diameter borehole is then drilled through the confining layer.

Well screens (Figure 4.55) for use in environmental sampling must be constructed of a noncorrosive and nonreactive material that will not impact the chemistry of the groundwater. The most commonly accepted materials for well screens by regulatory agencies are stainless steel and polyvinylchloride (PVC). Of these two materials, stainless steel is preferred because it is relatively inert and is very durable; however, it is also very expensive. For this reason, stainless steel well screens are most commonly installed at federal installations where groundwater monitoring will be performed for decades. In contrast, PVC is relatively inexpensive, and as a result is commonly selected for well screens in the private sector. However, it should be noted that some studies have shown that PVC may release and absorb trace amounts of organic constituents after prolonged exposure. If one chooses to install a PVC screen in a new groundwater monitoring well, it is important to get concurrence from the lead regulatory agency overseeing the investigation.

The most commonly used monitoring well screen slot sizes are 0.01 and 0.02 in. for fine- and medium-grain-sized formations. One can specially order well screens with larger slot sizes for coarser-grained formations. To determine the appropriate screen slot size for use in a particular formation, a sieve analysis must be performed either in the field or in a laboratory to determine the grain size distribution of the formation. With this information, slot size calculations can be made using procedures outlined by Sterrett (2007).

A sump is threaded to the bottom of a well screen for the purpose of catching any fine-grained soil that enters the well through the screen. If a sump is not used, soil accumulation will occur within the well screen and will eventually plug up the well. Common well sumps range in lengths from 1 to 3 ft but can be ordered in any size. Sumps should be made of the same material as the well casing, and they must be cleaned out on a regular basis to ensure that the screen remains clear.

A threaded riser pipe is attached to the well screen to complete the well to the ground surface. The riser pipe is cut near the ground surface to allow for either an aboveground or belowground completion. For an aboveground completion, the riser pipe is typically cut 1.5 to 2.0 ft above the ground surface. A locking protective steel casing is then grouted in place over the riser pipe. A concrete pad and guard posts are often positioned around an aboveground well completion for added protection (Figure 4.55).

For a belowground well completion, the riser pipe is cut several inches below the ground surface. A locking protective steel casing is then grouted in place over top of the riser pipe and completed flush with the ground surface (Figure 4.58). A watertight well cap should always be used in a belowground completion to prevent surface water from entering the well.

A sand pack should be built around the well screen to a level 2 to 3 ft above the top of the screen. The purpose of the sand pack is to reduce the amount of fine-grained formation soils entering the well screen. The sand pack is built above the top of the screen

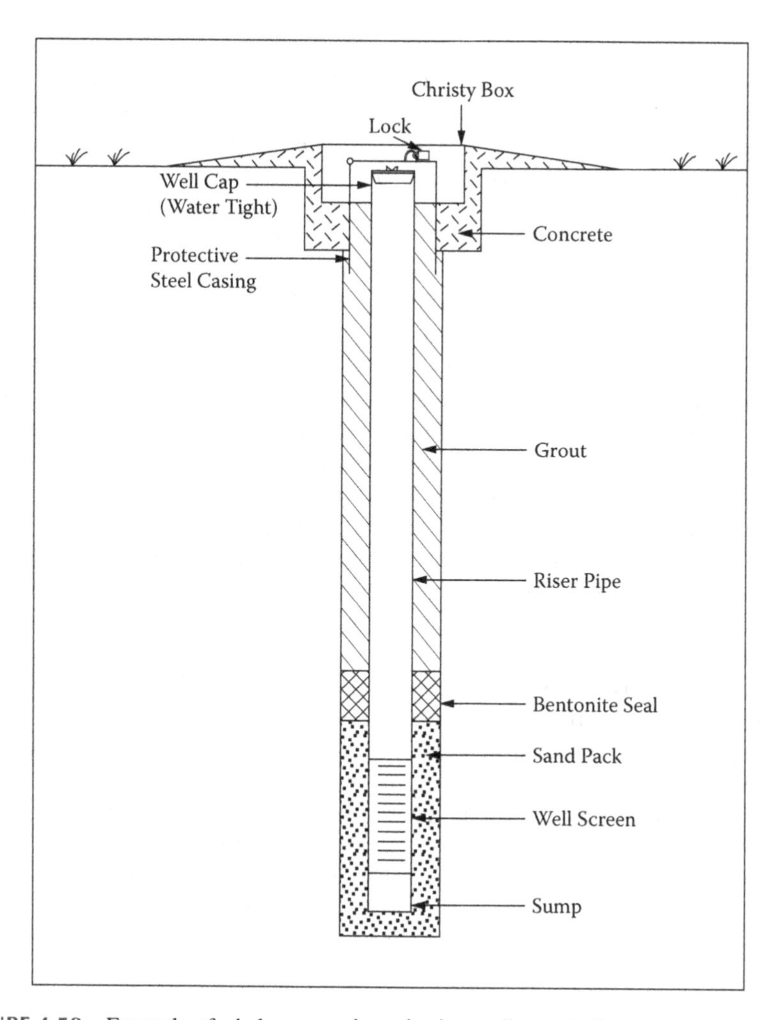

FIGURE 4.58 Example of a belowground monitoring well completion.

(From Byrnes, 1994. *Field Sampling Methods for Remedial Investigations.* Lewis Publishers, Boca Raton, FL.)

to allow for the potential settling of the sand pack over time. The grain size and size distribution of the sand pack used to build the well should be calculated using procedures outlined in Sterrett (2007), and should be compatible with the selected screen slot size. A 2- to 3-ft bentonite seal should be built over top of the sand pack using bentonite pellets. The remainder of the borehole should be filled with grout composed of Portland type I/II cement mixed with approximately 4 to 5% bentonite powder.

4.2.7.3 Well Development

Before collecting a groundwater sample from a well, it must first be properly developed and then purged. During well drilling, it is not uncommon for clayey soils to smear along the walls of the borehole. The development procedure restores the natural hydraulic conductivity and geochemical equilibrium of the aquifer near the well so that representative samples of formation water can be collected. In the development procedure, a well is surged and pumped long enough to clean out the well, and until the water stabilizes in pH, temperature, conductivity, turbidity, and dissolved oxygen.

Surging is the first step in the development procedure, which involves lowering a surge block down the well using a wire line from a drill rig or development truck. The surge block is repeatedly raised and lowered over 2- to 3-ft intervals starting at the top of the screened interval and moving downward. This procedure pulls soil particles that are finer than the well screen into the well so that they can later be removed. This procedure helps to compact the sand pack around the well screen, and should be performed several times during the development operation to ensure that most of the fine-grained soil particles have been removed from the immediate vicinity of the well screen. Following each surging step, a submersible pump or large bailer is used to remove turbid water and sediment from the well.

A submersible pump is used to develop wells that can yield water at rates greater than several gallons per minute. In low-yielding wells, a large bottom-filling bailer is commonly used. When a submersible pump is used, the intake is first set near the bottom of the well screen, then near the center, and finally near the top. The pump is left at each interval until the water has cleared, and the pH, temperature, conductivity, turbidity, and dissolved oxygen of the water have stabilized. For the most effective development, several different pumping rates are used at each development interval. The pumping rate for a submersible pump can be controlled by placing a restriction valve on the end of the discharge line, or by using a variable-speed pump.

For poor producing aquifers, a bottom-filling bailer is used for well development. Because the bailer method does not work as well as the submersible pump at pulling fine-grained soil from the surrounding formation, the surging procedure is that much more important when developing these wells. After surging, the bailer is lowered to the bottom of the well using a wire line from a drill rig or development truck. When the bailer is full, it is retrieved to the ground surface, and the water is transferred into drums or a water-holding tank. The bottom-filling bailer not only removes groundwater from the well, but also works effectively in removing any silt that has accumulated in the sump. In poor producing aquifers, it is tempting to add water to the well to assist the procedure; however, this should be avoided whenever possible because it can alter the groundwater chemistry. If a well will not develop without adding water, a field blank should be taken of the water added to the well, and it should be analyzed for the same parameters that the future groundwater samples are to be analyzed for. Development methods using air should be avoided because they have the potential to alter the groundwater chemistry, and can damage the integrity of the well.

At regular intervals during the development procedure, a sample of the development water should be collected for pH, temperature, conductivity, turbidity, and dissolved oxygen measurements. The results from these measurements are recorded on a Well Development Form (Chapter 5). A well is considered adequately developed when a minimum of three borehole volumes of water have been removed from the well, and when three consecutive readings are within the following limits: pH (± 0.1

units), temperature (± 3%), conductivity (± 3%), turbidity (± 10% for values >5 NTU, or if three turbidity readings are ≤5 NTUs, consider the values as stabilized), and dissolved oxygen (± 10% for values >0.5 mg/L, or if three dissolved oxygen values are ≤0.5 mg/L, consider the values as stabilized) (EPA 2017c). A borehole volume is calculated using the formula:

$$V = \pi r^2 l$$

where

V = volume
π = 3.14
r = radius of the borehole
l = thickness of the water column

If the foregoing physical parameters stabilize quickly, a minimum of three borehole volumes of water should still preferably be removed to consider the well adequately developed. In fine-grained formations it is not uncommon for it to take five or more borehole volumes for all the physical parameters to stabilize. Water from the well development process should be containerized and then sampled for analytical testing. One must come to an agreement with the lead regulatory agency on what analyses to run on the development water and whether or not any treatment of the water is required before its release. It is common for regulatory agencies to approve the release of this water to a sanitary or storm sewer drainage if specific release criteria are met.

Immediately after the completion of well development, the depth to groundwater should be measured (see Section 4.2.7.4.1), and recorded in a logbook. This information can later be used to help evaluate aquifer performance.

If a well is built using the appropriate screen slot size and sand pack, very little siltation should occur over time. In this instance, future sampling only requires a well to be purged (see Section 4.2.7.4) before sampling. However, if fine-grained sediment begins building up inside a well sump and screen over time, the well needs to be redeveloped.

4.2.7.4 Well Purging

After development, a well should be allowed to set for several days before purging and sampling. The standard accepted well-purging procedure is identical to that for well development, with the exception that surging is not performed and the pump is only required to be set at one position within the well screen. The objective of the purging procedure is to remove stagnant water from the well so that a representative water sample can be collected. The purging procedure should remove water throughout the screened interval to ensure that fresh formation water has replaced all stagnant water in the well. It is recommended that lower pumping rates be used during well purging than were used during well development. This will reduce the chances of additional silt being pulled into the well. At regular intervals during the purging procedure, field parameter measurements (e.g., pH, temperature, conductivity, turbidity, dissolved oxygen) should be collected. The results from these measurements are recorded on a Well Purging Form (see Section 5.2.3). Well purging should continue until the following field parameters have stabilized (EPA 2017c):

- pH (± 0.1 units)
- Temperature (± 3%)
- Conductivity (± 3%)
- Turbidity (± 10% for values >5 NTU, or if three turbidity readings are ≤5 NTUs, consider the values as stabilized)
- Dissolved oxygen (± 10% for values >0.5 mg/L, or if three dissolved oxygen values are ≤0.5 mg/L, consider the values as stabilized)

High purge rates should always be avoided, because purging a well dry causes formation water to cascade into the well, facilitating the loss of volatile organics.

Immediately after the completion of well purging, the depth to groundwater should be measured (see Section 4.2.7.4.1), and recorded on the Well Purging Form. This information can later be used to help evaluate aquifer performance. As a general rule, when using this standard well-purging method, one should attempt to sample the well before the static water level has had time to completely equilibrate.

For very-low-yielding wells (< 1 gpm), the foregoing purging procedure is not practical as it would take several days to complete. Rather, these wells should be bailed dry, or pumped dry, twice and then sampled before the static water level equilibrates.

The passive (no-purge) sampling method (see Section 4.2.7.5.1) and low-flow purging/sampling method (see Section 4.2.7.5.2) are also gaining acceptance by many regulatory agencies and are effective methods that eliminate or dramatically reduce the volume of purge water that needs to be disposed of.

There are three types of passive (no-purge) sampling method. These include methods that utilize a semi-permeable membrane through which chemicals diffuse or permeate, methods that simply allow chemicals to move through the body of the sampler by advection (horizontal flow of water) and dispersion or over time primarily by diffusion (movement from areas of higher concentration to areas of lower concentration), or samplers that contain a sorptive media where selected chemicals are sorbed onto the media for later extraction and analysis.

The low-flow purging/sampling method utilizes a pump to remove groundwater from the aquifer at a very slow rate. The pumping rate used by this method should be low enough that it creates a cone of depression of 0.3 ft or less (EPA 2017c). The low-flow purging/sampling method very often utilizes a bladder pump, compressor, pump control unit, and flow cell for monitoring field parameters. A submersible pump may be used instead of a bladder pump as long as it has an adjustable flow rate control and the proper pump size is selected. Some of the more significant advantages of the low-flow purging/sampling method over the standard well-purging method include the following:

- The volume of purge water generated can be reduced by as much as 95%.
- There is less disturbance to the water column, which leads to lower sample turbidity. This in turn reduces the variability in the sampling results.
- There is a reduced loss of volatile organics using this method because the formation water is less disturbed.
- Purging times are often reduced.

It is essential that one gets approval from the lead regulatory agency before beginning to use the passive (no-purge) or low-flow purging/sampling methods at a particular site. See Sections 4.2.7.5.1 and 4.2.7.5.2 for more details on the passive (no-purge) sampling and low-flow purging/sampling methods.

4.2.7.4.1 Depth to Groundwater Measurements/Groundwater Elevation Measurements

Collecting accurate depth to groundwater measurements from wells is critical for use in defining groundwater flow direction, groundwater gradient, aquifer hydraulic properties, extraction well area of influence, optimizing extraction/injection well pumping rates, etc. This section provides guidance on how to properly collect accurate depth to groundwater measurements using an electric tape and pressure transducer. This depth to groundwater data can then be used to calculate groundwater elevations. Before using either an electric tape or pressure transducer you must first verify that the instrument has been properly calibrated in accordance with guidance provided by the manufacturer.

4.2.7.4.1.1 Electric Tape Method An electric tape used to collect depth to groundwater measurements is typically composed of a probe, electric tape, reel, sounding device (beeper), test button, and volume control knob (Figure 4.59). Prior to collecting depth to groundwater measurements, a groundwater well should first be surveyed to define its elevation as well as its location coordinates (latitude and longitude). Well elevations are often surveyed to an accuracy of ±0.001 ft and latitude

FIGURE 4.59 Electric water level tape.

and longitude are often surveyed to an accuracy to five decimal places. However, it is important to check survey accuracy requirements for the particular county and state where your well is being installed. The elevation of a well is most often measured on the north side of the top of casing and is marked by putting a notch (or some other permanent marking) in the casing.

The following equipment and procedure can be used to measure the depth to water in a well using an electric tape:

Equipment

1. Electric water level tape
2. Field logbook
3. Permanent ink marker
4. Health and safety instruments
5. Health and safety clothing (see Chapter 10)
6. Waste container (e.g., 55-gal drum)
7. Plastic waste bag

Procedure

1. In preparation for collecting depth to water measurements, confirm that all necessary preparatory work has been completed, including obtaining property access agreements, meeting health and safety and equipment decontamination requirements, and checking the calibration of all health and safety and water level measurement instruments.
2. Unlock the well head and remove the cap from the well casing. Allow the water level to equilibrate for a short period of time (e.g., five to ten minutes).
3. Use health and safety instruments (e.g., photoionization, flame-ionization detector) to screen air quality at the well head. Based on results, adjust personal protective equipment if needed in accordance with project health and safety plan.
4. Turn the electric tape on and push the test button on the reel to verify the batteries are still working and the beeper sounds.
5. Slowly lower the probe and tape down the well until the beeper sounds, indicating that groundwater has been reached.
6. Holding the tape against the side of the casing where the elevation notch is located, slowly raise the tape upward until the beeper turns off.
7. Slowly lower the probe again until the beeper first sounds. Read the depth measurement off the tape to the closest 0.01 ft. Record the depth in a logbook or water level measurement form (see Chapter 5).
8. Repeat Steps 6 and 7 three times and record the results.
9. Calculate the average of the three measurements. Record the average depth measurement.
10. Raise the probe and tape to the ground surface.

11. Replace the well cap, lock the well head, and containerize any waste in a waste container. Before leaving the site, all waste containers should be sealed, labeled, and handled appropriately (see Chapter 11).
12. Decontaminate equipment in accordance with procedures outlined in Chapter 9.

When the elevation of the top of the well casing is known, one can calculate the groundwater elevation by simply subtracting the depth to water from the elevation of the top of casing.

In some cases the elevation is not measured at the top of casing, but rather a brass marker is cemented into the well pad instead, and the elevation is measured at the center of the brass marker (Figure 4.60). In this case, for an aboveground well completion, cut a small notch (or some other permanent marking) into the north side of the top of well casing where future depth to water measurements will be collected. Then carefully measure the casing stickup (at the notch) above the brass marker to an accuracy of ±0.01 ft. Add this measurement to the brass marker elevation to give the elevation at the top of well casing. Now you can calculate future groundwater elevations by simply subtracting the depth to water from the top of casing elevation. Figure 4.60 provides an example.

For a belowground well completion, where a brass marker is cemented at ground level next to the protective casing and the elevation is measured at the center of the brass marker (Figure 4.60), cut a small notch (or some other permanent marking) into the north side of the top of well casing where future depth to water measurements will be collected. Then carefully measure the depth that the top of casing (at the notch) is below the brass marker to an accuracy of ±0.01 ft and subtract this measurement from the brass marker elevation to give the elevation at the top of well casing. Now you can calculate future groundwater elevations by simply subtracting the depth to water from the top of casing elevation.

4.2.7.4.1.2 Pressure Transducer Method When pressure transducers are combined with a telemetry device (Figure 4.61), one can collect continuous or near-continuous water-level measurements from groundwater wells. This has led to improved understanding of aquifer hydraulic properties. In addition to water levels, many pressure transducers probes are also able to collect additional data, such as pH, temperature, specific conductivity, salinity, total dissolved solids, resistivity, and dissolved oxygen. The In Situ TROLL 600 shown in Figure 4.61 is one example.

The pressure transducer/water quality instrument probe shown on the left in Figure 4.61 is installed at a fixed depth in a well. The pressure transducer senses changes in pressure against a membrane in response to changes in the height, and thus in the weight, of the water column in the well above the transducer. As noted above, a variety of water quality measurements can be collected simultaneously with the pressure transducer data. In Figure 4.61 the device on the right is the top-of-well telemetry device. This device often contains a battery to power the instrument. Telemetry is the process of recording and transmitting the readings of an instrument. Batteries can sometimes last a year or more before needing replacement.

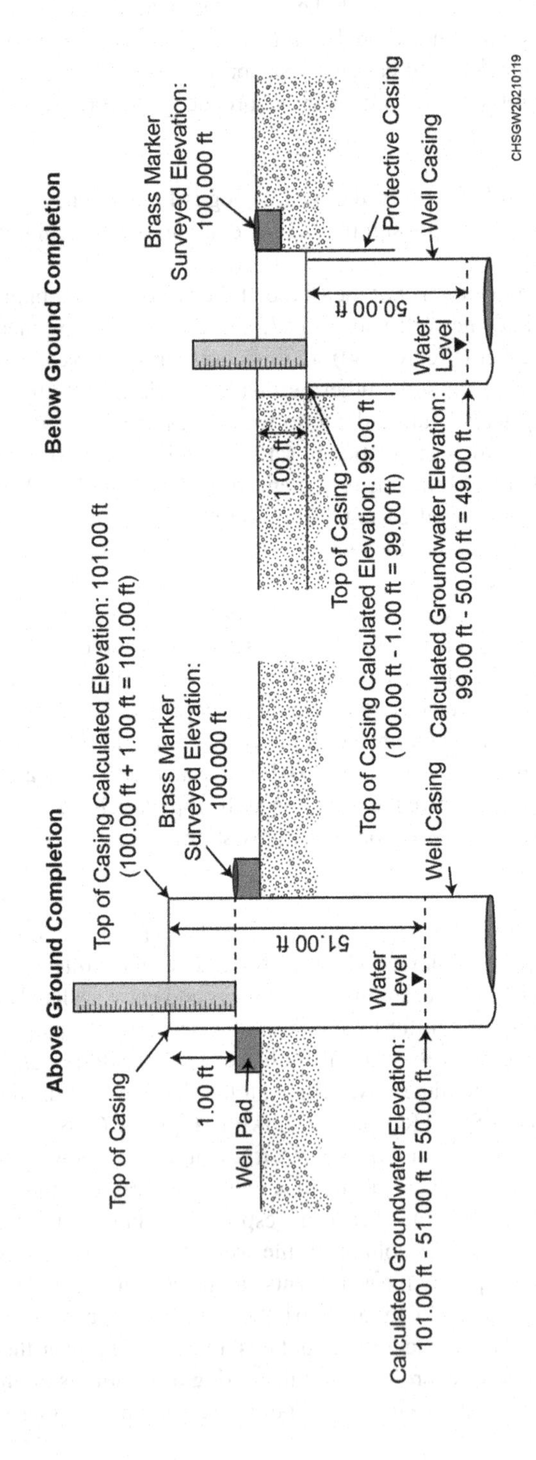

FIGURE 4.60 Calculating groundwater elevation at top of casing when brass marker is used.

FIGURE 4.61 In Situ TROLL 600 pressure transducer/water quality instrument.

When developing your sample design, the DQO process (see Section 3.2.5) will define the specific accuracy requirements for the water level measurement (and water quality) data being collected for your study. For a system such as the In Situ TROLL 600, a free App (VuSitu) can be downloaded to a tablet or mobile phone. After you download the app, you can connect to your device via Bluetooth to set up data logging. Some instruments have Bluetooth native to the actual instrument while others require a separate Bluetooth communication device to be connected to the instrument to communicate with the VuSitu app. A system can typically be programmed to record water level data in several different ways, such as:

- The change in the water level from the starting water level (e.g., −0.01 ft, +0.01 ft, −0.02 ft)
- The thickness of the water column above the transducer (e.g., 9.99 ft, 10.01 ft, 9.98 ft)
- The groundwater elevation (e.g., 149.99 ft, 150.01 ft, 149.98 ft).

This data can later be manipulated in the office to convert it to whatever format is most useful for preparing reports. At an environmental site, consider the possible effects of contaminants in the water that may corrode or otherwise degrade transducer components. Select components that are corrosion resistant, and easily decontaminated. Wells installed in areas subject to strong electromagnetic fields, such as near generators, motors, pumps, power supplies, or similar devices, may not be suitable candidates for some types of pressure transducers/water quality instruments.

One should consider running a few tests in the office such as checking the effects of temperature on the pressure transducer output and checking drift characteristics before taking it into the field. USGS (2004) provides details on how to run these and other tests. The manufacturer of your pressure transducer should also provide guidance on how to run these tests. Although most transducers are calibrated at the factory and come with the manufacturer's calibration specifications, an individual calibration check should still be done in the field as each transducer is installed.

To install a pressure transducer/water quality instrument in a monitoring well, implement the following steps:

Equipment

1. Pressure transducer/water quality instrument and telemetry device
2. Suspension line
3. Field logbook
4. Electric water level tape
5. Project specific sampling and analysis plan
6. Instrument user manual(s)
7. Tape measure
8. Zip-Ties
9. Permanent ink marker
10. Health and safety instruments
11. Health and safety clothing (see Chapter 10)
12. Waste container (e.g., 55-gal drum)
13. Plastic waste bag

Procedure

1. In preparation for installing a pressure transducer/water quality instrument, confirm that all necessary preparatory work has been completed, including obtaining property access agreements, meeting health and safety and equipment decontamination requirements, and checking the calibration of all health and safety and pressure transducer/water quality instruments.
2. Unlock the well head and remove the well cap from the well casing.
3. Use health and safety instruments (e.g., photoionization detector, flame-ionization detector) to screen air quality at well head. Based on results, adjust personal protective equipment if needed in accordance with project health and safety plan (see Chapter 10).
4. Using a suspension line, suspend the transducer/water quality instrument probe in the well from a stable fixed hanging point such as the well casing or the protective outer casing of the well. The depth at which the transducer is set will depend on a number of factors including whether or not a permanent sample pump is also to be installed in the well. Record the depth the transducer is set in a field logbook (see Section 5.2.1).
   a. For a monitoring well with a 20-ft well screen that is set near the top of the water table, if a sample pump is also installed in the well, the sample pump should be set near the base of the screened

interval. The transducer should then be set far enough above the pump intake that it will not be impacted by the turbulence created by the sample pump, and far enough below the static water level that it will remain below the cone of depression when the sample pump is running. If a sample pump is *not* installed in the well, the transducer should be set deep enough within the screened interval that it will remain below the maximum estimated natural drop in the water level.

   b. For a monitoring well with a 20-ft well screen set near the base of a thick aquifer (+100 ft), if a sample pump is to be installed, the sample pump should be set at the base of the well within the screened interval. The transducer should then be set at a depth that ensures the maximum pressure rating for the transducer is not exceeded. Also be sure the transducer is set far enough below the static water level so that it remains below the cone of depression when the sample pump is running. If a sample pump is *not* installed in the well, the transducer should be set at a depth that ensures the maximum pressure rating for the transducer is not exceeded and the transducer is set deep enough within the screened interval that it will remain below the maximum estimated natural drop in the water level.

5. Attach a Zip-Tie, or other method of marking, at the top of the suspension line. This mark will show if there has been any slippage in the line over time.
6. Collect a depth to water reading in the well using an electric tape (see Section 4.2.7.4.1.1) and record this measurement in a field logbook (see Section 5.2.1).
7. Turn the power on to the pressure transducer.
8. Check the calibration of the transducer in the well by lowering it and then raising it by known increments, then compare the incremental distance change with the transducer output change. Use a minimum of five points, in each direction, covering the operational range of the transducer.
   a. For example, if the transducer is set 10.00 ft below the static water level, and there is no pump in the well, then collect a measurement at a depth of 10.00, 12.00, 14.00, 16.00, 18.00, 16.00, 14.00, 12.00, and 10.00 ft below the static water level.
9. If the calibration check meets minimum data quality requirements specified in the sampling and analysis plan (Section 3.2.7), then begin recording water level measurements. For the next few days double check that everything is working properly.
10. If the calibration does *not* meet minimum data quality requirements specified in the sampling and analysis plan (Section 3.2.7), then refer to the instrument users' manual on how to make proper adjustment and then repeat Steps 6 and 7.

11. Once field data collection is complete, retrieve the transducer/water quality instrument probe from the well.
12. Replace the well cap, lock the well head, and containerize any waste in a waste container. Before leaving the site, all waste containers should be sealed, labeled, and handled appropriately (see Chapter 11).
13. Decontaminate equipment in accordance with procedures outlined in Chapter 9.
14. Back in the office, the collected data can be converted to whatever format is most useful to the project.

4.2.7.4.2 Depth to Bottom of Well Measurements

Collecting accurate depth to bottom of well measurements is important since this data can be used to support decisions such as the optimum depth for setting groundwater sample pumps, transducers, passive groundwater samplers, and in-well flow meters. Accurate depth to bottom of well measurements are also needed to make decisions like when a well needs to be re-developed to remove excess sediment buildup in the bottom of the well screen. This section provides guidance on how to properly collect depth to bottom of well measurements using a weighted downhole tape measure.

Depth to bottom of well measurements should be made to ±0.1 ft (EPA [Region 4] 2020a). Stainless steel weights can be purchased from vendors that will attach to the end of a downhole tape measure. Be certain to attach a weight to the tape measure that is heavy enough so that you will be able to feel when the weight has reached the bottom of the well. To collect a depth to bottom of well measurement perform the following steps.

Equipment

1. Downhole tape measure (long enough to reach the bottom of well)
2. Stainless steel weight
3. Field logbook
4. 25-ft tape measure (in tenths of a foot scale)
5. Permanent ink marker
6. Health and safety instruments
7. Health and safety clothing (see Chapter 10)
8. Waste container (e.g., 55-gal drum)
9. Plastic waste bag

Procedure

1. In preparation for collecting depth to bottom of well measurements, confirm that all necessary preparatory work has been completed, including obtaining property access agreements, meeting health and safety and equipment decontamination requirements, and checking the calibration of all health and safety instruments.
2. Unlock the well head and remove the well cap from the well casing.

3. Use health and safety instruments (e.g., photoionization, flame-ionization detector) to screen air quality at the well head. Based on results, adjust personal protective equipment if needed in accordance with project health and safety plan.
4. Clip the stainless steel weight on to the end of the downhole tape measure
5. Use a tape measure (in tenths of a foot scale) to measure (±0.01 ft) the distance that the attached weight extends beyond the end of the downhole tape measure. Record this measurement in the field logbook.
6. Lower the weighted tape measure down the well until you feel the weight has reached the bottom.
7. Holding the tape against the side of the well casing where the elevation notch is located, tug on the tape a few times and let it fall.
8. When you are confident that you have an accurate (±0.1 ft) depth to the bottom of the well, read the depth measurement at the elevation notch, add the length of the attached weight measured in Step 4, and record the reading in the field logbook.
9. Repeat Steps 7 and 8 three times, then calculate the average reading (±0.1 ft) and record this in the field logbook.
10. Retrieve the weighted tape measure to the ground surface.
11. Replace the well cap, lock the well head and containerize any waste in a waste container. Before leaving the site, all waste containers should be sealed, labeled, and handled appropriately (see Chapter 11).
12. Decontaminate equipment in accordance with procedures outlined in Chapter 9.

4.2.7.5 Groundwater Sampling after Well Installation

There are a number of effective methods for collecting groundwater samples after well installation. These include passive (no-purge) groundwater sampling methods (e.g., EON Dual Membrane Sampler, Snap Sampler®, HydraSleeve™, AGI Sample Module®), the low-flow groundwater sampling method, standard groundwater sampling methods (e.g., bailer, bomb sampler, bladder pump, submersible pump), and other unique methods such as the WestBay Multiple Port Sampling System. Flexible Liner Underground Technologies (FLUTe™) is another unique method typically used to collect groundwater samples in an open borehole. The optimum method to be used will depend upon the contaminants of concern being sampled for and the objectives of the sampling defined in the DQO process (see Section 3.2.5).

Passive groundwater sampling involves the collection of a water sample from a well without first purging the well. This is very cost-effective method of sampling groundwater since no purge water has to be handled and disposed of. Passive groundwater sampling is particularly useful in poor yielding formations or in wells that only have a few feet of water in them. Low-Flow groundwater sampling is used to collect groundwater samples from monitoring wells that are representative of ambient groundwater conditions using very low pumping rates. Standard groundwater sampling methods require standard well purging to be performed prior to sampling (see Section 4.2.7.4). More unique groundwater sampling methods (e.g., FLUTe™,

WestBay Multiple Port Sampling System) may provide some distinct advantages that other methods are not able to provide. Sections 4.2.7.5.4.1 and 4.2.7.5.4.2 provide details on what these advantages may be.

Sampling tools used to collect groundwater samples for laboratory analysis should be constructed of materials that are compatible with the media and the constituents that are being tested for. In other words, the sampler should *not* be constructed of a material that could cause a loss or gain in contaminant concentrations measured due to sorption, desorption, degradation, or corrosion (EPA 2002). Because stainless steel, Teflon, and glass are inert substances, sampling equipment made of these materials should be preferentially selected over equipment made of other materials. USGS (2003) notes that fluorocarbon polymers (Teflon®, Kynar®, and Tefzel®), stainless steel 316-grade, and borosilicate glass (laboratory grade) are acceptable for both inorganic and organic analyses; polypropylene, polyethylene, PVC, silicone, and nylon are only acceptable for inorganic analyses; and stainless steel 304-grade, and other metals (brass, iron, copper, aluminum, and galvanized and carbon steels) are only acceptable for organic analyses. Having a rigorous quality control sampling program (e.g., collecting rinsate blanks) in place will help ensure that the sample integrity is not being impacted by the sampling equipment (see Chapter 6).

Table 4.7 summarizes the effectiveness of each of the recommended procedures. A number "1" in the table indicates that a particular procedure is most effective in collecting samples for a particular laboratory analysis, sample type, or sample depth. A number "2" indicates that the procedure is acceptable but less preferred, whereas an empty cell indicates that the procedure is not recommended. For example, Table 4.7 indicates that the bladder pump and low-flow sampling are the most effective methods for collecting samples for volatile organic analysis. In contrast, the EON Dual Membrane Sampler®, Snap Sampler®, HydraSleeve™, AGI Sample Module®, bailer, bomb sampler, submersible pump, FLUTE™, and WestBay Multiple Port Sampling System are acceptable method for collecting samples for volatile organic analysis, but are less preferred.

When collecting groundwater samples, bottles for volatile organic analysis should always be the first to be filled. This is critical because volatile organics are continuously being lost to the atmosphere during the sampling procedure. The remaining sample bottles should be filled in an order consistent with their relative importance to the sampling program. If the analyses to be performed require preservation, preservatives should be added to sample bottles before sample collection. Collecting filtered and unfiltered groundwater samples for metals analysis should be considered, particularly for wells that remain cloudy or turbid throughout the development and purging procedure. Collecting samples in this way allows one to identify the dissolved concentration of metals. A pressurized filtration system utilizing a 0.45-μm millipore membrane filter is commonly used when filtering samples.

Shortly after a water sample has been collected from a well, it is common practice to take a final water level reading, and collect a final sample for pH, temperature, conductivity, turbidity, and dissolved oxygen readings.

4.2.7.5.1 Passive (No-Purge) Groundwater Sampling

Passive groundwater sampling involves the collection of a water sample from a well without first purging the well. This is a very cost-effective method of sampling

TABLE 4.7
Rating Table for Groundwater Sampling Methods

	Laboratory Analyses							Sample Type		Depth		
	Radionuclides	Volatiles	Semivolatiles	Metals	Pesticides	PCBs	TPH	Grab	Composite (Vertical)	Integrated	Shallow (0.0–50 ft)	Deep (> 50 ft)
EON Dual Membrane Sampler®		2	2	2				1	1	2	1	1
Snap Sampler®	2	2	2	2	2	2	2	1	1	2	1	1
HydraSleeve Sampler™	2	2	2	2	2	2	2	1	1	2	1	1
AGI Sample Module®		2	2		2			1			1	1
Low-Flow Sampler	1	1	1	1	1	1	1	1	2	1	1	1
Bailer	1	2	2	1	1	1	1	1		2	1	1
Bomb sampler	1	2	2	1	1	1	1	1	1	2	1	1
Bladder pump	1	1	1	1	1	1	1	1	2	1	1	1
Submersible pump	1	2	2	1	1	1	1	1	2	1	2	1
FLUTe™ sampling system	1	2	2	1	1	1	1	1	2	1	1	1
WestBay Multiple Port Sampling System	1	2	2	1	1	1	1	1	2	1	1	1

Note: 1 = preferred method; 2 = acceptable method; empty cell = method not recommended, a = Only recommended if well purging (standard or micropurging) is not possible for some site-specific reason.

groundwater since no purge water has to be handled and disposed of. In recent years much testing has been performed, which has shown that many passive groundwater sampling techniques are able to provide representative samples when compared to collecting groundwater samples using standard purge and sample methods. Regulatory agencies are more accepting of passive groundwater sampling for this reason. Passive groundwater sampling is particularly useful in poor yielding formations with slow recharge since purging with a pump is problematic. It is also useful in wells that only have a few feet of water in them above the bottom of the screen for the same reason.

USGS (2020) highlights that there are three primary types of passive groundwater samplers, including:

- *Equilibrium-membrane type*: These samplers use a semipermeable membrane through which chemicals diffuse or permeate. Permeation is the process of water or chemicals moving through openings in the membrane. The EON Dual Membrane Sampler® is an example of this type of passive sampler.
- *Equilibrium-thief type*: These samplers do not use a semipermeable membrane. Rather, chemical constituents simply move through the openings in the body of the sampler either initially through advection (horizontal flow of water) and dispersion or over time primarily by diffusion (movement from areas of higher concentration to areas of lower concentration). This sampler assumes chemical constituents reach equilibrium between the water in the sampler and the water in the well and are captured in the sampler when the sampler is closed. The Snap Sampler® and the HydraSleeve™ Sampler are examples of this type of passive sampler.
- *Accumulation-type*: These samplers contain sorptive media where selected chemical constituents are sorbed onto the media that the sampler contains for later extraction and analysis. The AGI Sample Module® is an example of this type of passive sampler.

The following sections provide information and procedures for implementing these different types of passive groundwater sampling methods.

4.2.7.5.1.1 EON Dual Membrane Sampler® EON's Dual Membrane™ Passive Diffusion Sampler (DMPDB™ sampler) (see Figure 4.62) is a type of passive diffusion bag (PDB) sampler that is used to collect groundwater samples from wells without purging or pumping. In addition to collecting samples for VOC analysis, this sampler is also able to collect samples for SVOCs, metals, anions, cations, 1,4-Dioxane, perfluoroalkyl and polyfluoroalkyl substances (PFAS), etc.

The DMPDB™ sampler is unique in that it utilizes two separate semipermeable membranes to increase the number and type of compounds that can be sampled. A sample chamber is formed inside a perforated tube by a small-pore hydrophobic membrane that is wrapped around the lower portion of the tube and a large-pore hydrophilic membrane around the upper length of the tube. In use, the chamber is filled with de-ionized water, sealed shut, and lowered on a suspension tether, into the saturated well screen, where natural groundwater flow intercepts the device.

When there are dissolved compounds in the groundwater surrounding the sampler, a concentration gradient exists between the groundwater surrounding the

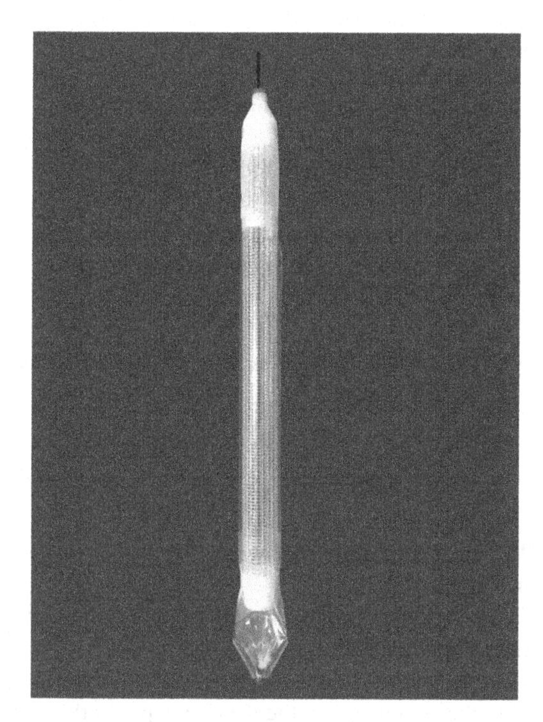

FIGURE 4.62 EON Dual Membrane™ Passive Diffusion Sampler.

sampler and the deionized water inside. The concentration gradient causes molecules to diffuse through one or the other of the membranes, depending on the compound and molecular characteristics, and into the deionized water where diffusion within the sampler causes the molecules for each compound to be uniformly dispersed.

Large and ionic molecules diffuse through the upper membrane, and smaller, non-polar VOCs, through the lower membrane. Because the diffusion process is driven by a concentration gradient, the sampler concentration adjusts to reflect changes in the surrounding groundwater concentrations. After a minimum equilibration period of three weeks, the sampler maintains a dynamic equilibrium with the groundwater, always representing the surrounding concentrations of the last few days of residence.

When the DMPDB sampler is removed from the well, the lower membrane, which serves as a sample reservoir, is pierced with a "juice-box" straw and the contents are decanted into laboratory bottles for standard laboratory analysis.

This sampler is 1.75-in. in diameter and comes in lengths up to 28 –in. which provide a sample volume of 650 mL. Multiple samplers can be strung together in a single well, if chemistry is believed to be stratified. Larger diameter DMPDB samplers are available for greater sample volume in larger wells (EON Products Inc. 2020).

Advantages of this sampling method include:

- Reduces sample acquisition costs over standard purging and sampling methods
- Samples for a broader range of chemicals than standard PDB samplers
- Sampler maintains dynamic equilibrium with the groundwater

- Provides a depth-specific concentration profile
- No purge water is generated
- Sampler can be used in wells as small as 2-in. in diameter

Disadvantages of this sampling method include:

- Requires continuous natural groundwater flow through well screen
- DMPDB™ sampler must be allowed to equilibrate for a minimum of three weeks.

The following equipment and procedure can be used to collect groundwater samples using the DMPDB™ sampler.

Equipment

1. DMPDB™ sampler
2. ASTM Type I or II deionized water
3. Teflon funnel
4. Project specific sampling and analysis plan (see Section 3.2.7)
5. Stainless steel connection rings
6. Weighted suspension tether (e.g., braided polypropylene) suitable for lowering the sampler(s) to the desired depth in the well
7. Tripod and reel (optional)
8. Plastic sheeting ground cover (optional)
9. Tape measure (in tenths of a foot scale) with stainless steel weight attached
10. Water level detector
11. Stainless steel snips or razor blade
12. Zip-Ties
13. Sample bottles
14. Sample preservatives
15. Sample labels
16. Cooler packed with Blue Ice®
17. Trip blank (only required for volatile organic analysis)
18. Coolant blank
19. Field logbook
20. Chain-of-custody forms
21. Chain-of-custody seals
22. Permanent ink marker
23. Health and safety instruments
24. Health and safety clothing (see Chapter 10)
25. Waste container (e.g., 55-gal drum)
26. Sampling table (optional)
27. Plastic waste bags

Procedure

1. In preparation for sampling, confirm that all necessary preparatory work has been completed, including obtaining property access agreements, meeting health and safety and equipment decontamination requirements, and checking the calibration of all health and safety instruments.
2. Just prior to mobilizing to the field perform the following tasks:
   a. Refer to project-specific sampling and analysis plan (see Section 3.2.6) regarding proper positioning of DMPDB™ sampler(s) within well screened interval.
   b. Assuming weight at end of suspension tether will rest at bottom of well, use tape measure to identify where along tether DMPDB™ sampler(s) should be attached.
   c. Use Zip Ties and stainless steel connection rings to attach top and bottom of DMPDB™ sampler(s) to suspension tether (Figure 4.63).
3. Unlock the well head and remove the well cap from the well casing.
4. Use health and safety instruments (e.g., photoionization detector and flame-ionization detector) to screen air quality at well head. Based on results, adjust personal protective equipment if needed in accordance with project health and safety plan (see Chapter 10).
5. Measure depth to groundwater (±0.01 ft) from top of casing using water level detector (see Section 4.2.7.4.1) and record the results in a field logbook.
6. Measure depth to bottom of well (±0.1 ft) from top of casing (see Section 4.2.7.4.2) using tape measure with stainless steel weight attached and record the results in a field logbook.
7. If you chose to use tripod and reel to lowering DMPDB™ sampler down well, then tie end of tether to reel, then reel in tether.
8. Remove red plug (Figure 4.63) from DMPDB™ sampler(s) and use Teflon funnel to fill DMPDB™ semi-permeable membranes with ASTM Type I or II deionized water. Replace red plug.
9. Slowly lower DMPDB™ sampler down well until weight at end of tether rests at bottom of well. Be careful not to create turbulence in groundwater.
10. Attach top of the tether to stainless steel connection ring, then attach to bottom of well cap.
11. Record in field logbook date and time when DMPDB™ sampler is in place.
12. Leave DMPDB™ sampler in place for length of time (minimum of three weeks) specified in sampling and analysis plan (see Section 3.2.6).
13. When it is time to retrieve DMPDB™ sampler, lift well cap just enough to allow water level detector tape down well. Measure depth to groundwater (±0.01 ft) from top of casing (see Section 4.2.7.4.1) and record the results in a field logbook.

FIGURE 4.63 EON Dual Membrane™ Passive Diffusion Sampler attached to tether line.

14. Retrieve DMPDB™ sampler being careful of not to create turbulence in groundwater. Record date and time when the sampler was retrieved in the field logbook.
15. Pierce lower membrane of sampler, which serves as sample reservoir, with a "juice-box" straw and decant water into sample bottles for laboratory analysis.
16. After each sample bottle is capped, attach a completed sample label and custody seal (see Section 5.2) and immediately place it into a sample cooler. Samples for chemical analysis should be packed in Blue Ice® along with coolant blank and trip blank (if analyzing for VOCs).
17. See Chapter 5 for details on preparing sample bottles and coolers for sample shipment.
18. Replace the well cap, lock the well head, and containerize any waste in a waste container. Before leaving the site, all waste containers should be sealed, labeled, and handled appropriately (see Chapter 11).
19. Decontaminate sampling equipment in accordance with procedures outlined in Chapter 9.

4.2.7.5.1.2 Snap Sampler® The QED Environmental Systems, Inc. Snap Sampler® (Figure 4.64) is an equilibrium-thief passive groundwater sampling method that assumes most (if not all) well screens exhibit ambient flow under natural groundwater gradients and therefore well screen sections exchange formation water without needing to be purged. The Snap Sampler® utilizes a unique double-end-opening bottle that snaps closed while the sample bottle is still below the water table. This eliminates the chance of air getting into the bottle and it removes the error introduced by the need to pour water from the sampler into sample bottles at the ground surface.

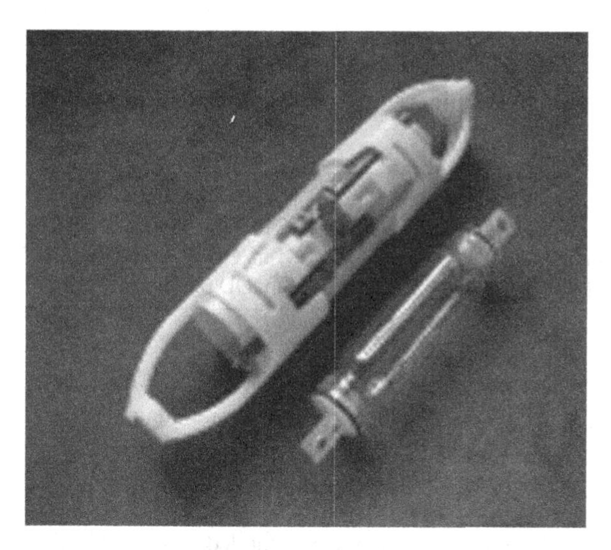

FIGURE 4.64 Snap Sampler®.

This sample collection method is the same for any user because the sample is captured downhole the same way every time, without impact from how the sampler fills the sample bottle. Positioning the sampler at the same depth in a well each time will optimize the consistency and comparability of the resulting analytical data over time. For this reason, dedicated sampling devices with dedicated trigger lines should be considered whenever possible. Trigger lines used to "snap" the sample bottles closed when it is time to collect a sample may be manual, with a mechanical wire connection from the ground surface to the sampler; or pneumatic, with an airline from ground surface to the sampler. The Snap Sampler® should not be set resting on the bottom of the well since any sediment that may be present will be disturbed when the sampler is triggered.

After the sampler has been deployed within the screened section of a well (with the "Snap Caps" in the open position), the sampler is left to equilibrate. Vroblesky (2001a) and Vroblesky (2001b) concluded that a one- to two-week equilibration period is needed for a well to return to steady-state conditions. When it is time to retrieve the sampler, one triggers the release of the "Snap Caps" that seal the bottle. Once the closed sample bottle is retrieved from the well, septa screw caps are placed over top of the "Snap Caps" to protect the seal. The bottle is then labeled in preparation for being shipped to the laboratory for analysis. As many as six Snap Samplers® (each containing a separate sample bottle) can be linked together to provide the volume of water needed to run a suite of analyses. The Snap Sampler® comes in three sizes as identified in Table 4.8.

Advantages of this sampling method include the following:

- Samples can be run for any COCs (e.g., VOCs, SVOCs, metals, radionuclides, pesticides, etc.)
- Provides a depth-specific concentration profile

TABLE 4.8
Dimensions of Snap Samplers® with Bottle Installed in Module

Sampler	Dimensions
SNAP-40 (with 40 ml glass bottle)	1.7 in. OD (43 mm), 7.8 in. (198 mm) Length
SNAP-125-350 (with 125 ml HDPE bottle)	1.88 in. OD (48 mm), 10.5 in. (277 mm) Length
SNAP-250 (with 250 ml HDPE bottle)	1.85 in. OD (47 mm), 14 in. (356 mm) Length
SNAP-125-350 (with 350 ml HDPE bottle)	3.4 in. OD (86 mm), 10.5 in. (277 mm) Length

- No purge water is generated
- Sampler can be used in wells as small as 2-in. in diameter

Disadvantages of this sampling method include the following:

- Higher initial capital expense when using dedicated systems
- Relies on assumption that there is continuous natural groundwater flow through well screen
- Limited volume of water is collected so there may not be enough water to run large list of COCs
- Increased training needs for sampling personnel

The following reference provides a thorough equipment list and step-by-step procedure for properly deploying and retrieving the Snap Sampler® from a groundwater well. This procedure was prepared by QED Environmental Systems, Inc. and can be downloaded from the WEB page cited below. For this reason, a step-by-step procedure is not provided in this book.

- QED, 2019, Snap Sampler® Zero – Passive Sampling System, Standard Operating Procedure for the Snap Sampler Passive Groundwater Sampling Method, November www.snapsampler.com/local/uploads/content/files/Snap_Sampler_SOP.pdf.

4.2.7.5.1.3 HydraSleeve™ Sampler The HydraSleeve™ (Figure 4.65) is an equilibrium-thief passive groundwater sampling method that assumes most (if not all) well screens exhibit ambient flow under natural groundwater gradients and therefore well screen sections exchange formation water without needing to be purged. Samples collected with this method can be analyzed for a full suite of analyses (e.g., VOCs, SVOCs, metals, radionuclides, pesticides, PCBs, explosive compounds, etc.). One limitation of this sampler is the volume of water that it can collect. GeoInsight (2019) cautions that the HydraSleeve™ Sampler is *not* designed to sample pure product (e.g., LNAPL or DNAPL).

The HydraSleeve™ consists of a closed-bottom sleeve of low-density polyethylene with a reed valve at the top. The sleeve is available in a variety of sizes, with lengths ranging from 30 to 66 in., diameters ranging from 1.4 to 2.7 in., and volumes ranging

FIGURE 4.65 HydraSleeve™ equilibrium-thief passive groundwater sampling method.

from 600 mL to 2.5 L. There are two holes at the top of the sleeve to attach the tether used to lower the sampler down the well. There are also two holes at the bottom of the sleeve where a reusable stainless steel weight is attached with a clip. This sampler is deployed in a collapsed state to the desired sampling depth within the well screened interval and is left for a period of time (hours to days) to equilibrate. It then fills with groundwater through the reed-type check valve when pulled upward at approximately 1 ft per second (in one continuous upward pull) through the sampling interval. When the sleeve is full, the reed valve closes and the sampler can be retrieved without the overlying groundwater disturbing the sample. At the ground surface, groundwater is transferred from the sleeve into sample bottles by puncturing the sleeve with the pointed discharge tube. The HydraSleeve™ is a single-use (disposable) sampler that is not intended for reuse, so there are no decontamination requirements for the sampler itself (EPA Region 4 2017d).

Advantages of this sampling method include:

- Samples can be run for any COCs (e.g., VOCs, SVOCs, metals, radionuclides, pesticides, etc.)
- Provides a depth-specific concentration profile
- No purge water is generated
- Sampler can be used in wells as small as 2-in. in diameter

Disadvantages of this sampling method include:

- Relies on assumption that there is continuous natural groundwater flow through well screen
- Limited volume of water is collected so that there may not have enough water to run a large list of COCs

The following equipment and procedure can be used to collect groundwater samples using the HydraSleeve™ sampler.

Equipment

1. HydraSleeve™ sampler
2. HydraSleeve™ spring clip to hold top of sampler open until sample is retrieved
3. HydraSleeve™ stainless steel weight and clip for bottom of sampler
4. HydraSleeve™ discharge tube
5. Project-specific sampling and analysis plan (see Section 3.2.7)
6. Tether to lower sampler to sampling depth
7. Tripod and reel (optional)
8. Zip-Ties
9. Tape measure (in tenths of a foot scale) with stainless steel weight attached
10. Water level detector
11. Stainless steel snips or razor blade
12. Sample bottles
13. Sample preservatives
14. Sample labels
15. Cooler packed with Blue Ice®
16. Trip blank (only required for volatile organic analysis)
17. Coolant blank
18. Field logbook
19. Chain-of-custody forms
20. Chain-of-custody seals
21. Permanent ink marker
22. Health and safety instruments
23. Health and safety clothing (see Chapter 10)
24. Waste container (e.g., 55-gal drum)
25. Sampling table
26. Plastic waste bags

Procedure

1. In preparation for sampling, confirm that all necessary preparatory work has been completed, including obtaining property access agreements, meeting health and safety and equipment decontamination requirements, and checking the calibration of all health and safety instruments.
2. Unlock the well head and remove the well cap from the well casing.

3. Use health and safety instruments (e.g., photoionization detector and flame-ionization detector) to screen air quality at the well head. Based on results, adjust personal protective equipment if needed in accordance with project health and safety plan.
4. Measure depth to groundwater (±0.01 ft) from top of casing using a water level detector (see Section 4.2.7.4.1) and record the results in a field logbook.
5. Measure depth to bottom of well (±0.1 ft) from top of casing (see Section 4.2.7.4.2) using a tape measure with stainless steel weight attached and record the results in a field logbook.
6. Remove the HydraSleeve™ from its packaging, unfold it, and hold it by its top.
7. Crimp the top of the HydraSleeve™ by folding the hard polyethylene reinforcing strips at the holes.
8. Attach the spring clip to the holes in top of HydraSleeve™ to ensure that the top will remain open until the sampler is retrieved.
9. Fold the flaps with the two holes at the bottom of the HydraSleeve™ together to align the holes and slide the weight clip through the holes.
10. Attach the stainless steel weight to the bottom of the weight clip to ensure that the HydraSleeve™ will descend to sampling depth.
11. Based on results from Steps 4 and 5 and requirements specified in sampling and analysis plan (see Section 3.2.7), identify the proper depth to lower the HydraSleeve™ down well. Measure and cut length of tether using stainless steel snips or razor blade that will allow sampling depth to be reached plus an extra 10 ft.
12. Tie one end of the tether to the spring clip at the top of HydraSleeve™. Mark the other end of the tether with the correct depth to lower HydraSleeve™ down well. This may be done by attaching a small Zip-Tie to tether.
13. If you chose to use a tripod and reel, then tie the end of tether to the reel of the tripod, and reel in the tether. If using hand-over-hand method instead, tie the end of tether to your wrist and wind the tether between your thumbs with your arms outstreched.
14. Slowly lower the HydraSleeve™ down well to proper sampling depth. Be careful not to create any turbulence in the groundwater. Hydrostatic pressure in water column will keep self-sealing check valve at top of HydraSleeve™ closed, and ensure it retains its flat, empty profile for an indefinite period prior to recovery.
15. Tie knot to secure the tether to the bottom of well cap to hold HydraSleeve™ at proper depth until it is time to retrieve sampler.
16. Record in field logbook the date and time when HydraSleeve™ is in place.
17. Leave the HydraSleeve™ in place for the length of time specified in sampling and analysis plan (see Section 3.2.7).
18. When it is time to retrieve HydraSleeve™, release the tether from bottom of the well cap.

19. Measure depth to groundwater (±0.01 ft) from top of casing using a water level detector (see Section 4.2.7.4.1) and record the results in a field logbook.
20. In one smooth motion, pull the tether upward the length of the sampler (30- to 60-in.) at a rate of ~1 ft per second. This will open top check valve and allow HydraSleeve™ to fill.
21. Carefully retrieve HydraSleeve™. Record in field logbook the date and time when the sampler reached the ground surface.
22. Dispose the small volume of water trapped in Hydrasleeve™ above check valve by pinching it off at the top under the stiffeners (above the check valve).
23. Remove Hydrasleeve™ discharge tube from its sleeve. Holding the Hydrasleeve™ at check valve, puncture HydraSleeve 3–4 in. below the reinforcement strips with the pointed end of the discharge tube.
24. Discharge water from Hydrasleeve™ into sample bottles. Control the rate of discharge by raising or lowering bottom of sleeve.
25. After each sample bottle is filled and capped, attach a completed sample label and custody seal (see Section 5.2) and immediately place it into a sample cooler. Samples for chemical analysis should be packed in Blue Ice®.
26. See Chapter 5 for details on preparing sample bottles and coolers for sample shipment.
27. Replace the well cap, lock the well head, and containerize any waste in a waste container. Before leaving the site, all waste containers should be sealed, labeled, and handled appropriately (see Chapter 11).

4.2.7.5.1.4 AGI Sample Module® The Amplified Geochemical Imaging, LLC (AGI) Sample Module (previously known as the GORE Module) was developed to sample soil gas and groundwater for a variety of VOC and SVOC compounds. This section is focused on how the AGI Sample Module can be used to support groundwater characterization. See Section 4.2.2.4 for using this module to support soil gas surveying.

The AGI Sample Module is an accumulation-type passive groundwater sampler and it relies on diffusion and sorption to accumulate analytes onto resins in the sampler (USGS 2020). The AGI Sample Module is approximately 0.25 in. in diameter and 13 in. in length and consists of a tube of GORE-TEX® membrane which is expanded polytetrafluoroethylene that is chemically-inert, vapor-permeable, and waterproof (Figure 4.66). Housed inside the membrane tubing are several packets of hydrophobic sorbents that have an affinity for a broad range of volatile and semi-volatile organic compounds. Reportedly, the AGI Sample Module can be used to detect chlorinated solvents, fuel-related compounds, oxygenates, 1,4-dioxane, some explosives, chemical warfare agent breakdown compounds, pesticides, and polycyclic aromatic hydrocarbons (US Army Corps of Engineers 2014).

With this method, the analyte must first partition from solution into the vapor phase. Once in the vapor phase, the molecule can then diffuse through the membrane while liquid water is prevented from passing through the (waterproof) membrane.

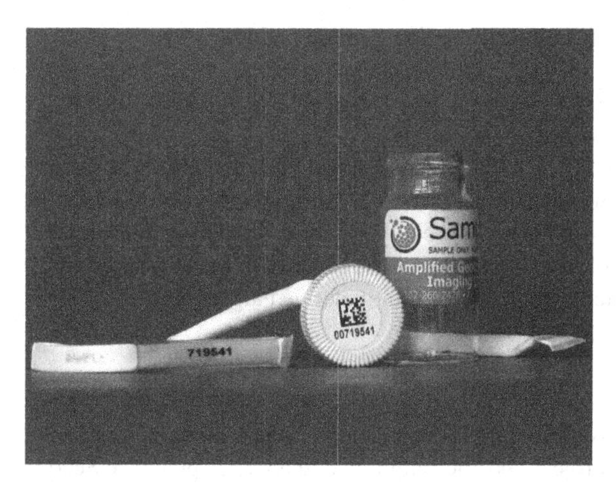

FIGURE 4.66 AGI sample module.

Once the analyte passes through the membrane, it is then sorbed by the adsorbent contained in the sampler. In groundwater monitoring wells, the sampler is deployed by tying it to a tether of the desired length (to reach the selected sampling depth), placing suitable weights on the end of the tether, and lowering it down the well. Any well or piezometer with a diameter greater than 0.25 in. can accept this sampler. The sampler immediately begins to collect the analytes, and typical sampling times range from 15 minutes to 4 hours (US Army Corps of Engineers 2014). Depending upon the flow dynamics in the well, high-resolution vertical profiling can be achieved in some cases by simply placing the modules at multiple sampling depths within the well.

While there is no limit to the depth at which an AGI Sample Module can be used, it should be noted that when the depth of the module below the water table exceeds 32 ft, analytes with higher aqueous solubility and lower Henry's Law constants are biased low. In this instance, methyl *tert*-butyl ether is lost entirely and 1,2-dichloroethane, 1,1,2-trichloroethane, and 1,1,2,2-tetrachloroethane are biased low by about 40% (US Army Corps of Engineers 2014).

Once retrieved from the well, the module is shipped to the manufacturer's laboratory. Analyses are performed by GC/MS, using either EPA SW-846 Method 8260C for VOCs or 8270 for SVOCs, which were modified for thermal desorption (US Army Corps of Engineers 2014). These analyses produce a mass flux or a total mass of organic compounds sorbed over the time of deployment. In other words, the concentration of contaminants in the groundwater is not measured directly but must be calculated using an experimentally derived algorithm.

Advantages of this sampling method include:

- Screens for a broad range of VOCs and SVOCs
- No purge water is generated
- Can be used to sample multiple depths within the well
- Sample does not require low-temperature storage following sample collection

Disadvantages of this sampling method include:

- Relies on assumption that there is a continuous natural groundwater flow through well screen
- Sampler is exposed to other analytes in water column above the sampling depth during deployment and retrieval (although this exposure is brief)
- Primary limitation, especially for some regulatory agencies, is that the concentration of contaminants in groundwater is not measured directly but must be calculated using experimentally derived algorithm.

Equipment

1. AGI Sample Module contained in sealed glass vial
2. Project-specific sampling and analysis plan (see Section 3.2.7)
3. Water level detector
4. Tape measure (in tenths of a foot scale) with stainless steel weight attached
5. Tether with stainless steel weight attached
6. Stainless steel scissors (or razor blade)
7. Field logbook
8. Tripod and reel (optional)
9. Zip Ties
10. Sample labels
11. Cooler
12. Trip blank
13. Chain-of-custody forms
14. Chain-of-custody seals
15. Permanent ink marker
16. Health and safety instruments
17. Health and safety clothing (see Chapter 10)
18. Waste container (e.g., 55-gal drum)
19. Plastic waste bags

Procedure

1. In preparation for sampling, confirm that all necessary preparatory work has been completed, including obtaining property access agreements, meeting health and safety and equipment decontamination requirements, and checking the calibration of all health and safety instruments.
2. Unlock the well head and remove the well cap from the well casing.
3. Use health and safety instruments (e.g., photoionization detector and flame-ionization detector) to screen air quality at the well head. Based on results, adjust personal protective equipment if needed in accordance with project health and safety plan.

4. Measure depth to groundwater (±0.01 ft) from top of casing using a water level detector (see Section 4.2.7.4.1) and record the results in a field logbook.
5. Measure depth to bottom of well (±0.1 ft) from top of casing (see Section 4.2.7.4.2) using a tape measure with stainless steel weight attached and record the results in a field logbook.
6. Based on results from Steps 4 and 5 and requirements specified in sampling and analysis plan (see Section 3.2.7), identify the depth at which the AGI Sample Module is to be placed.
7. Measure and cut a length of tether using stainless steel scissors (or razor blade) that will allow sampling depth to be reached. Attach stainless steel weight to one end of the tether.
8. If you chose to use tripod and reel to lowering AGI Sample Module down well, then tie the other end of the tether to reel, then reel in the tether.
9. Remove the AGI Sample Module from the sealed glass vial. Run the tip of a Zip Tie first through the loop end of the AGI Sample Module, then through the suspension tether at the correct sampling depth, then through the locking head of the Zip Tie, and tighten. (Note: Multiple AGI Sample Modules can be attached to the tether for vertical profiling if desired).
10. Slowly lower the weighted tether, with AGI Sample Module attached, down well until weight at the end of the tether rests at the bottom of well. Be careful not to create turbulence in groundwater.
11. Attach top of the tether to bottom of the well cap to maintain correct sampling depth.
12. Record in the field logbook the date and time when AGI Sample Module is in place.
13. Leave AGI Sample Module in place for the length of time specified in sampling and analysis plan (see Section 3.2.7).
14. When it is time to retrieve AGI Sample Module, lift the well cap just enough to allow the water level detector tape down well. Measure depth to groundwater (±0.01 ft) from top of casing (see Section 4.2.7.4.1) and record the results in a field logbook.
15. Retrieve AGI Sample Module being careful of not to create turbulence in groundwater. Record in field logbook the date and time when the sampler was retrieved.
16. Use stainless steel scissors (or razor blade) to free the AGI Sample Module from the tether and transfer the module back into the glass vial it came in and screw on the cap.
17. Attach a sample label and custody seal to the glass vial and immediately place it into a sample cooler.
18. See Chapter 5 for details on preparing sample bottles and coolers for sample shipment.

19. Replace the well cap, lock the well head, and containerize any waste in a waste container. Before leaving the site, all waste containers should be sealed, labeled, and handled appropriately (see Chapter 11).
20. Decontaminate sampling equipment in accordance with procedures outlined in Chapter 9.

4.2.7.5.2 Low-Flow Groundwater Sampling

The purpose of Low-Flow groundwater sampling is to collect groundwater samples from monitoring wells that are representative of ambient groundwater conditions in the aquifer using very low pumping rates. This is accomplished by setting the intake velocity of the sampling pump to a flow rate that limits drawdown inside the well to 0.3 ft or less (EPA 2017c). The advantages of Low-Flow sampling are that it minimizes the disturbance of sediment in the bottom of the well producing a sample with low turbidity; minimizes aeration of the groundwater during sample collection; reduces the volume of groundwater needing to be purged (and disposed of) as compared to conventional groundwater purging and sampling methods (see Section 4.2.7.4); and samples can be analyzed for a full suite of chemical and radiological contaminants.

Low-Flow groundwater sampling is best performed with an adjustable rate bladder pump that is constructed of stainless steel and Teflon® components. The bladder ensures that drive air (or gas) does not contact the sample, which avoids degassing or contamination of the sample. The components of a bladder pump system are shown in Figure 4.67 and include a bladder pump (e.g., Solinst Model 407), compressor (e.g., Solinst 12 Volt Compressor), pump control unit (e.g., Solinst Model 464 Pneumatic

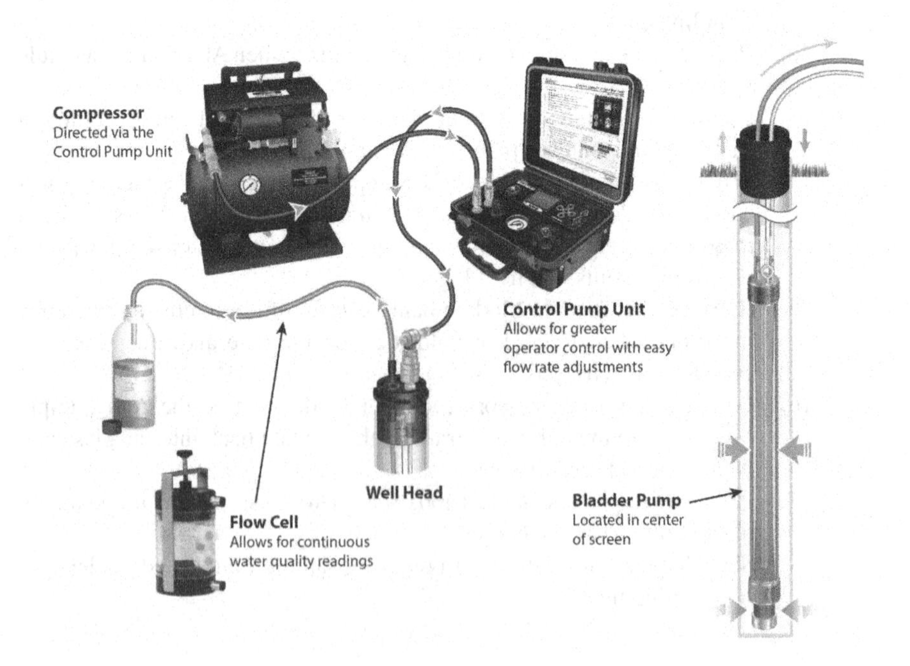

FIGURE 4.67 Components of a Solinst bladder pump system.

Pump Controller), and flow cell for monitoring field parameters. Bladder pumps come in different sizes but are typically 1.66 in. (42 mm) in diameter and often 2 ft (0.6 m) or 4 ft (1.2 m) in length.

If a bladder pump intake is set 100 ft below the ground surface, approximately 50 psi of pressure would be needed to bring a water sample to the ground surface (Solinst 2019). The drive gas is typically air or nitrogen delivered through either an air compressor or a gas cylinder. This type of pump can collect a groundwater sample from as deep as 500 ft below the ground surface. When a bladder pump is used to collect samples for VOC analysis, the pump should be set so that one pulse will deliver a water volume that is sufficient to fill a 40 mL VOC vial. The project-specific sampling and analysis plan (see Section 3.2.7) will define the optimum depth at which the bladder pump intake should be set.

While Low-Flow groundwater sampling could be performed using an adjustable rate peristaltic (suction) pump, EPA (2017c) notes that peristaltic pumps are to be used with caution when collecting samples for VOC and dissolved gas (e.g., methane, carbon dioxide) analyses. If a peristaltic pump is used, EPA (2017c) notes the inside diameter of the rotor head tubing needs to match the inside diameter of the tubing installed in the monitoring well since squeezing/pinching the pump's tubing will result in a pressure change that can result in the loss of VOCs and dissolved gases. USGS (2006, emphasis added) states that "*...peristaltic pumps, can operate at a very low pumping rate: however, using negative pressure to lift the sample can result in the loss of volatile analytes*". The use of electrical submersible pumps to support Low-Flow sampling also comes with a caution. USGS (2006) notes that

> Operating variable-speed, electrical submersible pumps at low flow rates may result in heating of the sample as it flows around and through the pump; this also can result in sample degassing and VOC loss, in addition to changes in other temperature-sensitive analytes.

The use of 1/4 in. or 3/8 in. (inside diameter) Teflon® or Teflon®-lined tubing to support Low-Flow sampling is recommended since it will help ensure that the tubing remains liquid filled when operating at very low pumping rates. A flow cell should be used to assist with monitoring field parameters and define when they have stabilized. Many flow cells are available that can measure pH, temperature, specific conductance, turbidity, Oxidation/Reduction Potential, dissolved oxygen, and calculated values such as total dissolved solids (TDS), salinity, and oxygen saturation.

When selecting the optimum Low-Flow pump intake depth, for well screens longer than 5 ft, one should consider initially collecting depth-discrete Low-Flow groundwater samples at multiple depths throughout the screened interval of the well to identify the interval showing the highest contaminant concentrations. These results should then be used in combination with data collected during well drilling to select the optimum interval to set the Low-Flow pump intake, which may include:

- Depth-discrete groundwater and soil contaminant concentrations
- Lithology log
- Borehole geophysical logging results (see Section 4.2.3)
- Grain size analysis
- Hydraulic testing

The most permeable zone within the screened interval is often selected for the Low-Flow pump intake since the majority of contaminant mass will be transported through this interval. If LNAPLs or DNAPLs are suspected of being present, then this would also be important to decide on where to set the pump intake.

When preparing to collect a groundwater sample from a well using the Low-Flow method, one must first purge the well. During well purging, EPA (2017c) and EPA (2017d) recommend monitoring the following indicator field parameters at intervals of five minutes or greater: pH, temperature, specific conductance, turbidity, and dissolved oxygen. See Section 4.2.7.4 for criteria used to define when well stabilization has occurred and groundwater sampling can begin.

It should be noted that EPA (2017d) does *not* recommend oxidation/reduction potential measurements to be used to help define well stabilization since these measurements often do not demonstrate enough stability. At the U.S. Department of Energy's Hanford Site the author's projects are no longer collecting oxidation/reduction potential measurements due to data reliability and usability issues.

Advantages of this sampling method include:

- Minimizes disturbance of sediment in the bottom of the well, thereby producing a sample with low turbidity
- Minimizes aeration of the groundwater during sample collection
- Greatly reduces the amount of purge water that needs to be disposed of when compared to conventional purging and sampling methods
- Increases reproducibility of data when a dedicated system is used

Disadvantages of this sampling method include:

- Sampling from non-dedicated systems requires greater setup time
- Higher initial capital expense when using dedicated systems
- Increased training needs for sampling personnel

A step-by-step procedure for using a Solinst bladder pump system has been prepared by Solinst and can be downloaded from the WEB page cited below. For this reason, a step-by-step procedure is not provided in this book.

- www.solinst.com/products/groundwater-samplers/464-pneumatic-pump-control-units/operating-instructions/electronic-control-unit-user-guide/464-user-guide.pdf

4.2.7.5.3 Standard Groundwater Sampling

Standard groundwater sampling methods that have been used for decades include the bailer, bomb sampler, bladder pump, and submersible pump. The following sections provide details on the components that make up these samplers, the advantages of these samplers over others, and standard operating procedures for implementation.

4.2.7.5.3.1 Bailer Method The bailer method is the simplest of all the groundwater sampling methods. A standard bailer is composed of a bailer body, which is available in various lengths and diameters, a pouring spout, and a bottom check valve, which contains a check ball (see Figures 4.46 and 4.68). As the bailer is lowered into groundwater, water flows into the sampler through the bottom check valve. When the sampler is retrieved, the check ball seals the bottom of the bailer, which prevents water from escaping. Water is poured from the bailer through the pouring spout into sample bottles.

Some common modifications to the bailer include the use of extension couples to increase the length of the bailer, and the use of a controlled flow bottom assembly. The bottom assembly allows the bailer to be emptied through the bottom of the sampler, which reduces the opportunity for volatilization.

Some of the major advantages of the bailer are that it is easy to operate, portable, available in many sizes, is relatively inexpensive, and can be purchased as presterilize disposable samplers. The disadvantage is that aeration of the sample can be a problem when transferring water into sample bottles.

The bailer should be lowered down a well using a monofilament line, such as a common fishing line. The selected line should be cut to a length long enough to reach the groundwater, and it must be strong enough to lift the weight of the bailer when it is full of water. The line should be decontaminated in the same manner as the sampling bailer (see Chapter 9). The two most effective means of lowering a bailer and sampling line down a well are using the hand-over-hand or tripod and reel methods.

To implement the hand-over-hand method, one end of the sampling line is tied to the top of the bailer, and the other end is tied to the sampler's wrist. With the sampler's arms fully extended horizontally, the slack in the line is removed by winding it between the sampler's thumbs. To lower the bailer down the well, the sampler simply allows the line to unwind. To retrieve the bailer, the line is rewound. This method works very effectively for sampling wells that are less than 25 ft in depth, and when a small 2-in.-diameter bailer is used.

To implement the tripod and reel method, one end of the sampling line is tied to the top of the bailer, and the other end is tied to a reel (Figure 4.68). To lower the bailer down the well, the line is allowed to unwind from the reel. The handle on the reel is then used to rewind the line when the bailer is retrieved. This method is most effective in collecting samples from a depth less than 50 ft, and can handle a bailer as large as 4 in. in diameter.

For most sampling programs, four people are sufficient for the purging and sampling procedure. Two are needed for field testing, sample collection, labeling, and documentation; a third is needed for health and safety; and a fourth is needed for miscellaneous tasks such as managing wastewater containers and equipment decontamination.

The following equipment and procedure can be used to collect groundwater samples for chemical or radiological analysis:

Equipment

1. Bailer
2. Monofilament line
3. Tripod and reel (not needed if the hand-over-hand method is selected)

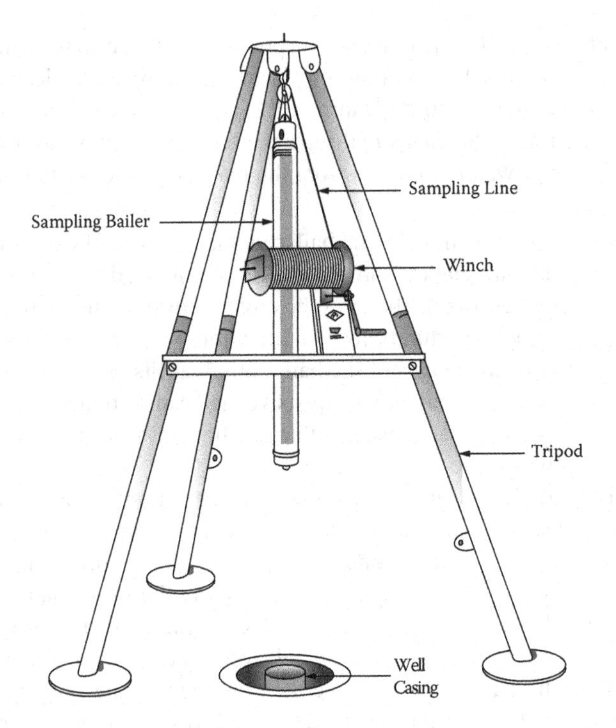

FIGURE 4.68 Tripod and reel used to assist groundwater sampling using a bailer.

(From Byrnes, 1994. *Field Sampling Methods for Remedial Investigations.* Lewis Publishers, Boca Raton, FL.)

4. Water level detector
5. Sample bottles
6. Sample preservatives
7. pH, temperature, conductivity, turbidity, and dissolved oxygen meter
8. Wide-mouth glass jar
9. Sample labels
10. Cooler packed with Blue Ice® (Blue Ice® is not required for radiological analysis.)
11. Trip blank (only required for volatile organic analyses)
12. Coolant blank (not required for radiological analysis)
13. Sample logbook
14. Chain-of-custody forms
15. Chain-of-custody seals
16. Permanent ink marker
17. Health and safety instruments
18. Chemical and/or radiological field screening instruments
19. Health and safety clothing
20. Sampling table
21. Waste container (e.g., 55-gal drum)
22. Plastic waste bags

Sampling Procedure

1. In preparation for sampling, confirm that all necessary preparatory work has been completed, including obtaining property access agreements, meeting health and safety and equipment decontamination requirements, and checking the calibration of all health and safety and chemical and/or radiological field screening instruments.
2. Before sampling, a groundwater well must be properly developed and purged (see Sections 4.2.7.3 and 4.2.7.4). If a well has been previously developed, there is no need to repeat this procedure unless accumulated sediment is blocking the well screen, or the well was inadequately developed the first time. When using this method, a well must be purged each time before it is sampled.
3. When well purging is complete, collect a final water level measurement (Section 4.2.7.4.1), and record this information on the Well Purging and Sampling Form (see Section 5.2.3).
4. Cut a length of monofilament line long enough to reach the water table. Tie one end of the line to the top of the bailer, and the other end to either the sampling reel or sampler's wrist, depending on whether the tripod and reel or hand-overhand method is used.
5. Lower the bailer down the well to a depth just above the water table. At this point, slowly lower the bailer through the water just deep enough to fill it with water. When the bailer is full, slowly raise it out of the water, and then retrieve it quickly to the ground surface. Collecting a groundwater sample in this manner creates little disturbance of the water column, which in turn reduces the loss of volatile organics and minimizes the turbidity of the water sample.
6. Transfer the water from the bailer carefully into the appropriate sample bottles. If the analyses to be performed require the sample to be preserved, this should be performed before filling the sample bottles. Begin filling bottles for volatile organic analysis first. Bottles for other analyses should be filled in an order consistent with their relative importance to the sampling program.
7. After each bottle is capped, attach a sample label and custody seal to the sample bottle and immediately place it into a sample cooler. Samples for chemical analysis should be packed in Blue Ice®.
8. See Chapter 5 for details on preparing sample bottles and coolers for sample shipment.
9. Fill a glass jar with sample water and collect and record a final pH, temperature, conductivity, turbidity, and dissolved oxygen measurement, in addition to collecting a final water level measurement (Section 4.2.7.4.1) from the well. Record this information on the Well Purging and Sampling Form (Section 5.2.3).
10. Replace the well cap, lock the well, and containerize any waste in a waste container. Before leaving the site, all waste containers should be sealed, labeled, and handled appropriately (see Chapter 11).

11. Transfer any other sampling-related wastes (e.g., gloves and foil) into a plastic waste bag.

4.2.7.5.3.2 Bomb Sampler Method The bomb sampler method provides the advantage of being able to collect a grab sample from a specific depth interval. A standard bomb sampler is composed of a sampling tube, center rod, and a support ring (see Figure 4.49). With this method, a support line is used to lower the sampler to the desired sampling depth, whereas a sampling line is used to open and close the sampler inlet via the center rod. Within the body of the bomb sampler, a spring keeps the center rod in the closed position when lowering the sampler to the desired sampling depth. This prevents water from entering the sampler. When the desired sampling depth is reached, the sampling line is lifted against the pressure of the spring, which allows water to enter the sampling tube. When the sampling line is released, the center rod drops to reseal the sampling tube. After the sampler is retrieved, water is transferred into a sample bottle by placing the bottle beneath the center rod and lifting up on the sampling line.

Some of the major advantages of the bomb sampler are that it is effective for depth interval sampling, easy to operate, portable, available in many sizes, and is relatively inexpensive. The disadvantage is that aeration of the sample can be a problem when transferring water into sample bottles.

The support and sampling line should be monofilament, such as common fishing line, and should be discarded between wells. The selected line should be cut to a length long enough to reach the desired sampling depth, and it must be strong enough to lift the weight of the sampler when it is full of water. The sampling line should be decontaminated in the same manner as other sampling equipment before sampling (see Chapter 9).

The most effective means of lowering a bomb sampler down a well is using the tripod and reel method. To implement this method, one end of the sampling line is tied to the top of the sampler, and the other end is tied to the reel. To lower the sampler down the well, the line is allowed to unwind from the reel. The handle on the reel is then used to rewind the line when retrieving the sampler.

For most sampling programs, four people are sufficient for the purging and sampling procedure. Two are needed for field testing, sample collection, labeling, and documentation; a third is needed for health and safety; and a fourth is needed for miscellaneous tasks such as managing wastewater containers and equipment decontamination.

The following equipment and procedure can be used to collect groundwater samples for chemical and/or radiological analysis:

Equipment

1. Bomb sampler
2. Monofilament line
3. Tripod and reel
4. Water level detector
5. Sample bottles

6. Sample preservatives
7. pH, temperature, conductivity, turbidity, and dissolved oxygen meters
8. Wide-mouth glass jar
9. Sample labels
10. Cooler packed with Blue Ice® (Blue Ice® is not required for radiological analysis.)
11. Trip blank (only required for volatile organic analyses)
12. Coolant blank (not required for radiological analysis)
13. Sample logbook
14. Chain-of-custody forms
15. Chain-of-custody seals
16. Permanent ink marker
17. Health and safety instruments
18. Chemical and/or radiological field screening instruments
19. Health and safety clothing
20. Sampling table
21. Waste container (e.g., 55-gal drum)
22. Plastic waste bags

Sampling Procedure

1. In preparation for sampling, confirm that all necessary preparatory work has been completed, including obtaining property access agreements, meeting health and safety and equipment decontamination requirements, and checking the calibration of all health and safety and chemical or radiological field screening instruments.
2. Before sampling, a groundwater well must be properly developed and purged (see Sections 4.2.7.3 and 4.2.7.4). If a well has been previously developed, there is no need to repeat this procedure unless accumulated sediment is blocking the well screen, or the well was inadequately developed the first time. A well must be purged each time before it is sampled.
3. When well purging is complete, collect a final water level measurement (Section 4.2.7.4.1), and record this information on the Well Purging and Sampling Form (see Section 5.2.3).
4. Cut two lengths of monofilament line long enough to reach the sampling interval. Tie one end of one line to the top of the bomb sampler and the other end to the sampling reel. The second line is tied to the top of the center rod.
5. Using the support line, the bomb sampler is lowered down the well to a depth just above the water table, and then slowly lowered through the water to the sampling interval.
6. When the sampling interval is reached, hold the support line steady while lifting up on the sampling line. When the sampler is full of water, release the sampling line. Using the support line, slowly raise the sampler just above the water surface, and then retrieve it quickly to the

ground surface. Collecting a groundwater sample in this manner creates little disturbance of the water column, which in turn reduces the loss of volatile organics and minimizes the turbidity of the water sample.

7. Transfer water from the sampler carefully into sample bottles by placing the bottle beneath the center rod and lifting up on the center rod. If analyses to be performed require the sample to be preserved, do so before filling the sample bottles. Begin filling bottles for volatile organic analysis first. Bottles for other analyses should be filled in an order consistent with their relative importance to the sampling program.
8. After each bottle is capped, attach a sample label and custody seal and immediately place it into a sample cooler. Samples for chemical analysis should be packed in Blue Ice®.
9. See Chapter 5 for details on preparing sample bottles and coolers for sample shipment.
10. Fill a glass jar with sample water and collect and record a final pH, temperature, conductivity, turbidity, and dissolved oxygen measurement, in addition to collecting a final water level measurement (Section 4.2.7.4.1). Record this information on a Well Purging and Sampling Form (Section 5.2.3).
11. Replace the well cap, lock the well, and containerize any waste in a waste container. Before leaving the site, all waste containers should be sealed, labeled, and handled appropriately (see Chapter 11).
12. Transfer any other sampling-related wastes (e.g., gloves and foil) into a plastic waste bag.

4.2.7.5.3.3 Bladder Pump A bladder pump combined with an air compressor (or compressed gas cylinder [carbon dioxide, nitrogen, or air]) and an electronic controller can be used as a very effective well-purging and sampling tool. The bladder pump is composed of a stainless steel pump body, bottom and top check ball, fill tube, Teflon bladder, outer sleeve, and discharge and air supply line (Figure 4.69). The advantages of this sampling tool are that it can be used effectively as either a dedicated or a portable pump. It is a very clean system in which the compressed air does not contact the sample water, it is effective in collecting samples as deep as 250 ft, and it is regarded highly by the EPA as an effective tool to collect samples for all parameters, including volatile organic compounds. The disadvantage of this method is the low pumping rate.

This pump operates on an alternating fill and discharge cycle. During the fill cycle, the bottom check ball allows the bladder to fill with water, while the upper check ball prevents any liquid in the discharge line from dropping back into the pump. During the discharge cycle, the bottom check ball seats as compressed gas squeezes the bladder, which in turn forces water up the discharge line. Both the air supply and discharge lines should preferably be made of Teflon or should be Teflon lined.

The bladder pump is powered by a source of compressed air such as an electric-powered compressor or a gas cylinder (containing carbon dioxide, nitrogen, or air). As a general rule, 0.5 psi is required per foot of well depth, plus 10 psi for pressure that is lost in the pressure line. An electronic controller is used to control the timing

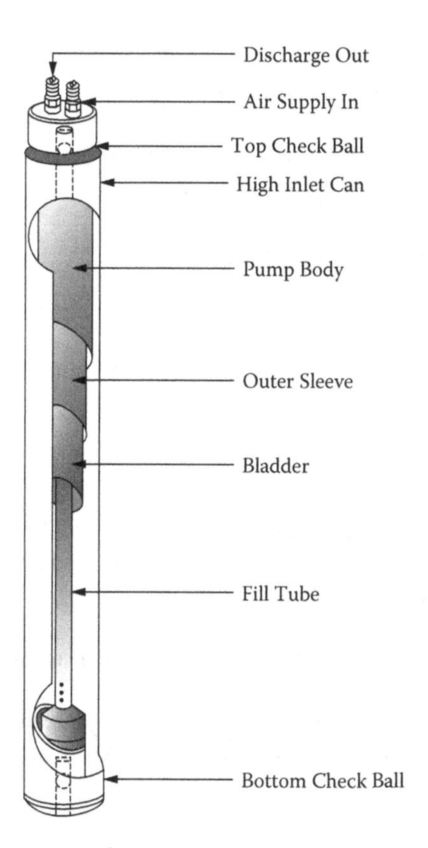

FIGURE 4.69 Bladder pump used to collect groundwater samples.

(From Byrnes, 1994. *Field Sampling Methods for Remedial Investigations.* Lewis Publishers, Boca Raton, FL.)

of the fill and discharge cycle of the bladder pump. The controller also monitors and controls the air pressure.

A 1.6-in.-diameter bladder pump is able to pump water at a rate between 1.0 and 1.5 gpm at a depth of 25 ft. At a depth of 100 ft the pumping rate typically drops to around 0.5 gpm. These pumps are ideal for use in 2-in.-diameter wells, but can also be used in larger-diameter wells. Bladder pumps are also available in a 3.0-in.-diameter size, which is ideal for larger-diameter wells. This larger pump is able to pump water at a rate between 3 and 4 gpm at a depth of 25 ft, and 1 to 2 gpm at 100 ft.

When the wells at a site are to be sampled regularly for an extended period of time, it is recommended that one bladder pump be installed in each groundwater well. Although this is an expensive recommendation for any site, a significant amount of sampling time and money will be saved in the long run by not having to decontaminate the pump and discharge line between wells. Also, equipment blank samples are not needed if a dedicated sampling system is used.

For most sampling programs, four people are sufficient for the purging and sampling procedure. Two are needed for field testing, sample collection, labeling, and documentation; a third is needed for health and safety; and a fourth is needed for miscellaneous tasks such as managing wastewater containers, and equipment decontamination.

The following equipment and procedure can be used to collect a groundwater sample for chemical and/or radiological analysis:

Equipment

1. Bladder pump
2. Discharge and air supply line
3. Air compressor or compressed gas (containing carbon dioxide, nitrogen, or air) bottle
4. Electronic controller
5. Water level probe
6. Sample bottles
7. Sample preservatives
8. pH, temperature, conductivity, turbidity, and dissolved oxygen meters
9. Wide-mouth glass jar
10. Sample labels
11. Cooler packed with Blue Ice® (Blue Ice® is not required for radiological analysis.)
12. Trip blank (only required for volatile organic analyses)
13. Coolant blank (not required for radiological analysis)
14. Sample logbook
15. Chain-of-custody forms
16. Chain-of-custody seals
17. Permanent ink marker
18. Health and safety instruments
19. Chemical and/or radiological field screening instruments
20. Health and safety clothing
21. Sampling table
22. Waste container (e.g., 55-gal drum)
23. Plastic waste bags

Sampling Procedure

1. In preparation for sampling, confirm that all necessary preparatory work has been completed, including obtaining property access agreements, meeting health and safety and equipment decontamination requirements, and checking the calibration of all health and safety and chemical and/or radiological field screening instruments.
2. Before sampling, a groundwater well must be properly developed and purged (see Sections 4.2.7.3 and 4.2.7.4). If a well has been previously developed, there is no need to repeat this procedure unless accumulated sediment is blocking the well screen, or the well was inadequately developed the first time. A well must be purged each time before it is sampled.

3. When well purging is complete, collect a final water level measurement (Section 4.2.7.4.1), and record this information on the Well Purging and Sampling Form (see Section 5.2.3).
4. Reduce the pumping rate (if needed) so that a slow but steady flow of water flows from the discharge line.
5. Begin sampling by filling bottles for volatile organic analysis first. Bottles for other analyses should be filled in an order consistent with their relative importance to the sampling program. If the analyses to be performed require the sample to be preserved, this should be performed before filling the sample bottle.
6. After the bottle is capped, attach a sample label and custody seal to the sample bottle and immediately place it into a sample cooler. Samples for chemical analysis should be packed in Blue Ice®.
7. See Chapter 5 for details on preparing sample bottles and coolers for sample shipment.
8. Use a flow cell or fill a glass jar with sample water and collect and record final pH, temperature, conductivity, turbidity, and dissolved oxygen measurement, in addition to collecting a final water level measurement (Section 4.2.7.4.1). Record this information on a Well Purging and Sampling Form (Section 5.2.3).
9. Replace the well cap, lock the well, and containerize any waste in a waste container. Before leaving the site, all waste containers should be sealed, labeled, and handled appropriately (see Chapter 11).
10. Transfer any other sampling-related wastes (e.g., gloves and foil) into a plastic waste bag.

4.2.7.5.3.4 Submersible Pump Submersible pumps tend to be used in wells greater than 100 ft in depth. These systems utilize a submersible pump, a discharge line, an electronic flow rate control system, and a power generator (Figure 4.70).

The submersible pump utilizes an electric motor that rotates a number of impellers, which in turn force water up the discharge line. These pumps are effective in collecting samples hundreds of feet in depth, and depending on the size of pump, they are able to pump at rates as low as <1 gpm to 100 gpm or more. These high pumping rates should never be used for monitoring well purging and sampling but rather are designed for remediation extraction wells. The relatively light weight of the smaller pumps allows the flexibility of permanently installing the pump for repeated sampling of the same well, or moving the pump between wells after being thoroughly decontaminated.

The advantage of the electronic flow rate control system is that the sampler can control the discharge rate of the pump without throttling the discharge line. Throttling is not desired when sampling, because it facilitates the loss of volatile organics. Another advantage is that this pump is available in sizes small enough for use in 2-in.-diameter wells.

When the monitoring wells at a site are to be sampled routinely, it is recommended that dedicated pumps be installed in each well. This is an expensive recommendation for a large site; however, a significant amount of sampling time and money will be saved in the long run by not having to decontaminate the pump and discharge line between wells, nor will equipment blanks be needed each time the well is sampled.

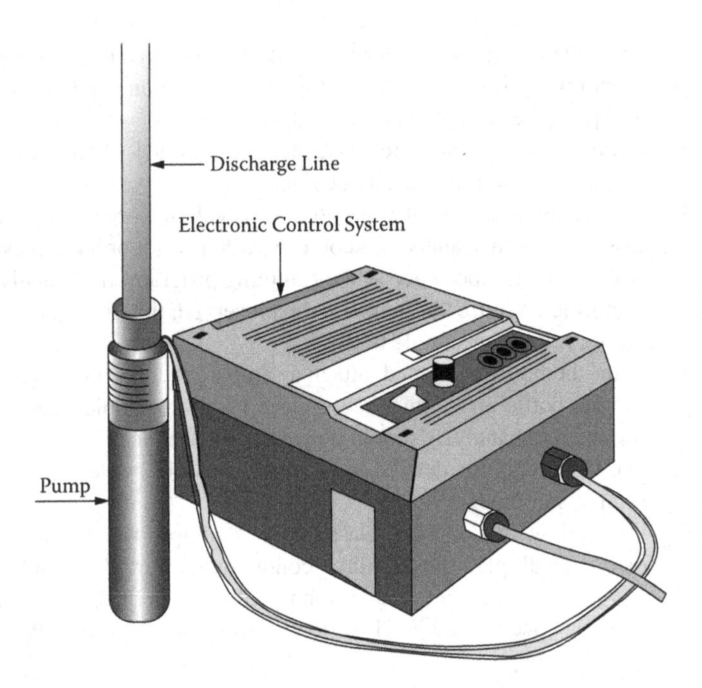

FIGURE 4.70 Submersible pump used to collect groundwater samples.

(From Byrnes, 1994. *Field Sampling Methods for Remedial Investigations.* Lewis Publishers, Boca Raton, FL.)

For most sampling programs, four people are sufficient for the purging and sampling procedure. Two are needed for field testing, sample collection, labeling, and documentation; a third is needed for health and safety; and a fourth is needed for miscellaneous tasks such as managing wastewater drums, and equipment decontamination.

The following equipment and procedure can be used to collect a groundwater sample for chemical and/or radiological analysis:

Equipment

1. Submersible pump
2. Discharge line
3. Generator or electrical power source
4. Electronic flow rate controller
5. Water level probe
6. Sample bottles
7. Sample preservatives
8. pH, temperature, conductivity, turbidity, and dissolved oxygen meters
9. Wide-mouth glass jar
10. Sample labels
11. Cooler packed with Blue Ice® (Blue Ice® is not required for radiological analysis.)

12. Trip blank (only required for volatile organic analyses)
13. Coolant blank (not required for radiological analysis)
14. Sample logbook
15. Chain-of-custody forms
16. Chain-of-custody seals
17. Permanent ink marker
18. Health and safety instruments
19. Chemical and/or radiological field screening instruments
20. Health and safety clothing
21. Sampling table
22. Waste container (e.g., 55-gal drum)
23. Plastic waste bags

Sampling Procedure

1. In preparation for sampling, confirm that all necessary preparatory work has been completed, including obtaining property access agreements, meeting health and safety and equipment decontamination requirements, and checking the calibration of all health and safety and chemical and/or radiological field screening instruments.
2. Before sampling, a groundwater well must be properly developed and purged (see Sections 4.2.7.3 and 4.2.7.4). If a well has been previously developed, there is no need to repeat this procedure unless accumulated sediment is blocking the well screen, or the well was inadequately developed the first time. A well must be purged each time before it is sampled.
3. When well purging is complete, collect a final water level measurement (Section 4.2.7.4.1), and record this information on the Well Purging and Sampling Form (see Section 5.2.3).
4. Reduce the pumping rate to <1 gpm so that a slow but steady flow of water flows from the discharge line.
5. Begin sampling by filling bottles for volatile organic analysis first. Bottles for other analyses should be filled in an order consistent with their relative importance to the sampling program. If the analyses to be performed require the sample to be preserved, this should be performed before filling the sample bottle.
6. After the bottle is capped, attach a sample label and custody seal to the sample bottle and immediately place it into a sample cooler. Samples for chemical analysis should be packed in Blue Ice®.
7. See Chapter 5 for details on preparing sample bottles and coolers for sample shipment.
8. Use a flow meter or fill a glass jar with sample water and collect and record a final pH, temperature, conductivity, turbidity, and dissolved oxygen measurement, in addition to collecting a final water level measurement (Section 4.2.7.4.1). Record this information on a Well Purging and Sampling Form (Section 5.2.3).

9. Replace the well cap, lock the well, and containerize any waste in a waste container. Before leaving the site, all waste containers should be sealed, labeled, and handled appropriately (see Chapter 11).
10. Transfer any other sampling-related wastes (e.g., gloves and foil) into a plastic waste bag.

4.2.7.5.4 Other Groundwater Sampling Methods

This section addresses more complex groundwater sampling systems that are designed to either collect depth-discrete groundwater samples in an open borehole or from a monitoring well that has multiple screened intervals. Two of the better known systems are the FLUTe Sampling System and the WestBay Multiple Port Sampling System. These two systems are discussed in the following sections. It is important to note that before a multi-screened groundwater monitoring well can be installed to support the WestBay Multiple Port Sampling System, regulatory agencies must be consulted and a waiver will likely be required to be signed before proceeding.

4.2.7.5.4.1 Flexible Liner Underground Technologies (FLUTe™) Sampling System FLUTe™ offers many methods that can be used to support groundwater investigations. This method uses a flexible tubular liner to seal the walls of an open borehole to keep it open and to help maintain the original aquifer hydraulic conditions. This method may be used in boreholes drilled through either unconsolidated sediment or bedrock.

The tubular FLUTe™ liner is constructed of a nylon fabric with a urethane coating. FLUTe™ liners are typically installed immediately after a borehole is completed and developed. Once a liner has been installed in an open borehole, it is filled with water to a level above the water table (e.g., 10 ft). The difference between the elevation of the water level in the liner and the elevation of the water table is called the "excess head" and is what forces the liner against the walls of the borehole and forms a seal. The WEB page www.flut.com provides a video and additional details on how the liner is installed and retrieved from a borehole.

Special covers can be placed over the outside of the FLUTe™ liner. One of those is the nonaqueous phase liquid (NAPL) cover (Figure 4.71). This is a thin hydrophobic cover with dye stripes on the outside of the material (similar to the Ribbon NAPL Sampler discussed in Section 4.2.3.1). When installed in a borehole, if the cover contacts a NAPL, the NAPL will be wicked into the hydrophobic cover. If that liquid is a solvent, the NAPL will dissolve the dye stripes on the exterior of the cover and carry the dye into the hydrophobic material. That dye transport produces a strong stain of the white interior surface of the cover material. The NAPL cover does not usually need more than an hour in the borehole to react. Since the FLUTe™ liner is removed from the borehole by inversion, the NAPL cover does not contact any other portion of the borehole when it is retrieved. The cover material is removed from the liner after it has been retrieved from the borehole to observe any stains caused by contact with a NAPL. A tape measure can then be used to define the depth at which NAPL is found within the formation.

Since the NAPL cover does not respond to the dissolved phase of the NAPLs, an activated carbon felt strip is available for the interior surface of the NAPL FLUTe

FIGURE 4.71 Nonaqueous phase liquid cover for FLUTe™ liner.

cover. This activated felt system is called FLUTe Activated Carbon Technique. A diffusion barrier between the carbon felt and the liner isolates the carbon felt so that it can only wick contaminants from the borehole wall pore space or fractures. The activated carbon felt wicks the contaminants by diffusion and therefore the carbon must typically be left in place for one to two weeks in order to gain a reasonable replica of the dissolved contaminant distribution in the borehole wall pore space. The activated carbon will contain a replica of the relative contaminant distribution in the borehole wall. Upon retrieval the activated carbon felt strip is sectioned for analysis. The volatile organics are typically extracted from the activated carbon using methanol. The methanol is then analyzed using a gas chromatograph mass spectrometer to identify the volatile organic compounds that are present and the relative amounts that are present in the activated carbon. The equally divided carbon segments provide a continuous distribution of dissolved contaminants in both the formation pore space and in the fractures.

A transparent FLUTe™ liner may also be used which allows an optical tele-viewer to see the borehole wall and to detect color changes behind the liner. This transparent liner is particularly useful in viewing fractures in bedrock using an acoustic tele-viewer. The FLUTe liner can be used to map the location and flow rates of fractures intersecting the borehole, and collect aquifer hydraulic property data such as transmissivity and hydraulic head. See www.flut.com for additional detail and references for how this data is measured or calculated.

As shown in Figure 4.72, one or more sample ports can be installed through the flexible tubular liner to support the collection of groundwater samples. When collecting a groundwater sample, water first passes from the aquifer through the sample port into the "port to pump tube." The water in the "port to pump tube" flows past the first check valve and into the U-shaped sampling device. Two tubes connected to

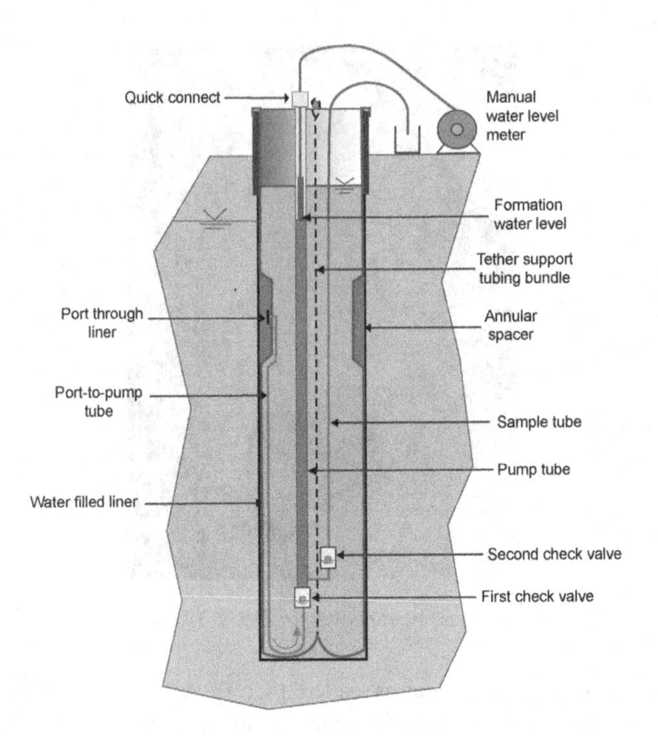

FIGURE 4.72 FLUTe™ groundwater sampling system.

the U-tube extend to the ground surface. Pressure is applied to the larger diameter "pump tube" using compressed nitrogen gas which forces the water in the U-tube to the ground surface through the smaller diameter "sample tube." Before filling sample bottles, the system is first purged multiple times from each port in order to draw samples a significant distance from the borehole. Sample bottles can then be filled from the water exiting the "sample tube" during a subsequent lower pressurization of the pump tube. With use of a manifold all sample ports can be purged simultaneously. The pressure history at each port can be monitored with dedicated downhole transducers or movable air coupled transducers at the surface. The head at each port is also measured with in electric water level meter in the pump tube. The Water FLUTe can be removed from the borehole for decommissioning without drilling.

For additional details on this sampling method visit www.flut.com.

4.2.7.5.4.2 WestBay Multiple Port Sampling System The WestBay Multiple Port Sampling System® is a no-purge/low-flow sampling system that is capable of sampling and monitoring a number of depth-discrete intervals within the same well. The WestBay Multiple Port Sampling System® is a modular PVC pipe system that is inserted into a well casing. Some of the modular components incorporate port openings, which are positioned adjacent to the well screen, and allow for groundwater samples to be collected along with pressure measurements. Other components of the system incorporate water-inflated packers that are located between screened sections of the well. Each screened interval may be equipped with as many as three monitoring ports and one pumping port. The monitoring ports provide for low-flow groundwater sampling and for attachment of a transducer for long-term water level monitoring. The pumping ports are designed for higher-flow groundwater sampling. The Westbay design allows for monitoring multiple screen sections within a well.

To install a WestBay Multiple Port Sampling System®, a large diameter borehole is first drilled using a drilling method that minimizes impact to the chemistry and hydraulic properties of the aquifer (e.g., avoid mud rotary drilling). Conventional blank well casing and multiple screen sections are then lowered down the borehole to screen those sections of the aquifer of interest. One should consider collecting depth-discrete groundwater samples during well drilling, followed by quick turnaround laboratory analysis, to support the selection of the optimum intervals within the aquifer to screen (see Section 4.2.7.2). Also the distance between screen sections must be enough to allow placement of Westbay packers to isolate each screen zone. Wells are then completed by installing a sand filter pack outside the multiple screened sections of the casing, installing bentonite seals outside the blank casing sections separating the screened intervals, and then grouting the remainder of the blank casing to the ground surface. As a caution, before proceeding with this type of multi-screen well completion, check with applicable regulatory agencies since a special variance will likely be required. Each section of the well screen must be thoroughly developed (see Section 4.2.7.3), followed by the installation of the WestBay Multiple Port Sampling System®.

After the WestBay Multiple Port Sampling System® has been installed in the well, quality control tests should be performed to confirm the hydraulic integrity of the Westbay Multiple Port casing and packer seals, and confirm the proper operation of the port valves and monitoring probes. Deionized (DI) water is placed inside the Westbay casing to reduce buoyancy of the casing in the well and to partially equilibrate formation pressures at deeper ports. Solar panels and 12-volt batteries are often installed to provide power for the data logger and transducer systems.

Groundwater can be sampled using a monitoring port by lowering a sampling device into a well that is attached to as many as four 1-L containers. The containers are attached to a vacuum pump before each sampling run, and the pressure in each container is lowered to approximately 2 psi. The empty containers are then lowered to the monitoring port at the screen interval to be sampled. When attached to the port, the sample containers draw water from the well into the containers. The sample containers fill at a rate of approximately 4 L/h (1 gal/h) (LANL 2007). The pumping ports will fill sample containers at a higher rate and are operated using compressed gas to open or close the pumping port sliding valve. When this slide valve is opened, it allows a much larger opening to the screen than do the monitoring ports. Opening a pumping port allows formation water to enter the Westbay casing where a submersible pump is used to pump the water to the surface.

The following Los Alamos National Laboratory (LANL) reference provides the results from a seven-year study using the WestBay Multiple Port Sampling System®.

- LANL, 2007, Evaluation of Sampling Systems for Multiple Completion Regional Aquifer Wells at Los Alamos National Laboratory, EP2007-0486, August.

This reference also compares the WestBay Multiple Port Sampling System® to several other sampling systems such as the FLUTe Sampling System (see Section 4.2.7.5.4.1). The FLUTe system was rated higher than the WestBay Multiple Port Sampling System® in several categories including cost and complexity of installation/removal.

A procedure for sampling groundwater using the WestBay Multiple Port Sampling System® was developed by LANL and is found in the following reference:

- LANL, 2008, "Groundwater Sampling Using the Westbay MP System®," Los Alamos National Laboratory Standard Operating Procedure SOP-5225, Los Alamos, New Mexico, October. https://eprr.lanl.gov/?page=2&q=ENV-WQH-SOP-050.3&search_field=all

4.2.7.5.5 Defining Vertical Contaminant Concentration Profile in Completed Wells

To optimize the recovery of contaminants from an aquifer, it is important to consider defining the vertical contaminant concentration profile. This profile is best defined by collecting depth-discrete groundwater samples during well drilling using the methods described in Section 4.2.7.2.1. If this profile sampling was not performed during drilling, it can be performed after well installation but is more challenging to obtain the vertical contaminant concentration profile. Some projects collect depth-discrete groundwater samples within the screened interval of completed wells using the low-flow groundwater sampling method described in Section 4.2.7.5.2. However, this approach may not provide representative results due to the mixing of water from multiple zones within the well screened interval. This vertical flow mixing within the well screened interval occurs as a result of hydraulic or density gradients. Detailed guidance on how to best define the vertical contaminant concentration profile in the screened section of a completed monitoring or extraction well is provided in the following reference and it entails performing depth-discrete flowmeter or spinner log surveying in combination with depth-discrete groundwater sampling:

- Sukop, M.C., 2000, *Estimation of Vertical Concentration Profiles from Existing Wells*, Vol. 38, No. 6, Ground Water, November – December.

Sukop (2000) notes that for extraction wells, a spinner log survey should be performed at pumping rates close to those expected to be used under normal operating conditions. This reference goes on to clarify for monitoring wells, the spinner log survey should be performed at pumping rates low enough to avoid aquifer solids mobilization and excessive drawdown, yet high enough to provide a quality spinner log. Depth-discrete groundwater samples should be collected while the well is pumped at the same rate used for spinner logging. Depth-discrete groundwater samples should be collected at depths corresponding to breaks in the slope of the spinner log. If the spinner log slope is constant, the depth-discrete groundwater samples should be collected at regular depth intervals (Sukop 2000).

When collecting depth-discrete groundwater samples from monitoring wells under low flow conditions, the installation and removal of a sampling device may cause mixing that temporarily distorts the true vertical distribution of contaminants within the well. For that reason, it is recommended that at least one well volume of water be purged after the sampler has reached the intended depth before sample collection. Sukop (2000) clarifies that the collected samples do not provide a direct measure of aquifer contaminant concentrations but by combining the flow- and concentration-weighted depth-discrete sample results with the flowmeter log it is possible to estimate the contaminant concentrations at the sampling intervals using a water/mass-balance approach.

4.2.7.6 Tracer Testing

This section provides the reader with a general overview of how tracer testing can be used to support groundwater investigations. Tracer testing can provide valuable information about groundwater flow direction, groundwater flow velocity, heterogeneity in an aquifer, and physical characteristics of an aquifer such as porosity and hydraulic conductivity. This information is often used to help refine site conceptual site models and numerical models used to describe subsurface flow and transport conditions. This section focuses on tracer tests that are performed between groundwater wells and is focused on natural gradient tests. Forced-gradient tracer testing is also commonly performed but it is outside the scope of this chapter. A few references that may help the reader whether a forced-gradient tracer test needs to be performed include:

- Marco Bianchi, Chunmiao Zheng, Geoffrey R. Tick, Steven M. Gorelick, 2011, *Investigation of Small-Scale Preferential Flow with a Forced-Gradient Tracer Test*, Ground Water, Vol. 49, No. 4, July–August.
- Einat Magal, Noam Weisbrod, Alexander Yakirevich, Daniel Kurtzman, and Yoseph Yechiele, 2010, *Line-Source Multi-Tracer Test for Assessing High Groundwater Velocity*, Ground Water, Vol. 48, No. 6, November–December.
- William E. Sanford, Peter G. Cook, Neville I. Robinson, and Douglas Weatherill, 2006, *Tracer Mass Recovery in Fractured Aquifers Estimated from Multiple Well Tests*, Ground Water, Vol. 44, No. 4, July–August.

A tracer test typically involves the injection of a tracer solution into a groundwater well and then tracking the natural migration of the tracer solution in downgradient observation wells. Ideal tracer testing wells would be in areas requiring remedial action and would include wells spaced from tens to a few hundred feet apart (depending on anticipated groundwater flow velocity and arrival times). The selected wells should be easy to access with plenty of room for tracer testing equipment and personnel, and they should have easy access to water for preparing the tracer solutions. One should expect that a tracer test will take a few days to a few months to implement. Tracer arrival measurements need to continue until a clearly defined peak concentration is identified at each downgradient tracer observation well. However, it should be noted that it may be the center of mass of the tracer that is important to capture rather than the peak concentrations. Since tracer test curves typically show a steep rising limb followed by a longer tail limb, the center of mass might not be at the peak concentration.

Many types of tracers have been used over the years, a few of which include:

- Fluorescent dyes (e.g., fluorescein, rhodamine WT, rhodamine B)
- Inorganic ions (e.g., bromide)
- Heat
- Strong electrolytes (e.g., sodium chloride, potassium chloride)
- Short half-life radionuclides (e.g., tritium)

Fluorescent dyes, bromide, and heat have been used as tracers at the U.S. Department of Energy Hanford Site and their specific applications are the focus of this section. These are all "conservative tracers," which means that they are nonreactive and stay entirely in the phase in which they are injected. "Partitioning tracers," "adsorption tracers," and "reactive tracers" are sometimes needed to meet certain study objectives. Partitioning tracers are compounds that have some affinity for multiple phases (e.g., aqueous and non-aqueous phase liquid) and therefore partition between two or more phases during the test. Adsorption tracers are those tracers that have certain adsorption properties, while reactive tracers are those that undergo a chemical reaction during the test. While the focus of this section is conservative tracers, refer to the following Idaho National Engineering and Environmental Laboratory (INEEL) document for guidance on the use of partitioning, adsorption, and reactive tracers.

- INEEL, 2004, Tracers and Tracer Testing: Design, Implementation, and Interpretation Methods, INEEL/EXT-03-01466, January.

When using a fluorescent dye to support a tracer test, the fluorescence intensity is affected into varying degrees by certain physical and chemical factors, such as concentration, temperature, pH, photochemical decay, and fluorescence quenching (USGS 1986). USGS (1986) notes that while there are a great number of dyes that are commercially available, only a few exhibit the combination of properties essential for water tracing. The two dyes recommended by USGS (1986) are rhodamine WT and pontacyl pink (also known as intracid rhodamine B, pontacyl brilliant pink B, and acid red 52). These dyes are generally good tracers because they are:

- Water soluble
- Strongly fluorescent
- Fluorescent in a part of the spectrum not common to materials found in groundwater
- Harmless in low concentrations
- Inexpensive
- Reasonably stable in a normal water environment

A few additional advantages of using fluorescent dyes as conservative tracers include the fact that they are easy to measure in the field, have been used extensively as tracers in the past, are detectable at very low concentrations, are generally not reactive with soil particles, and are readily available.

Bromide has been used successfully as a conservative tracer at the Hanford Site. Some of the advantages of using bromide include the following: it rarely sorbs to soil particles; the toxicity of bromide is low; it is biologically stable; and there are numerous analytical methods that can be used to test for bromide. Heat has also been used successfully at Hanford as a conservative tracer. Two advantages of using heat as a tracer are that it is simple and cost-effective.

At the Integrated Field Research Challenge (IFRC) site in the Hanford Site 300 Area, a tracer test was conducted using both bromide and heat tracers. The bromide tracer data from this test was needed to help improve the calibration of a groundwater flow model which was challenged by the highly dynamic nature of the flow field. This study concluded that bromide tracer data provided valuable constraints for calibration of hydraulic conductivity and boundary conditions under highly dynamic flow conditions. However, the bromide concentrations collected were obtained from fully screened observation wells, so there was no depth-discrete resolution to use for vertical characterization. For this reason, to supplement the bromide tracer study a heat tracer test was implemented by installing special electrical-resistivity tomography electrodes in 28 wells in addition to thermistors and cables in regular intervals. The thermistors and cables allowed for the continuous recording of groundwater temperature at different depth intervals. Heat tracer testing was selected because it was relatively simple and inexpensive to implement. The study concluded that "*heat can be used as a cost-effective proxy for solute tracers for calibration of the hydraulic conductivity distribution, especially in the vertical direction*" (Water Resources Research 2012, emphasis added). Full details on this tracer test are provided in Water Resources Research (2012).

It is important to note that while the concentration of bromide in natural waters is typically very low this is not always the case. The Hanford Site Solid Waste Landfill/Non-Radioactive Dangerous Waste Landfill (SWL/NRDWL) study was *not* able to use bromide as a tracer because background concentrations of bromide in the aquifer were high (DOE 2017).

Strong electrolytes, such as sodium chloride or potassium chloride, can be used as conservative tracers. In this case, a known amount and concentration of a selected electrolyte is injected into a tracer injection well. Electrical conductivity or water chemistry measurements are then collected at frequent intervals in a downgradient tracer observation well. When considering using electrolytes as a tracer it is important to determine if it will be prone to biological uptake or reactions with water or sediment that would cause it to drop out of the dissolved state.

Tritium (hydrogen-3) is a radioactive isotope of hydrogen. Tritium is an effective tracer because it rarely occurs naturally in the environment. Tritium is a normal byproduct from the operation of nuclear reactors. Other characteristics that make tritium an effective tracer include the following: it does not interact with the aquifer materials as it moves downgradient from the injection well; it can be measured at low levels; and it has a short half-life of just over 12 years. Since regulatory agencies can be sensitive to the use of tritium, this tracer is often used at nuclear sites where longer half-life radionuclides are already present within the aquifer.

4.2.7.6.1 Steps for Developing a Successful Tracer Test

Four chronological steps should be performed to ensure that a tracer test is appropriate for a specific project. This is a modified version of guidance provided by the Idaho National Engineering and Environmental Laboratory (INEEL). The following sections provide the purpose of each of the four steps and a short summary of the activities that need to be performed. For more detailed guidance on this topic, refer to the following reference:

- Idaho National Engineering and Environmental Laboratory (INEEL), 2004, Tracers and Tracer Testing: Design, Implementation, and Interpretation Methods, INEEL/EXT-03-01466, January

4.2.7.6.1.1 Step 1: Define tracer test objective The purpose of this step is to clearly define all of the primary objectives of the tracer study, the properties of the aquifer that need to be estimated, and the scale of the study. Based on these, one needs to determine if a tracer test is able to provide the data that is needed, at that scale, and if a tracer test is the best method to be used to obtain this data. If the answer is "yes," then proceed to the following Step 2 to select the most appropriate tracer to use for the test. If the answer is "no," then stop and evaluate other options for obtaining this data.

4.2.7.6.1.2 Step 2: Implementation Strategy and Tracer Selection and Testing The purpose of this step is to develop an implementation strategy for the tracer test. This step will define the number and type(s) of tracer(s) to be used in the study; the mass of tracer required to run the test; the test duration; how and where the tracer will be introduced into the aquifer; details of the sampling locations; and the sampling frequency and analyses to be performed. These details need to be documented in a field test plan that includes maps showing tracer injection wells and tracer observation wells locations, as well as tables identifying the analytical performance requirements.

It is important to carefully select the most appropriate tracer(s) for a study that will meet all of the specific study objectives. The selected tracer(s) should be:

- Nontoxic
- Available and affordable
- Soluble for injection
- Capable of being measured at low detection limits
- Negligible effect on transport properties (density, viscosity, pH, etc.)
- Stable
- Low natural background concentrations

The INEEL (2004) reference cited earlier in Section 4.2.7.6 provides details on a variety of tracers to consider along with the advantages of some tracers over others. The following references provide additional information that can assist with the selection of an optimum tracer for a study:

- S. Gerenday, J.F. Clark, Department of Earth of Science; J. Hansen, I. Fischer, J. Koreny, HDR Engineering, Inc., 2020, "*Sulfur Hexafluoride and Potassium Bromide as Groundwater Tracers for Managed Aquifer Recharge*." Available at: https://ngwa.onlinelibrary.wiley.com/doi/abs/10.1111/gwat.12983
- DOE 2017, Central Plateau Groundwater Tracer Study Work Plan, SGW-60511, Rev. 0, Richland, WA, August.

- Rubin, Y., and S.S. Hubbard (2005), Hydrogeophysics, Water Sci. Technol. Libr., vol. 50, 523 pp., Springer, New York.
- EPA, 1987, Handbook – Groundwater, Chapter 7, EPA/625/6-87/016, March.
- USGS 1986, Techniques of Water-Resources Investigations of the United States Geological Survey, Chapter A12, Fluorometric Procedures for Dye Tracing, Book 3.
- EPA 1985, An Introduction to Groundwater Tracers, PB86-100591, National Technical Information Service, Tucson, Arizona, March.

4.2.7.6.1.3 Step 3: Field implementation This step is the implementation of the field test plan. If new groundwater wells are required to be installed to support the tracer test, it is important that the drilling method selected does not negatively impact the data to be collected. Prior to field implementation one must first identify:

- How the tracer is going to be prepared (in the laboratory or in the field)
- How the tracer will be handled in the field and how it will be emplaced in the well
- What steps will be taken to ensure that the tracer is the same concentration over the entire screen interval

The following are chronological steps for field implementation:

- Drilling, construction, and development of wells supporting tracer test
- Injection of the tracer
- Collecting and analyzing groundwater samples during tracer test
- Field documentation

4.2.7.6.1.4 Step 4: Test interpretation This step involves the interpretation of tracer test results. A basic plot of the concentration of a tracer as a function of time is called a breakthrough curve (Figure 4.73). If the tracer observation well is directly downgradient of the tracer injection well, then the first arrival of a conservative tracer in the tracer observation well provides the maximum velocity of groundwater flow between those two points. The arrival time of the center of mass of the tracer provides an approximation of the average velocity of a molecule of groundwater traveling between the two points. The interpretation of tracer test results is beyond the scope of this book. The following references provide detailed guidance on tracer data interpretation:

- Rubin, Y., and S.S. Hubbard (2005), Hydrogeophysics, Water Sci. Technol. Libr., vol. 50, 523 pp., Springer, New York
- Idaho National Engineering and Environmental Laboratory (INEEL), 2004, Tracers and Tracer Testing: Design, Implementation, and Interpretation Methods, INEEL/EXT-03-01466, January
- EPA, 1987, Handbook – Groundwater, Chapter 7, EPA/625/6-87/016, March

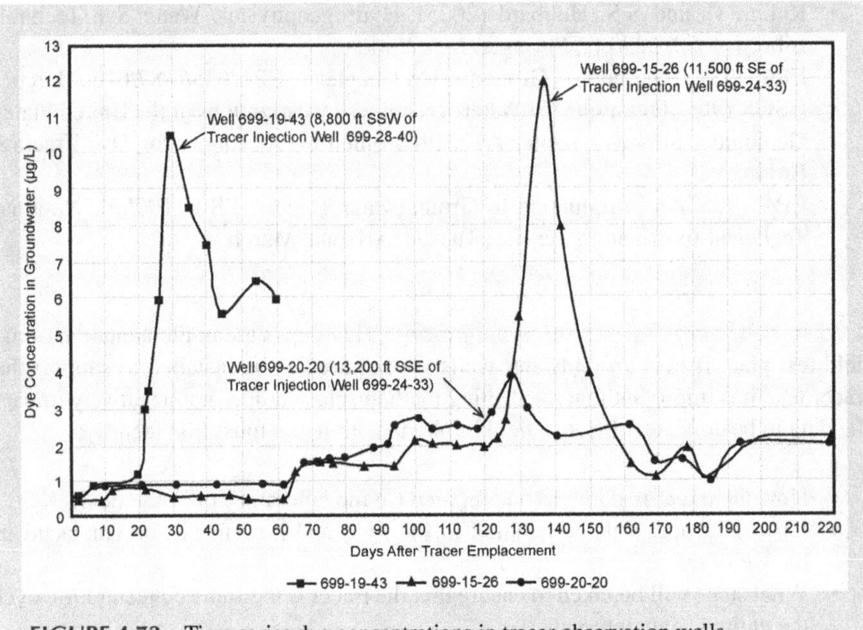

FIGURE 4.73 Time series dye concentrations in tracer observation wells.

4.2.7.6.2 Tracer Test Case Study

As shown in Figure 4.74, the Hanford Site Study Area 1 is located west of the SWL/NRDWL. The tracer study performed at this location was designed to obtain the groundwater flow velocity data needed to confirm the suspected presence of a high-conductivity paleochannel. This suspected paleochannel was providing unique challenges to the groundwater flow and transport modeling in the area. This study utilized one upgradient tracer injection well (699-28-40, formally known as 699-28-41) and one tracer observation well (699-19-43). Fluorescent dye (e.g., 100 lbs of sodium fluorescein in 100 gal of water) was injected into the injection well followed by the collection of groundwater samples in the downgradient tracer observation well.

The following year a second tracer study was performed at Hanford Site Study Area 2 in a similar manner southeast of the SWL/NRDWL. As shown in Figure 4.75, this second study utilized one tracer injection well (699-24-33) and two tracer observation wells (699-15-26 and 699-20-20). Fluorescent dye was injected into the injection well followed by the collection of groundwater samples in the two observation wells.

Groundwater samples collected during these two studies were analyzed using a fluorometer to track the arrival of the tracer. The graph in Figure 4.73 shows the length of time it took to see the first arrival and peak concentration of the fluorescent dye tracer in the three tracer observation wells that were used in the two studies. Table 4.9 summarizes the results from these tests.

The maximum observed groundwater velocities in these studies were based on the timing of the first tracer arrival at the observation wells, which ranged from 52 to 134 m/day (170 to 440 ft/day). The observed groundwater velocity for the peak

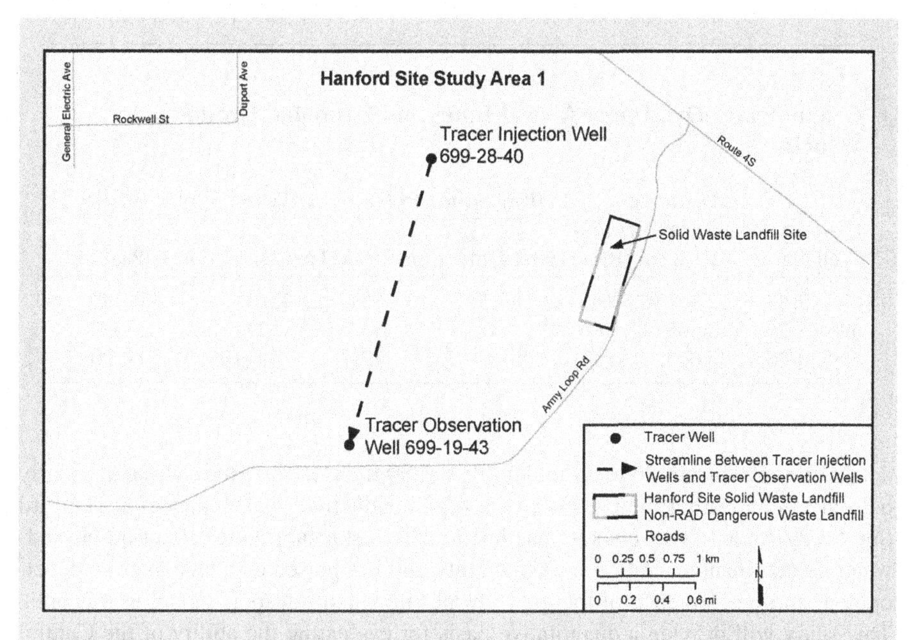

FIGURE 4.74 Hanford Site Study Area 1.

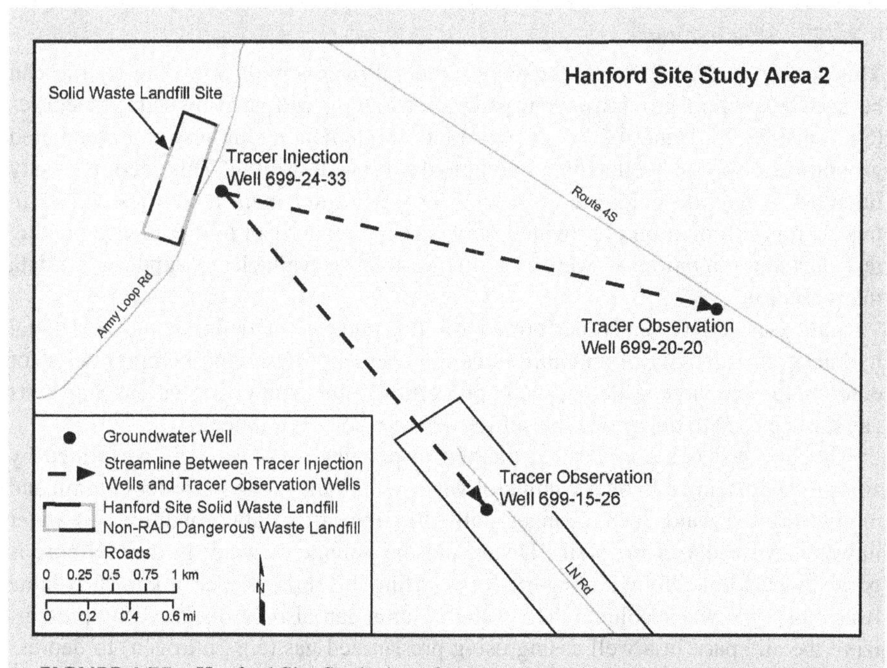

FIGURE 4.75 Hanford Site Study Area 2.

TABLE 4.9
Groundwater Dye Tracer Arrival Times and Estimated Groundwater Velocity

Observation Well Name	Distance from Injection Well (m [ft])	Travel Time in Days		Velocity (m/d [ft/d])	
		First Detection	Peak	First Detection	Peak
699-19-43	2,862 (8,800)	20	31	134 (440)	85 (280)
699-15-26	3,505 (11,500)	67	134	52 (170)	26 (85)
699-20-20	4,023 (13,200)	67	129	59 (195)	30 (100)

detection ranged from 26 to 85 m/day (85 to 280 ft/day). This study was particularly helpful in confirming the presence of a high-conductivity paleochannel at Hanford Site Study Area 1. This channel has historically had a substantial effect on groundwater contaminant migration in this vicinity and has presented a challenge to development and performance of the groundwater flow and transport model in this area. This study will provide a quantitative basis for evaluating the ability of the Central Plateau Groundwater Model to predict groundwater flow and contaminant migration in this portion of the Hanford Site (DOE 2017). Full study details are provided in DOE (2017).

4.2.7.7 Slug Testing

This section provides the reader with a general overview of how slug testing can be used to support groundwater investigations along with step-by-step procedures (Sections 4.2.7.7.1 and 4.2.7.7.2) on how to implement a slug test in a completed groundwater well as well as in a borehole during well drilling. This section is only intended to provide guidance on how to properly implement a slug test. Later in this section information is provided on where the reader can find guidance on slug test data interpretation as well as software that is available to support the data interpretation.

Slug tests are typically performed for the purpose of defining the horizontal hydraulic conductivity of an aquifer and are useful for designing pumping tests for determining the large-scale hydraulic properties of the aquifer formation. Slug tests can also be used to determine the sufficiency of well development.

The slug test is a simple field method in practice. A slug test is performed by abruptly displacing a known volume of water within the wellbore water column and monitoring the water-level changes until they return to static conditions as water moves into (or out of the well). Displacing the volume of water in the wellbore is typically performed by withdrawing (or inserting) a solid cylinder of known volume from (into) the water column. The water column can also be displaced by pressurizing the air space in a well casing using pressurized gas (e.g., nitrogen) to depress the water column to a new equilibrium level and then instantaneously releasing the pressure. Following pressure release, changes in water level are monitored until they return to static conditions. During pressurization, both the water level pressure and the pressurized gas are monitored when depressing the water column. The advantage of pneumatic tests is much greater control and variation in the imposed stress level.

Since the hydraulic properties determined by a slug test are only representative of the formation in the vicinity of the well, a series of slug tests should be considered in closely spaced wells whenever possible. Conducting slug tests in a large number of closely spaced wells screened in the same portion of the aquifer can provide valuable information on the spatial distribution of the hydraulic conductivity. Since hydraulic responses to slug tests can be affected by many factors, the results should always be compared to known values for similar geologic media to determine if they are reasonable.

Slug tests should be performed in properly designed and thoroughly developed groundwater wells. Without thorough well development, fine-grained soils may remain smeared on the outside walls of a borehole or fine-grained material may remain in the formation near the borehole as a result of the drilling process. If this material is not removed, the slug testing data will represent an artificially low hydraulic conductivity. Slug tests may also be performed within a borehole during well drilling. However, designing and implementing the correct test configuration and then properly developing the test interval is more challenging during well drilling.

A pressure transducer and telemetry device (see Section 4.2.7.4.1.2) should be used to collect water level measurements during these tests. The importance in using a pressure transducer is to measure water levels at a high measurement rate to capture the rapidly changing water levels. This is very important for determining when time zero occurs (i.e., the start of the test) and for determining the volume displaced at the beginning of the test, particularly when testing in high permeability zones. For deeper wells, a drill rig or drilling support vehicle is needed to assist the lowering and raising of the slug rod and/or other support equipment (e.g., tubing string, short temporary screen, packer, and purge pump) down the well. For shallower wells the slug rod can be raised/lowered manually by hand.

For guidance on the design, performance, and interpretation of slug test data, refer to:

- Butler, J.J., Jr., 2020. *The Design, Performance, and Analysis of Slug Tests (2nd ed.)*, Lewis Publishers, New York, 266p. [errata: sheet (pdf), spreadsheet (xlsx)]
- S.S. Papadopulos & Associates, Inc., 2021, Fundamentals of the Execution and Interpretation of Slug Tests, Notes Prepared for the University of Waterloo Hydrogeology Field School, Waterloo, Ontario
- Butler, J.J., Jr., and E.J. Garnett, 2000, Simple procedures for analysis of slug tests in formations of high hydraulic conductivity using spreadsheet and scientific graphics software, Kansas Geological Survey Open-File Rept. 2000-40, Lawrence, Kansas

Below are a few software programs and/or spreadsheets that are available to support the analysis of slug test results:

- *AQTESOLV™ Version 4.5*
- Kansas Geological Survey (KGS) Model
- USGS AIRSLUG Version 1.1
- USGS AQTESTSS Version 1.2
- Geoprobe® Slug Test Analysis Software – Version 2.0

4.2.7.7.1 Slug Testing Procedure for Completed Groundwater Well

The following is a slug test procedure for a completed and thoroughly developed groundwater well (see Section 4.2.7.3). The below procedure assumes that a slug rod is used to support the slug test and is modified after the procedure presented in the following EPA standard operating procedure:

- EPA, 1994, Slug Tests, SOP 2046, Rev. 0, October

Running a slug test using pressurized air is more complex and is not addressed in this book. A procedure for implementing a pneumatic slug test can be found in the following references:

- Geoprobe®, 2014, Geoprobe® Pneumatic Slug Test Kit (GW1600), Installation and Operation Instructions for USB System, Instructional Bulletin No. MK3195, February.
- U.S. Geological Survey, 1995, Methods of Conducting Air-Pressurized Slug Tests and Computation of Type Curves for Estimating Transmissivity and Storativity, Open-File Report 95-424, Reston, Virginia.

Equipment

1. Drill rig or drilling support vehicle (with cable for raising/lowering the slug rod)
2. Slug rod (solid cylinder of known volume)
3. Pressure transducer and telemetry device (with user manual)
4. Project-specific sampling and analysis plan (see Section 3.2.7)
5. Water level detector
6. Tape measure (in tenths of a foot scale) with stainless steel weight attached
7. Zip-Ties
8. Plastic sheeting ground cover (optional)
9. Field logbook
10. Permanent ink marker
11. Health and safety instruments
12. Health and safety clothing (see Chapter 10)
13. Waste container (e.g., 55-gal drum)
14. Sampling table (optional)
15. Plastic waste bags

Procedure

1. In preparation for slug testing, confirm that all necessary preparatory work has been completed, including obtaining property access agreements, meeting health and safety and equipment decontamination requirements, and checking the calibration of all health and safety instruments.

2. Unlock the well head and remove the well cap from the well casing.
3. Use health and safety instruments (e.g., photoionization detector, flame-ionization detector) to screen air quality at well head. Based on the results, adjust personal protective equipment if needed in accordance with project health and safety plan (see Chapter 10).
4. Measure depth to groundwater (±0.01 ft) from top of casing using water level detector (see Section 4.2.7.4.1) and record the results in a field logbook.
5. Measure depth to bottom of well (±0.1 ft) from top of casing (see Section 4.2.7.4.2) using tape measure with stainless steel weight attached and record the results in a field logbook.
6. Connect the slug rod to a drill rig (or drilling support vehicle) cable and lower the slug rod down the well casing to just above the water table. Mark the cable at the top of the well casing with a Zip-Tie (or other marking methods).
7. Lower the slug rod to the submerged depth and mark the cable at the top of the well casing with a Zip-Tie (or other marking method). Remove the slug rod from the well.
8. Using the procedure in Section 4.2.7.4.1.2, install pressure transducer and telemetry device. Begin collecting static water level measurements at ±0.01 ft.
   a. Refer to project-specific sampling and analysis plan (see Section 3.2.6) for frequency at which to collect water level measurements during the slug test. It is critical to collect as many measurements as possible in the early part of the slug test. Refer to the transducer user manual for guidance on adjusting measurement frequency.
   b. Set transducer 1–2 ft below the lowest depth that the slug rod will be dropped.
9. Record in a field logbook the volume of the slug rod being used in the slug test. This volume will be provided by the manufacturer of the slug rod. If the volume of the slug rod is not known, refer to the following reference which provides a volume for different size slug rods.
   a. Cunningham, W.L. and C.W. Schalk (comps.), 2011, *Groundwater Technical Procedures of the U.S. Geological Survey*, "GWPD 17 – Conducting an Instantaneous Change in Head (Slug) Test with a Mechanical Slug and Submersible Pressure Transducer," p. 145-151, U.S. Geological Survey, Reston, Virginia. Available at: https://pubs.usgs.gov/tm/1a1/pdf/GWPD17.pdf.
10. Lower the slug rod to just above the water table.
11. Perform falling head test by instantaneously lowering the full length of the slug rod below the static water level, then allow the displaced water column to stabilize as water level measurements continue to be collected by transducer and telemetry device. Continue collecting water level data 10 to 15 minutes after water level has stabilized. Record the date and time the slug test was initiated in a field logbook along with the date and time when water level stabilized.

12. Perform rising head test by instantaneously raising the full length of the slug rod so that it hangs a few feet above the static water level, then allow the displaced water column to stabilize as water level measurements continue to be collected by transducer and telemetry device. Record the date and time when the slug rod was raised in the field logbook along with the date and time when water level stabilized.
13. If a review of transducer data confirms the slug test data are acceptable, proceed to Step 14. If the slug test data are unacceptable repeat Steps 11 through 13.
    a. Note: It is not uncommon for the project-specific sampling and analysis plan (see Section 3.2.7) to require three (or more) falling head and/or rising head tests be performed in a well. Two tests with the same slug rod volume, and one other test with a different slug rod volume. This would require Steps 9 and 10 to be repeated prior to proceeding to Step 12. Repeated tests will help demonstrate reproducibility and assess adequacy of well development and skin effects.
14. Retrieve slug rod and pressure transducer from well. Collect final water level measurement using water level detector.
15. Replace the well cap, lock the well head, and containerize any waste in a waste container. Before leaving the site, all waste containers should be sealed, labeled, and handled appropriately (see Chapter 11).
16. Decontaminate equipment in accordance with procedures outlined in Chapter 9.

4.2.7.7.2 Slug Testing Procedure during Well Drilling

The following is a procedure for performing a slug test during well drilling. This procedure assumes that a slug rod is used to support the slug test and it assumes the drilling method is advancing temporary drill casing to keep the borehole open as when drilling with air rotary method. This method is modified after the procedure presented in the following EPA standard operating procedure:

- EPA, 1994, Slug Tests, SOP 2046, Rev. 0, October

Equipment

1. Drill rig (with cable for raising and equipment)
2. Tubing string (with diameter large enough for slug rod) with short (2 to 3 ft) temporary screen
3. Inflatable well packer
4. Slug rod (solid cylinder of known volume)
5. Pressure transducer and telemetry device (with user manual)
6. Development pump and surge block
7. Flow cell to monitoring field parameters (e.g., pH, temperature, conductivity, turbidity, dissolved oxygen)
8. Project-specific sampling and analysis plan (see Section 3.2.7)

9. Water level detector
10. Zip-Ties
11. Plastic sheeting ground cover (optional)
12. Field logbook
13. Permanent ink marker
14. Health and safety instruments
15. Health and safety clothing (see Chapter 10)
16. Waste container (e.g., 55-gal drum)
17. Plastic waste bags

Procedure

1. Prior to the start of borehole drilling, confirm that all necessary preparatory work has been completed, including obtaining property access agreements, meeting health and safety and equipment decontamination requirements, and checking the calibration of all health and safety instruments.
2. Have the driller advance the borehole to the targeted slug testing interval.
   a. Use health and safety instruments (e.g., photoionization detector, flame-ionization detector) to screen air quality during drilling. Based on the results, adjust personal protective equipment if needed in accordance with project health and safety plan (see Chapter 10).
3. Remove drill stem and bit from borehole.
4. Lower tubing string (with diameter large enough for slug rod) with short (2 to 3 ft) temporary screen and packer to bottom of borehole.
5. Pull back drill casing a few feet to expose the entire temporary screen to the formation.
6. Inflate packer to provide seal between tubing string and drill casing to isolate the slug testing interval.
7. Measure depth to groundwater (±0.01 ft) inside the tubing string using water level detector (see Section 4.2.7.4.1) and record the results in a field logbook.
8. Lower surge block down tubing string to the temporary screen section. Thoroughly surge the screened section to remove fines. Remove surge block and lower development pump. Pump until indicator field parameters (pH, temperature, specific conductance, turbidity, and dissolved oxygen) stabilize (see Section 4.2.7.3). Multiple rounds of surging may be required.
9. Remove well development pump from borehole.
10. Connect slug rod to drill rig cable and lower slug rod down the well casing to just above water table. Mark the cable at the top of the temporary drill casing with a Zip-Tie (or other marking method).
11. Lower the slug rod to the submerged depth and mark the cable at the top of the temporary drill casing with a Zip-Tie (or other marking method). Remove the slug rod from the well.

12. Lower a pressure transducer down the inside of the tubing string to a depth 1–2 ft below the lowest depth that the slug rod will be dropped, at least several feet below the expected depth of the slug rod at full immersion. Using the procedure in Section 4.2.7.4.1.2, install pressure transducer and telemetry device. Begin collecting static water level measurements at ±0.01 ft.
    a. Refer to project-specific sampling and analysis plan (see Section 3.2.7) for frequency at which to collect water level measurements during the slug test. It is critical to collect as many measurements as possible in the early part of the slug test. Refer to the transducer user manual for guidance on adjusting measurement frequency.
13. Record in a field logbook the volume of the slug rod being used in the slug test. This volume will be provided by the manufacturer of the slug rod. If the volume of the slug rod is not known, refer to the reference provided earlier in this section which provides a volume for different size slug rods.
14. Monitor baseline water level with the transducer and allow water levels to stabilize.
15. Lower the slug rod to just above the water table.
16. Perform falling head test by instantaneously lowering the full length of the slug rod below the static water level, then allow the displaced water column to stabilize as water level measurements continue to be collected by a transducer and telemetry device. Continue collecting water level data 10 to 15 minutes after water level has stabilized. Record the date and time the slug test was initiated in field logbook along with date and time when water level stabilized.
17. Perform rising head test by instantaneously raising the full length of the slug rod so that it hangs a few feet above the static water level, then allow the displaced water column to stabilize as water level measurements continue to be collected by transducer and telemetry device. Record the date and time the slug rod was raised in field logbook along with date and time when water level stabilized.
18. If a review of transducer data confirms the slug test data are acceptable, proceed to Step 19. If the slug test data are unacceptable, repeat Steps 16 through 18.
    a. Note: It is not uncommon for the project-specific sampling and analysis plan (see Section 3.2.6) to require multiple falling head and/or rising head tests be performed in a borehole using multiple slug rod volumes. This would require Steps 16 through 18 to be repeated prior to proceeding to Step 19.
19. Deflate packer. Retrieve packer, slug rod, and pressure transducer from borehole.
20. Remove tubing string and screen from borehole.
21. Continue drilling to the next slug testing interval in borehole and repeat Steps 3 through 20.

22. When the last slug test has been completed in the borehole, containerize any waste in a waste container. Before leaving the site, all waste containers should be sealed, labeled, and handled appropriately (see Chapter 11).
23. Decontaminate equipment in accordance with procedures outlined in Chapter 9.

4.2.8 Drum and Waste Container Sampling

In an effort to reduce the spread of contamination at a site, soil cuttings, decontamination water, well development and purging water, personal protective equipment, and other types of investigation-derived waste material should be containerized, labeled, and handled as potentially contaminated waste. The following subsections provide the reader with guidance for collecting soil, sludge, and water samples from waste drums and other containers to support waste disposal.

Waste containers of known or unknown origin must be sampled to determine their content before disposal. The only exception to this rule is when there is thoroughly documented process knowledge of the container contents. In this case, the container is designated on the basis of process knowledge, and appropriate references are cited. For those containers that require sampling, whatever process knowledge is available should be used to assist in identifying which analyses need to be run on the waste samples. The analyses to be run must include determining whether or not the waste is an RCRA characteristic waste by assessing its toxicity, corrosivity, ignitability, and reactivity (Section 2.1.2.2). One must also determine if any RCRA-listed waste codes apply to the waste, and if there is any reason to suspect the waste contains PCB, asbestos, or is radioactive. An understanding of all of these components is needed in order to properly designate and dispose of the waste to a properly licensed disposal facility.

Much greater health and safety concerns are related to the sampling of containers of unknown origin because the contents may be highly toxic, corrosive, reactive, explosive, or radioactive. To address health and safety concerns, it is common practice to pierce the lid of containers of unknown origin with a spike by means of an unmanned mechanical device. By piercing the drum, the volatile organics in the head space will be released, which in turn reduces the chances of an explosion. The piercing device and other tools used to open these drums should be made from a "nonsparking" material. Using robotic devices, supplied air, chemical safety suits, or other health and safety precautions is essential whenever sampling these types of containers.

4.2.8.1 Soil Sampling from Drums and Waste Containers

The methods recommended for collecting soil samples from drums or waste containers for chemical or radiological analysis are the hand auger and core barrel (slide hammer) methods discussed earlier in Sections 4.2.4.1.2 and 4.2.4.1.3. These methods are recommended over other soil sampling methods because they can be used effectively to collect soil samples throughout the depth of a drum or waste container. Refer to Table 4.1 for a summary of the advantages and limitations of these two sampling methods. When many drums or waste containers need to be characterized, one should consider sampling the drum showing the highest contamination based on field screening measurements, or using a statistical sampling approach. Use the DQO process (Chapter 3, Section 3.2) to assist in defining the optimum sampling design, and the required number and types of samples to collect.

The hand auger and core barrel (slide hammer) sampling methods for collecting samples from drums or waste containers are identical to those described in Sections 4.2.4.1.2 and 4.2.4.1.3 except that in Step 2 the plastic sheeting should be placed around the outside of the drum or container being sampled to help prevent the spread of contamination.

4.2.8.2 Sludge Sampling from Waste Drums and Waste Containers

The AMS sludge sampler presented below and the Coliwasa method presented in Section 4.2.8.3.1 are the preferred methods for collecting sludge samples from drums and waste containers. The AMS sludge sampler is the preferred sampler when the sludge is composed of a higher percentage of solid material as compared to liquid. On the other hand, the Coliwasa method is preferred when the sludge is composed of a higher percentage of liquid as compared to solid material.

4.2.8.2.1 AMS Sludge Sampler Method

The tool recommended for collecting sludge samples from drums or containers for chemical or radiological analysis is the AMS sludge sampler (Figure 4.76). The AMS sludge sampler is recommended over other sampling methods because the tip of the sampler incorporates a one-way valve that allows the sludge to enter the sampler, but prevents it from escaping by closing as the sampler is withdrawn from the material. The sampler is constructed so that it can accept sample liners. The sampler is typically attached to a shaft that has a T-bar handle. The sampler is advanced by placing a downward pressure on the T-bar handle while rotating. For more information on the AMS sludge sampler, see www.ams-samplers.com. Note that the Coliwasa method described in Section 4.2.8.3.1 may also be used to collect sludge samples from drums or other waste containers.

For most sampling programs, four people are sufficient for this sampling procedure. Two are needed for sample collection, lithology description, labeling, and documentation; a third is needed for health and safety; and a fourth is needed for miscellaneous tasks such as waste management and equipment decontamination.

FIGURE 4.76 AMS sludge sampler used to collect sludge samples from drums and waste containers.

The following equipment and procedure can be used to collect sludge samples for chemical and/or radiological analysis:

Equipment

1. AMS sludge sampler
2. Sample liners
3. Stainless steel bowl
4. Stainless steel spoon
5. Sample jars
6. Sample labels
7. Cooler packed with Blue Ice® (Blue Ice® is not required for radiological analysis.)
8. Trip blank (only required for volatile organic analyses)
9. Coolant blank (not required for radiological analysis)
10. Sample logbook
11. Chain-of-custody forms
12. Chain-of-custody seals
13. Permanent ink marker
14. Health and safety instruments
15. Chemical and/or radiological field screening instruments
16. Health and safety clothing
17. Waste container (e.g., 55-gal drum)
18. Sampling table
19. Plastic sheeting
20. Plastic waste bags

Sampling Procedure

1. In preparation for sampling, confirm that all necessary preparatory work has been completed, including obtaining property access agreements, meeting health and safety and equipment decontamination requirements, and checking the calibration of all health and safety and chemical and/or radiological field screening instruments.
2. Lay plastic sheeting flat on the ground surface around the vicinity of the drum (or waste container) to be sampled. The purpose of this sheeting is to help prevent the spread of contamination. If the drum is of unknown origin, consider piercing first to avoid potential for explosion as described in Section 4.2.8.
3. Begin collecting a sludge sample from the drum (or waste container) by applying a downward pressure on the sampler while rotating it clockwise. When the sampler is full of sludge, remove it from the container. If a sample liner was used, remove the liner from the sludge sampler and place Teflon caps over the ends of the liner. If a sample liner was not used, spoon sludge from the sampler directly into a sample jar.

4. After the liner or sample jar is capped, attach a sample label and custody seal and immediately place it into a sample cooler. Samples for chemical analysis should be packed in Blue Ice®.
5. Scan the drum (or waste container) where the sample was collected using chemical and/or radiological field screening instruments, and record the results in a bound logbook.
6. See Chapter 5 for details on preparing sample jars and coolers for sample shipment.
7. Transfer any sludge left over from the sampling back into the drum (or waste container) from which it was derived.
8. Transfer any other sampling-related wastes (e.g., gloves and foil) into a plastic waste bag.

4.2.8.3 Liquid Sampling from Waste Drums and Waste Containers

The Coliwasa method discussed in the following subsection is the preferred method for collecting liquid samples from drums and waste containers.

4.2.8.3.1 Coliwasa Method

The Coliwasa is an effective tool for collecting grab and composite samples of liquid or sludge samples from drums or other waste containers. This sampler is composed of a vertical sampling tube, a piston suction plug, and a handle (Figure 4.77). The inlet for the sampler is located at the base of the sample tube. To collect a grab sample, the bottom of the sampler is lowered to the desired sampling depth. The suction plug is then raised to draw the sample into the sampling tube. When the sampler is full, it is removed from the drum, and its contents are transferred into a sample bottle. On the other hand, to collect a composite sample, the bottom of the sampler is positioned at more than one sampling depth. At each depth, the suction plug is raised to fill a portion of the sampling tube. When the sampler is full, it is removed from the drum and its contents are transferred into a sample bottle.

For most sampling programs, four people are sufficient for this sampling procedure. Two are needed for field testing, sample collection, labeling, and documentation; a third is needed for health and safety; and a fourth is needed for miscellaneous tasks such as waste management and equipment decontamination.

The following equipment and procedure can be used to collect liquid or sludge samples for chemical and/or radiological analysis:

Equipment

1. Coliwasa sampler
2. Sample bottles
3. Sample preservatives
4. pH, temperature, conductivity, and dissolved oxygen meters
5. Sample labels
6. Cooler packed with Blue Ice® (Blue Ice® is not required for radiological analysis.)
7. Trip blank (only required for volatile organic analyses)

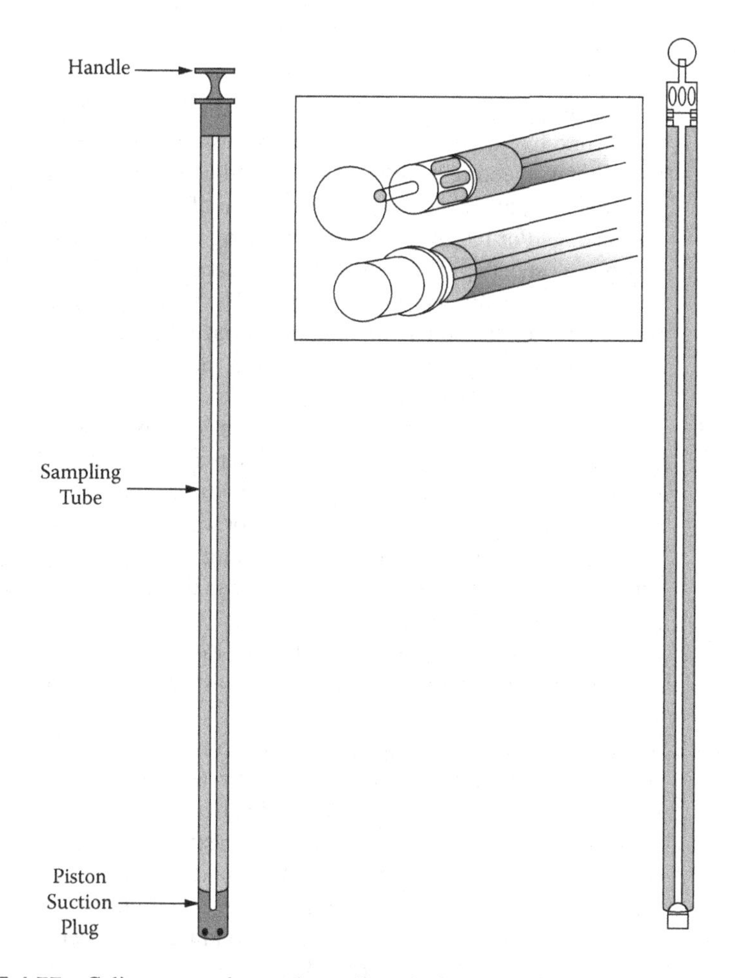

FIGURE 4.77 Coliwasa sampler used to collect liquid and sludge samples from drums and waste containers.

(From Byrnes, 1994. *Field Sampling Methods for Remedial Investigations.* Lewis Publishers, Boca Raton, FL.)

8. Coolant blank (not required for radiological analysis)
9. Sample logbook
10. Chain-of-custody forms
11. Chain-of-custody seals
12. Permanent ink marker
13. Health and safety instruments
14. Chemical and/or radiological field screening instruments
15. Health and safety clothing
16. Sampling table
17. Plastic waste bags

Sampling Procedure

1. In preparation for sampling, confirm that all necessary preparatory work has been completed, including obtaining property access agreements, meeting health and safety and equipment decontamination requirements, and checking the calibration of all health and safety and chemical and/or radiological field screening instruments.
2. Lay plastic sheeting flat on the ground surface around the vicinity of the drum or waste container to be sampled. The purpose of this sheeting is to help prevent the spread of contamination. If the drum is of unknown origin, consider piercing first to avoid potential for explosion as described in Section 4.2.8.
3. To collect a grab sample, lower the Coliwasa sampler into the drum (or waste container) to the selected depth interval to be sampled. Raise the suction plug to draw the liquid/sludge into the sampling tube. When the sampler is full, remove it from the drum (or waste container) and transfer its contents into a sample bottle by holding the open bottle below the sampler and lowering the suction plug. Scan the sample using chemical and/or radiological screening instruments, and record the results in a bound logbook. Screw on the bottle cap. Return any liquid remaining in the Coliwasa sampler to the drum (or waste container) from which it was derived.
4. To collect a composite sample, lower the Coliwasa sampler into the drum (or waste container) to the first depth interval to be sampled. Raise the suction plug partially to draw the liquid/sludge into the sampling tube. Move the intake of the Coliwasa to the next sampling interval. Raise the suction plug partially to draw the liquid/sludge into the sampling tube. Repeat this procedure until the sampler is full. Remove the sampler from the drum (or waste container) and transfer its contents into a sample bottle by holding the open bottle below the sampler and lowering the suction plug. Scan the sample using chemical and/or radiological screening instruments, and record the results in a bound logbook. Screw on the bottle cap. Return any liquid remaining in the Coliwasa sampler to the drum (or waste container) from which it was derived.
5. After the bottle is capped, attach a sample label and custody seal to the sample bottle and immediately place it into a sample cooler. Samples for chemical analysis should be packed in Blue Ice®. Transfer any leftover liquid back into the drum (or waste container) from which it was derived.
6. Fill a clean glass jar with the liquid waste from the drum (or waste container) and measure the pH, temperature, conductivity, and dissolved oxygen. Record this information in a sample logbook.
7. See Chapter 5 for details on preparing sample bottles and coolers for sample shipment.
8. Transfer any sampling-related wastes (e.g., gloves and foil) into a plastic waste bag. See Chapter 11 for guidance on managing this waste.

4.2.9 Buildings Material Sampling

The following subsections provide procedures for collecting swipe samples, concrete samples, and paint samples.

4.2.9.1 Swipe Sampling

Swipe samples are typically collected from material surfaces for the purpose of determining the amount of removable radioactivity. It is important to define the amount of removable radioactivity from building surfaces and equipment because if not controlled, this activity can easily be spread outside the radiologically controlled area and has the potential to expose workers or other receptors to contaminants through the inhalation and ingestion pathways.

- To prevent the spread of contamination from a radiological controlled area, 10 CFR 835.1101 requires that material and equipment not be released from the study area if removable surface contamination levels on accessible surfaces exceed the values specified in 10 CFR 835 Appendix D.

Once a swipe sample has been collected, it is most frequently analyzed for gross alpha and beta/gamma activity using a swipe counter. A swipe counter is gas-filled detector or scintillator. A swipe counter is used in part because handheld detectors are not always effective in measuring alpha and beta/gamma contamination in the field. The swipe counter has a better geometry and a better counting efficiency than field instruments.

The following procedure identifies how a swipe sample should be collected to define the amount of removable activity. For most sampling programs, three people are sufficient for implementing this sampling procedure. Two are needed for sample collection, labeling, recording sampling location, and documentation, and one is needed for health and safety.

The following equipment and procedure should be used to collect swipe samples for determining the amount of removable radioactivity from a surface:

Equipment:

1. Swipe counter and dry filter papers
2. 100-cm^2 (10 × 10 cm) plastic template
3. Plastic ziplock bags
4. Sample labels
5. Sample logbook
6. Chain-of-custody forms
7. Chain-of-custody seals
8. Permanent ink marker
9. Health and safety instruments
10. Chemical and/or radiological field screening instruments
11. Health and safety clothing
12. Sampling table

13. Waste container (e.g., 55-gal drum)
14. Plastic waste bags

Sampling Procedure:

1. In preparation for sampling, confirm that all necessary preparatory work has been completed, including obtaining property access agreements, meeting health and safety and equipment decontamination requirements, and checking the calibration of all health and safety and radiological field screening instruments.
2. Place the 100-cm^2 plastic template over the location where the swipe sample is to be collected.
3. Lay a piece of swipe counter dry filter paper inside the 100-cm^2 plastic template. Place index finger over top of filter paper and apply moderate pressure. Using a circular motion, rub the filter paper throughout the entire sampling area as shown in Figure 4.78.
4. Transfer the filter paper into a plastic ziplock bag.
5. After sealing the bag, attach a sample label and custody seal to the bag.
6. See Chapter 5 for details on preparing samples for shipment.
7. Transfer any sampling-related wastes (e.g., gloves and foil) into a plastic waste bag.
8. Survey the coordinates of the sampling point to preserve the sampling location. See Chapter 11 for guidance on managing this waste.

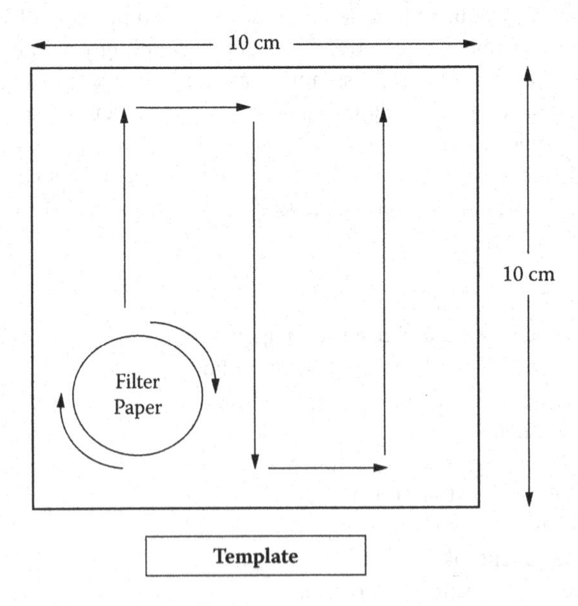

FIGURE 4.78 Swipe sampling method.

(From Byrnes, 2001. *Sampling and Surveying Radiological Environments.* Lewis Publishers, Boca Raton, FL.)

4.2.9.2 Concrete Sampling

Concrete samples are often collected from the floors, walls, and ceilings of a building to be decommissioned. The purpose of this sampling is often to determine the chemical and/or radiological composition for waste disposal purposes. The concrete drilling method and concrete coring methods presented in the following text should be used to collect concrete samples for these purposes.

4.2.9.2.1 Surface Concrete Sampling

Concrete samples are commonly collected from the surfaces of building walls, floors, ceilings, and other surfaces to support building characterization activities. These surface concrete samples are most often collected to help assess contaminant levels that building workers are being exposed to, or they can also be used to help support waste disposal decisions.

Surface concrete samples are typically collected using a hand drill and a 0.5-in.-diameter (or larger) drill bit, where multiple shallow holes (of approximately 0.25 in. depth) are drilled from either hotspots identified by chemical or radiological screening surveys (when wanting to assess maximum health hazard exposure), or from statistically selected locations (when wanting to identify mean concentrations of contaminants in concrete material for disposal purposes). The drill cuttings are collected in sample bottles and are then sent to a laboratory for analysis. The number of shallow holes required to be drilled at each sampling location is dependent on the sample volume required by the laboratory for the analyses to be performed.

The following procedure identifies how a surface concrete sample should be collected. For most sampling programs, four people are sufficient for implementing this sampling procedure. Two are needed for sample collection, labeling, recording sampling location, and documentation; a third is needed for health and safety; and a fourth is needed for miscellaneous tasks such as waste management and equipment decontamination.

The following procedure should be used to collect concrete samples for chemical and/or radiological analysis:

Equipment

1. Hand drill and drill bit (0.5-in. or larger)
2. Stainless steel spoon and bowl
3. Aluminum foil
4. 12 in. × 12 in. × 0.5 in. piece of plywood (or other material)
5. Ruler
6. Zip-Ties
7. Sample jars
8. Sample labels
9. Sample logbook
10. Chain-of-custody forms
11. Chain-of-custody seals
12. Permanent ink marker
13. Health and safety instruments
14. Chemical and/or radiological field screening instruments

15. Health and safety clothing
16. Sampling table
17. Waste container (e.g., 55-gal drum)
18. Plastic waste bags

Sampling Procedure

1. In preparation for sampling, confirm that all necessary preparatory work has been completed, including obtaining property access agreements, meeting health and safety and equipment decontamination requirements, and checking the calibration of all health and safety and chemical and/or radiological field screening instruments.
2. Use a ruler and a Zip-Tie to mark the maximum drilling depth of 0.25 in. on the drill bit.
3. Tightly wrap aluminum foil around the 12 in. × 12 in. × 0.5 in. piece of plywood (or other material) (shiny side down). The entire surface of the block should be covered by foil.
4. To collect a grab sample, drill as many 0.25-in.-deep holes at the selected sampling location as needed to provide the minimum sample volume of concrete cuttings required by the laboratory for analysis. When collecting samples from walls and ceilings, hold the foil-covered block beneath each hole during drilling to collect the drill cuttings. Transfer the cuttings into a sample jar using a stainless steel spoon.
5. To collect a composite sample, drill multiple 0.25-in.-deep holes at the first sampling location while holding the foil-covered block beneath each hole to collect the drill cuttings. Use a stainless steel spoon to transfer the cuttings to a stainless steel bowl. Move to the next sampling location. Repeat the preceding procedure. Once all of the locations have been sampled, use a stainless steel spoon to homogenize the sample before filling a sample jar.
6. After the jar is capped, attach a sample label and custody seal to the jar and immediately place it into a sample cooler. Samples for chemical analysis should be packed in Blue Ice®.
7. See Chapter 5 for details on preparing sample jars and coolers for sample shipment.
8. Transfer any drill cuttings left over from the sampling into a waste container. Before leaving the site, all waste containers should be sealed, labeled, and handled appropriately (see Chapter 11).
9. Transfer any other sampling-related wastes (e.g., gloves and foil) into a plastic waste bag.
10. Have a professional surveyor survey the coordinates of the sampling point to preserve the exact sampling location.

4.2.9.2.2 Concrete Coring Method

Concrete core samples are commonly collected from building walls, floors, ceilings, and other surfaces to estimate the concentration of contaminants in the concrete if the building were to be demolished. When evaluating waste disposal options for demolished building materials, it is essential that concrete core samples be collected throughout the entire thickness of building walls, floors, ceilings, and other surfaces for laboratory analysis. This is because the results from scanning surveys (see Sections 4.1.6.1 and 4.1.6.2) and surface concrete sampling (Section 4.2.9.2.1) will overestimate the contamination levels if the building material is demolished. Core samples collected for this purpose are typically collected from the locations showing the highest contamination based on scanning surveys.

For most sampling programs, four people are sufficient for implementing this sampling procedure. Two are needed for sample collection, labeling, recording sampling location, and documentation; a third is needed for health and safety; and a fourth is needed for miscellaneous tasks such as waste management and equipment decontamination.

The following procedure should be used to collect concrete core samples for chemical or radiological analysis:

Equipment

1. Coring drill and coring barrel (length and diameter based on floor/wall thickness and required sample volume)
2. Hose and water supply for coring drill
3. Aluminum foil
4. Stainless steel chisel and hammer
5. Stainless steel spoon
6. Plastic ziplock bags
7. Sample bottles
8. Sample labels
9. Sample logbook
10. Chain-of-custody forms
11. Chain-of-custody seals
12. Permanent ink marker
13. Health and safety instruments
14. Chemical and/or radiological field screening instruments
15. Health and safety clothing
16. Sampling table
17. Waste container (e.g., 55-gal drum)
18. Plastic waste bags

Sampling Procedure

1. In preparation for sampling, confirm that all necessary preparatory work has been completed, including obtaining property access agreements, meeting health and safety and equipment decontamination requirements, and checking the calibration of all health and safety and chemical and/or radiological field screening instruments.
2. Connect water supply line to core drill, and begin drilling. Keep a steady downward pressure on the coring barrel. Raise the core barrel every few minutes to facilitate the removal of drill cuttings.

3. Once the cut has been completed, remove the coring barrel from the hole. Typically, the concrete core will remain in the core barrel. Tap the outside of the core barrel gently with a hammer until the core drops out.
4. To collect a grab sample, tightly wrap the concrete core in aluminum foil, then slide it into in a large plastic ziplock bag. After sealing the bag, attach a sample label and custody seal and immediately place it into a sample cooler.
5. Although not commonly performed, to collect a composite sample, core samples from multiple locations are placed in a stainless steel bowl. The concrete core samples are then broken into small pieces using a stainless steel chisel and hammer. A stainless steel spoon is then used to homogenize the sample and then transfer the material into a sample jar. After attaching a sample label and custody seal to the jar, immediately place it into a sample cooler.
6. Samples for chemical analysis should be packed in Blue Ice®.
7. See Chapter 5 for details on preparing coolers for sample shipment.
8. Transfer any drill cuttings left over from the sampling into a waste container. Before leaving the site, all waste containers should be sealed, labeled, and handled appropriately (see Chapter 11).
9. Transfer any other sampling-related wastes (e.g., gloves and foil) into a plastic waste bag.
10. Have a professional surveyor survey the coordinates of the sampling point to preserve the exact sampling location.

4.2.9.3 Paint Sampling

Paint samples are commonly collected from a variety of building surfaces including walls, piping, support beams, etc., in support of building decommissioning activities. Paint is most frequently sampled in older buildings because of concerns over lead and PCB content (which has the potential to create RCRA or Toxic Substance Control Act (TSCA) waste). At radioactive sites, paint (and various types of resins) is sometimes used to fix radiological contaminants on building surfaces.

The following procedure identifies how a paint sample should be collected from a building surface for analytical testing. For most sampling programs, four people are sufficient for implementing this sampling procedure. Two are needed for sample collection, labeling, recording sampling location, and documentation; a third is needed for health and safety; and a fourth is needed for miscellaneous tasks such as waste management and equipment decontamination.

The following procedure should be used to collect paint samples for chemical and/or radiological analysis:

Equipment

1. Stainless steel spatula
2. Stainless steel spoon
3. Stainless steel bowl

4. Aluminum foil
5. 12 in. × 12 in. × 0.5 in. piece of plywood (or other material)
6. Sample jars
7. Sample labels
8. Sample logbook
9. Chain-of-custody forms
10. Chain-of-custody seals
11. Permanent ink marker
12. Health and safety instruments
13. Chemical and/or radiological field screening instruments
14. Health and safety clothing
15. Sampling table
16. Waste container (e.g., 55-gal drum)
17. Plastic waste bags

Sampling Procedure

1. In preparation for sampling, confirm that all necessary preparatory work has been completed, including obtaining property access agreements, meeting health and safety and equipment decontamination requirements, and checking the calibration of all health and safety and chemical and/or radiological field screening instruments.
2. Tightly wrap an aluminum foil around the outside of the piece of plywood (or other material) so that no wood surface is exposed. The shiny side of the foil should face down. This block will be used to catch paint chips.
3. To collect a grab sample, use a stainless steel spatula to scrape the painted surface at a single location, directing the paint chips onto the block wrapped in aluminum foil. Continue this effort until the minimum sample volume required by the laboratory has been collected. Use the stainless steel spatula to transfer paint chips from the block into a sample jar.
4. To collect a composite sample, use a stainless steel spatula to scrape the painted surface of the first location, directing the paint chips onto the block wrapped in aluminum foil. Transfer the paint chips to a stainless steel bowl. Move to the next sampling location. Use a stainless steel spatula to scrape the painted surface at this location, directing the paint chips onto the block wrapped in aluminum foil. Transfer the paint chips to the stainless steel bowl. Continue this effort until the minimum sample volume required by the laboratory has been collected. Use a stainless steel spoon to homogenize the paint chips for the various sampling locations, and then transfer the chips into a sample jar.
5. After the jar is capped, attach a sample label and custody seal to the jar and immediately place it into a sample cooler. Samples for chemical analysis should be packed in Blue Ice®.

6. See Chapter 5 for details on preparing sample jars and coolers for sample shipment.
7. Transfer any paint chips left over from the sampling into a waste container. Before leaving the site, all waste containers should be sealed, labeled, and handled appropriately (see Chapter 11).
8. Transfer any other sampling-related wastes (e.g., gloves and foil) into a plastic waste bag.
9. Have a professional surveyor survey the coordinates of the sampling point to preserve the exact sampling location.

4.2.10 Pipe Surveying

The following subsections provide the reader with information regarding pipe surveying tools that are available to assist building decontamination and decommissioning and soil remediation activities. The primary objective behind performing a pipe survey is to determine if contaminated material is present in the pipe, determine if there are any cracks in the pipe where the contamination could leak into the surrounding soil, and determine where the pipe ultimately discharges. If contaminated material is found within a pipe, special tools are available that can clear it from the pipe.

Surveying tools can be advanced through a pipe using either a self-propelled crawler unit or fiberglass push cable. The crawler unit is currently commercially available in sizes small enough to investigate 6-in.-diameter pipes. Smaller-diameter pipes can be investigated by advancing tools using a fiberglass push cable. Some of the investigative tools that are currently available for pipe surveying include:

- High-resolution video cameras
- Radiation detectors
- Organic vapor analyzers
- Combustible gas indicators
- Temperature and pH meters

Pipe crawler units are commonly designed to meet site-specific requirements and conditions, such as:

- Pipe diameters
- Pipe bend angles
- Pipe surface texture
- Surveying distances
- Analytical instrumentation
- Site chemical or radiation contamination levels

Depending on the pipe geometry, it is often necessary to build the crawler unit in modules, which are linked together in a trainlike configuration. When a modular system is used, a separate module can be used for the motor, scanning instruments, and video cameras (Figure 4.79).

FIGURE 4.79 Versa Trax 100 pipe inspection system in modular configuration. (Courtesy of Inuktun Services LTD. Shilho Uzawa.)

Wheels or tracks are used to propel the crawler units. Wheeled units use four or six wheels, which are narrow, and chamfered to the approximate radius of the pipe. These wheels combined with a direct-current-powered motor provide traction to negotiate most sludge and mudlike environments (Figure 4.80).

Tracked systems can be advantageous for short-distance applications (<100 ft) in large-diameter pipes or ducts because of their ability to turn on their own center (Figures 4.81 and 4.82). This feature can be very useful when negotiating complex runs, particularly when there is a significant amount of debris. With this one exception, wheeled systems outperform tracked systems in most other applications.

The distance a system can travel is primarily dependent on the weight of the tractor package, and the size and weight of the telemetry cable. As a general rule, a tractor can only pull its own weight in cable and associated drag. Onboard multiplexing or using fiber-optic telemetry transmission cable can help to reduce the drag. However, fiber-optic cable is very fragile and therefore is not recommended, because piping interrogation activities are commonly performed in rugged environments. Some wireless telemetry-controlled systems are now available.

Another approach that can be used for long-distance application (> 100 ft) is powering the crawler and the sensor instruments with onboard batteries. With this approach, the cables are only needed to provide trickle charge to the crawler package and to transmit information. This approach can achieve horizontal distances of 1,500 ft or more.

Some accessories that can be designed modularly and that can be used to expand the capability of the crawler unit include:

- Pan-and-tilt for vision or directional sensors
- Laser distance and scaling instrumentation
- Pipe-clearing tooling
- Inclinometry for measuring settling or slope

If a specific detector requires the pipe to be free of debris before performing the survey, a squeegee-like tool can be attached to a crawler unit. Tethers are attached to all devices to allow retrievability in the event the drive unit fails.

The manual push method is generally required for pipes smaller than 6 in. in diameter, and utilizes ⅛ to ⅝-in. solid flexible polyethylene rods to push and retrieve

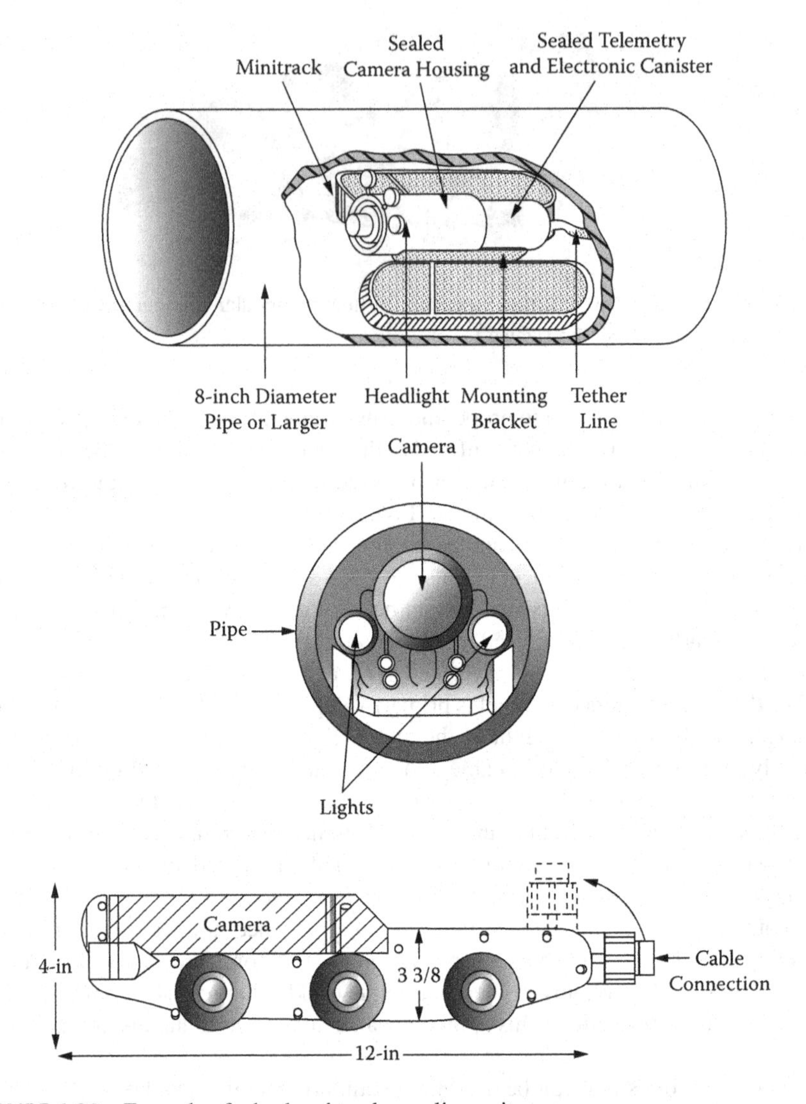

FIGURE 4.80 Example of wheel and track crawling unit.

(From Byrnes, 1994. *Field Sampling Methods for Remedial Investigations.* Lewis Publishers, Boca Raton, FL.)

sampling tools (Figure 4.83). At this time, as far as the author is aware, the longest push accomplished in mockup is approximately 700 ft, and approximately 350 ft in actual field application. The same investigative tools available for the pipe crawler are also available for the push method. To keep the sampling instrument or camera properly positioned in the pipe, instrument-bearing skids are used. For larger-diameter pipes (3- to 6-in.), wheeled "dogbone"-shaped skids are most effective, whereas skids with runners are used in smaller-diameter pipes. In an effort to maximize distance,

FIGURE 4.81 Side view of a track crawling unit made by Non-Entry System Ltd. Reprinted with permission from Non-Entry System Ltd.

FIGURE 4.82 Rear view of a track crawling unit made by Non-Entry System Ltd. Reprinted with permission from Non-Entry System Ltd.

small three-wheeled devices can be attached to the push-rod every several feet to reduce the friction of the cable rubbing against the pipe.

Software systems are currently being developed that can provide precise system-positioning information. The only requirement for this method is a physical reference point within the area of interest. This method can also be used to navigate equipment within an area remotely, and may replace more complex methods.

There is instrumentation now available that accepts real-time voltage or current inputs, translates them into reportable values, and displays the results in real time on the video screen. This device has a high-level/low-level alarming capability and

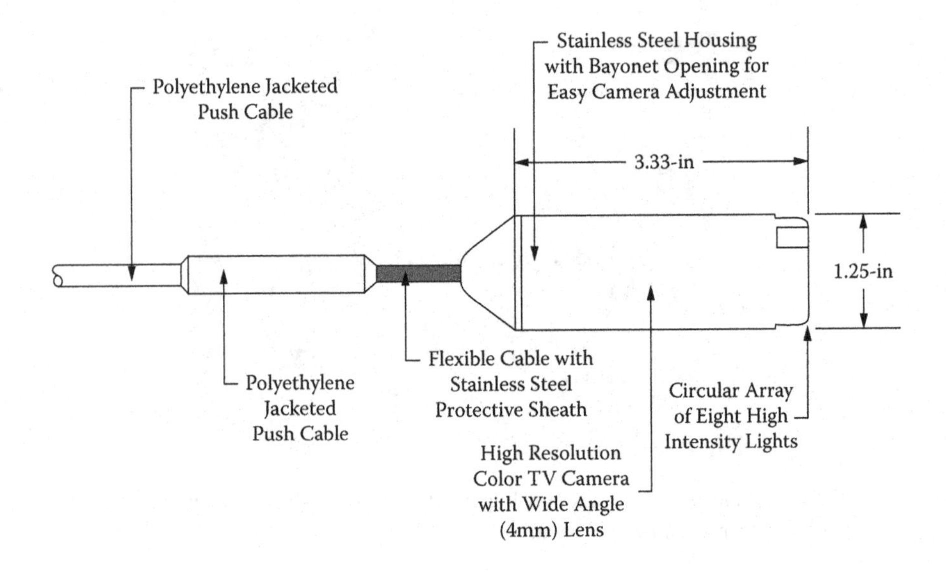

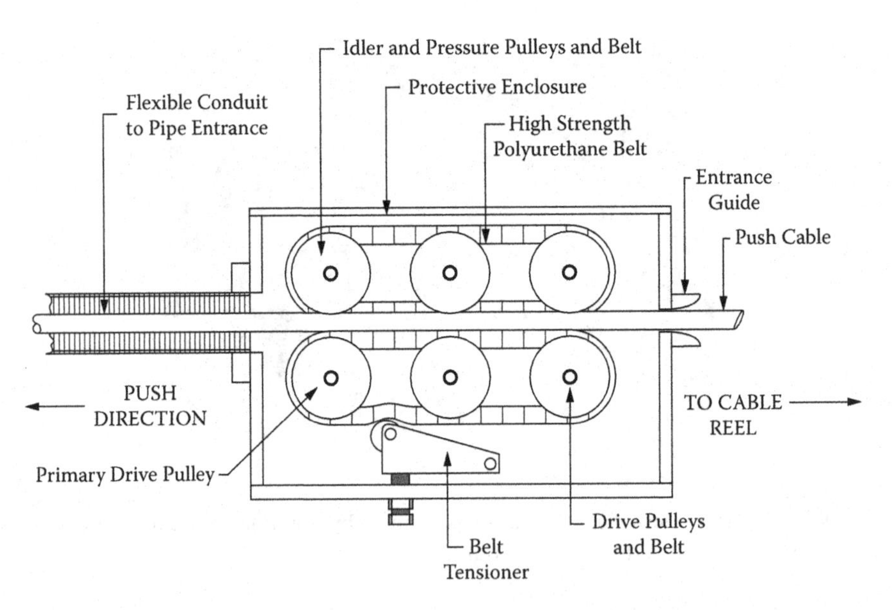

FIGURE 4.83 Small-diameter television camera and push cable mechanism.

(From Byrnes, 1994. *Field Sampling Methods for Remedial Investigations.* Lewis Publishers, Boca Raton, FL.)

will accept analog, log, nonlinear, or quadrature inputs. This instrument is particularly valuable for providing a single-source documentation package of raw data and acquired information.

For further information on pipe and remote surveying instrumentation, contact Pacific Northwest National Laboratory.

4.2.11 Remote Surveying

Remote surveying should be considered in highly contaminated environments or environments where there are potential explosives, or confined spaces. Using remote surveying instruments to collect samples in these types of hazardous environments will minimize worker exposure. The Mark VA-1 is an all-terrain platform that can be equipped in a multitude of tools to support hazardous environmental investigations. For details on other all-terrain platforms that are available, refer to the following web site: www.army-technology.com/contractors/mines/northrop-remotec/

4.2.11.1 ANDROS Mark VA-1

The ANDROS Mark VA-1 remote hazardous duty system is designed for deployment in areas where personnel are prohibited access, such as highly contaminated environments or environments where there are potential explosives, or confined spaces (Figure 4.84).

An earlier version of this robot, the ANDROS Mark VI, was initially deployed for remote collection of characterization data such as gross gamma radiation readings, video imaging, and smear samples. The Pacific Northwest National Laboratory deployed the ANDROS Mark VI robot into the 221-U Facility Railroad Tunnel where radiation levels were expected to be high. The robot traversed to the outer rollup door, collecting gross radiation data and videotaping the condition of the tunnel. Nine smear samples were taken from selected locations along the tunnel. For this deployment, the robot was configured with:

- Two camera and lighting systems (one color and one black and white)
- A real-time gross gamma detector

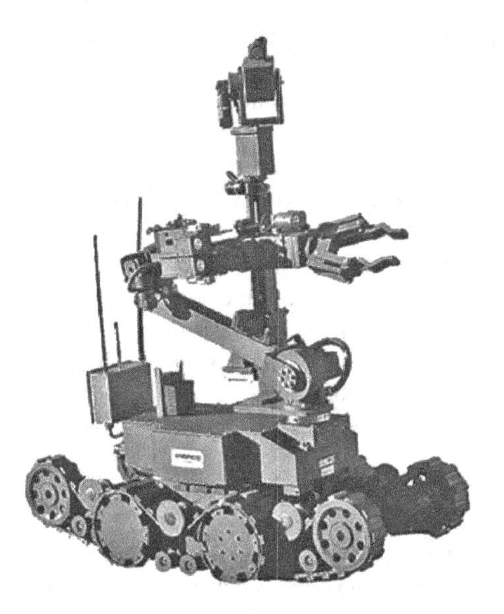

FIGURE 4.84 ANDROS Mark VA-1 remote hazardous duty system.

- Smear sample pads
- A lifting fixture (with camera and lights) for the robot, with a motorized cable playout and retrieval system
- An operator control station with video recording equipment

The deployment of the ANDROS Mark VI system in the railroad tunnel was a success, and the system has since been deployed at other locations.

The ANDROS Mark VA-1 robot comes equipped with a stationary arm camera, surveillance camera with 216:1 total zoom and pan/tilt capability, lighting system, multiple mission tool/sensor mounts, gripper arm with continuous rotation, and a track system that can traverse very rough terrain.

4.3 DEFINING BACKGROUND CONDITIONS

When performing an environmental investigation, it is not uncommon for the lead regulatory agency overseeing the work to require a comparison to be made between contaminant concentrations at the site and natural background concentrations. This is often required in addition to comparing concentrations against a regulatory threshold or a risk-based action level.

Background soil samples should be collected from the same soil formation and approximately the same depths as the soil samples collected from the contaminated site. For this reason, background soil samples should preferably be collected from locations within approximately 0.5 mi of the study area. Background soil samples must be collected from locations where the soil has not been disturbed and must be collected using the same type of sampling equipment used to collect the samples for the contaminated site. In an urban environment, background sampling may need to be performed at a local park, or private property that has not been disturbed. (Note that sampling on private property would require written authorization from the landowner.) Background soil samples should be analyzed for all of the parameters as the samples collected from the contaminated site. It is recommended that ten background samples be collected for analysis to define the background population distribution (e.g., normal and lognormal) for each of the contaminants at the site to assure that a reliable estimate of the population mean and variance can be calculated.

Background groundwater samples should be collected from a groundwater monitoring well located upgradient from the site. This background well must screen the same aquifer, and the same depth interval within the aquifer, as the wells located downgradient from the contaminated site. Groundwater samples collected from the background well should be analyzed for all of the same parameters as the wells downgradient of the contaminated site.

BIBLIOGRAPHY

Amplified Geochemical Imaging (AGI), 2016, *Passive Gas Environmental Site Assessment and Monitoring*, https://cdn.ymaws.com/www.envirobank.org/resource/resmgr/memberdevelopmentgroups/Presentation_with_Notes_10-1.pdf.

Boart Long Year, 2021, www.boartlongyear.com/insite/sonic-drilling-works/.

Butler, J.J., Jr., 2020. *The Design, Performance, and Analysis of Slug Tests (2nd ed.)*, Lewis Publishers, New York, 266p. [errata: sheet (pdf), spreadsheet (xlsx)].

Byrnes, M.E., 2001. *Sampling and Surveying Radiological Environments.* Boca Raton, FL: Lewis Publishers.

Byrnes, M.E., 1994. *Field Sampling Methods for Remedial Investigations.* Boca Raton, FL: Lewis Publishers.

California EPA, 2014, *Well Design and Construction for Monitoring Groundwater at Contaminated Sites*, Department of Toxic Substances Control, Geology Services Branch, Berkeley, California.

California EPA, 2012, *Guidelines for Planning and Implementing Groundwater Characterization of Contaminated Sites*, Department of Toxic Substance Control, June.

California State Water Resource Control Board, 2005, *GeoTracker™ Survey XYZ, Well Data and Site Map Guidelines & Restrictions, Electronic Deliverable Format and Data Dictionary*, Revision 6.1, California State Water Resources Control Board, Sacramento, California. Available at: www.waterboards.ca.gov/ust/electronic_submittal/docs/geotrackersurvey_xyz_4_14_05.pdf.

Cascade, 2019, *Sampling 102: 4 Ways to Collect Deep Groundwater Samples during Drilling, ISOFLOW*, August, www.cascade-env.com/resources/blogs/archive/sampling-102-4-ways-to-collect-deep-groundwater-samples-during-drilling/.

Castle Drilling Company, 2021, https://castledrill.com/an-introduction-to-reverse-circulation-drilling/.

Cunningham, W.L. and C.W. Schalk (comps.), 2011, *Groundwater Technical Procedures of the U.S. Geological Survey*, "GWPD 17 – Conducting an Instantaneous Change in Head (Slug) Test with a Mechanical Slug and Submersible Pressure Transducer," pp. 145–151, U.S. Geological Survey, Reston, Virginia. Available at: https://pubs.usgs.gov/tm/1a1/pdf/GWPD17.pdf.

DOE, 2020, *200-ZP-1 Operable Unit Ringold Formation Unit A Characterization Sampling and Analysis Plan*, DOE/RL-2019-23, Rev. 0, February, pp. 58–59.

DOE, 2017, Central Plateau Groundwater Tracer Study Work Plan, SGW-60511, Rev. 0, Richland, WA, August.

DOE, 1999, *In-Situ Object Counting System, Deactivation and Decommissioning Focus Area*, DOE/EM-0477, September.

En Novative Technologies, Inc. WEB page www.fieldenvironmental.com/assets/files/Literature/En%20Core%20Sampler%20Brochure.pdf.

EON Products Inc., 2020, *EON Dual Membrane Passive Diffusion Samplers (DMPDB™)*, https://store.eonpro.com/store/p/2768-EON-Dual-Membrane-Passive-Diffusion-Samplers.aspx.

EPA (Region 4), 2020a, *Groundwater Level and Well Depth Measurement*, LSASDPROC-105-R4, Laboratory Services and Applied Science Division, Athens, Georgia, Page 5 of 8.

EPA (Region 4), 2020b, *Waste Sampling*, LSASDPROC-302-R4, May.

EPA (Region 4), 2020c, *Soil Sampling Operating Procedure*, Laboratory Services and Applies Science Division, Athens, Georgia, LSASDPROC-300-R4, June, www.epa.gov/sites/production/files/2015-06/documents/Soil-Sampling.pdf.

EPA, 2020d, *Groundwater Level and Well Depth Measurement*, LSASDPROC-105-R4, Laboratory Services and Applied Science Division, Athens, Georgia, May.

EPA, 2019, Method TO-15A, *Determination of Volatile Organic Compounds (VOCs) in Air Collected in Specially Prepared Canisters and Analyzed by Gas Chromatography-Mass Spectrometry (GC-MS), September.*

EPA (Region 4), 2017a, *Field X-Ray Fluorescence Measurement*, SESDPROC-107-R4, Science and Ecosystem Support Division, Athens, Georgia, September.

EPA (Region 4), 2017b, *Superfund X-Ray Fluorescence Field Operations Guide*, SFDGUID-001-R0, Superfund Division, Atlanta, Georgia, July.

EPA (Region 1), 2017c, *Low Stress (Low Flow) Purging and Sampling Procedure for the Collection of Groundwater Samples From Monitoring Wells*, EQASOP-GW4, Quality Assurance Unit, North Chelmsford, MA, September.

EPA (Region 4), 2017d, *Groundwater Sampling*, SESDPROC-301-R4, Science and Ecosystem Support Division, Athens, Georgia, March.

EPA, 2016, *Quality Assurance Guidance Document 2.12, Monitoring $PM_{2.5}$ in Ambient Air Using Designated Reference or Class I Equivalent Methods*, EPA-454/B-16-001, January.

EPA (Region 4), 2016, *Ambient Air Sampling*, SESDPROC-303-R5, Athens, Georgia, March.

EPA (Region 4), 2013, *Design and Installation of Monitoring Wells*, SESDGUID-101-R1, Science and Ecosystem Support Division, Athens, Georgia, January.

EPA (Region 1), 2011, *Canister Sampling Standard Operating Procedure, ECASOP-Canister Sampling SOP5*, September.

EPA, 2007, *Field Portable X-Ray Fluorescence Spectrometry for the Determination of Elemental Concentrations in Soil and Sediment*, Method 6200, Revision 0, February.

EPA, 2005, *Groundwater Sampling and Monitoring with Direct Push Technologies, OSWER No. 9200.1-51, EPA540/R-04/005,* August.

EPA, 2002. *EPA Quality Assurance Guidance Document: Method Compendium, Field Standard Operating Procedures for the PM2.5 Performance Evaluation Program*, United States Environmental Protection Agency Office of Air Quality Planning and Standards, Revision No. 2, March.

EPA, 2002, *RCRA Waste Sampling Draft Technical Guidance*, EPA 530-D-02-002, August.

EPA, 1999a, *Compendium of Methods for the Determination of Toxic Organic Compounds in Ambient Air, Second Edition, Compendium Method TO-15, Determination of Volatile Organic Compounds (VOCs) in Air Collected in Specially-Prepared Canisters and Analyzed by Gas Chromatography/Mass Spectrometry (GC/MS),* EPA/625/R-96/010b, January.

EPA, 1999b, *Compendium of Methods for the Determination of Toxic Organic Compounds in Ambient Air, Second Edition, Compendium Method TO-17, Determination of Volatile Organic Compounds in Ambient Air Using Active Sampling Onto Sorbent Tubes,* EPA/625/R-96/010b, January.

EPA, 1999, *Compendium of Methods for the Determination of Inorganic Compounds in Ambient Air*, EPA/625/R-96/010a, June.

EPA, 1999, *Compendium of Methods for the Determination of Toxic Organic Compounds in Ambient Air* – Second Edition, EPA/625/R-96/010b, January.

EPA, 1999, *Compendium of Methods for the Determination of Toxic Organic Compounds in Ambient Air, Compendium Method TO-4A, Determination of Pesticides and Polychlorinated Biphenyls in Ambient Air Using High Volume Polyurethane Foam (PUF) Sampling Followed by Gas Chromatographic/Multi-Detector Detection (GC/MD)*, Second Edition, EPA/625/R-96/010b, January.

EPA, 1999, *Compendium of Methods for the Determination of Toxic Organic Compounds in Ambient Air, Compendium Method TO-13A, Determination of Polycyclic Aromatic Hydrocarbons (PAHs) in Ambient Air Using Gas Chromatography / Mass Spectrometry (GC/MS)*, EPA/625/R-96/010b, January.

EPA, 1999, *Compendium of Methods for the Determination of Toxic Organic Compounds in Ambient Air – Second Edition*, EPA/625/R-96/010b, January.

EPA, 1998, *Guidance for Using Continuous Monitors in $PM_{2.5}$ Monitoring Networks*, EPA/R-98-012, May.

EPA, 1998, *Environmental Technology Verification Report, Passive Soil Gas Sampler*, W.L. Gore & Associates, Inc. GORE-SORBER Screening Survey, EPA/600-98/095, August.

EPA, 1996, *SW-846, Test Methods for Evaluating Solid Waste, Physical/Chemical Methods,* Office of Solid Waste, Washington, D.C., www.epa.gov/epawaste/hazard/testmethods/sw846/online/index.htm.

EPA, 1994, *Slug Tests*, SOP 2046, Rev. 0, October.

EPA, 1987, *Handbook – Groundwater*, Chapter 7, EPA/625/6-87/016, March.

EPA 1985, *An Introduction To Groundwater Tracers*, PB86-100591, National Technical Information Service, Tucson, Arizona, March.

EPA, 1984, *Method (TO-1) for the Determination of Volatile Organic Compounds in Ambient Air Using Tenax® Adsorption and Gas Chromatography/Mass Spectrometry (GC/MS)*, Revision 1, April.

Flury, M. and N.N. Wai, 2003, *Dyes as tracers for vadose zone Hydrology, Reviews in Geophysics* 41(1): 1002.

Freeman, L.A., et al., 2004, *Use of Submersible Pressure Transducers in Water-Resources Investigations, Chapter A of Book 8, Instrumentation, Section A, Instruments for Measurement of Water Level*, U.S. Department of the Interior, U.S. Geological Survey. Available at https://pubs.usgs.gov/twri/twri8a3/#N10009

GeoInsight, 2019, *HydraSleeveTM Standard Operating Procedure: Sampling Groundwater with a HydraSleeve*, www.hydrasleeve.com/images/stories/support/HydraSleeve_SOP.pdf.

Geoprobe®, 2014, *Geoprobe® Pneumatic Slug Test Kit (GW1600), Installation and Operation Instructions for USB System*, Instructional Bulletin No. MK3195, February.

Geotech, 2020, *Geotech Bladder Pumps Installation and Operation Manual*, Rev 1/14/2020, Geotech Environmental Equipment, Inc., Denver, Colorado. Available at: www.geotechenv.com/Manuals/Geotech_Bladder_Pumps.pdf.

GeoVision, 2007, www.geovision.com/.

Gerenday, S., et al., 2020, *Sulfur Hexafluoride and Potassium Bromide as Groundwater Tracers for Managed Aquifer Recharge*, Department of Earth of Science, HDR Engineering, Inc., https://ngwa.onlinelibrary.wiley.com/doi/abs/10.1111/gwat.12983.

Hodny, J.W. et al., 2009, *Quantitative Passive Soil Gas and Air Sampling in Vapor Intrusion Investigations, Proceedings of Vapor Intrusion*, Air & Waste Management Association, San Diego, CA, January.

IAEA (International Atomic Energy Agency), 1991, *Airborne Gamma Ray Spectrometer Surveying, Technical Report Series No. 323,* Vienna, Austria.

Idaho National Engineering and Environmental Laboratory (INEEL), 2004, *Tracers and Tracer Testing: Design, Implementation, and Interpretation Methods*, INEEL/EXT-03-01466, January.

Interstate Technology and Regulatory Council (ITRC), 2007, *Protocol for Use of Five Passive Samplers to Sample for a Variety of Contaminants in Groundwater.* Washington, DC: The Interstate Technology and Regulatory Council, Diffusion Sampler Team. www.itrcweb.org/documents/DSP-5.pdf.

Keller, Carl, 2021, *Flexible Liner Applications*, CRC Press, in publication.

Los Alamos National Laboratory (LANL), 2013, *Westbay Pressure Transducer Installation, Removal, and Maintenance*, Rev. 0, SOP-5226, October.

Los Alamos National Laboratory (LANL), 2007, *Evaluation of Sampling Systems for Multiple Completion Regional Aquifer Wells at Los Alamos National Laboratory*, EP2007-0486, August.

Los Alamos National Laboratory (LANL), 2008, *Groundwater Sampling Using the Westbay MP System®*, Los Alamos National Laboratory Standard Operating Procedure SOP-5225, Los Alamos, New Mexico, October.

National Academy of Sciences, Engineering and Medicine, 2019, Manual on Subsurface Investigations, Washington, D.C., The National Academies Press, https://doi.org/10.17226/25379.

New Jersey Department of Environmental Protection, 2003, *Low-Flow Purging & Sampling Guidance*, Trenton, New Jersey, www.nj.gov/dep/srp/guidance/lowflow/.

Ohio EPA, 2014, *Establishing the Monitoring Well Fixed Survey Elevation Reference Point*, Division of Drinking and Ground Waters, February. Available at: www.epa.state.oh.us/Portals/28/documents/TGM-Supp2.pdf.

Ohio EPA, 2007, *Technical Guidance Manual for Ground Water Investigations*, Chapter 6, *Drilling and Subsurface Sampling, Division of Drinking and Ground Waters*, Columbus, Ohio, www.epa.state.oh.us/ddagw/

Ohio EPA, 2006, *Technical Guidance Manual for Groundwater Investigations, Chapter 4 Pumping and Slug Tests*, December, www.epa.state.oh.us/ddagw/.

Ohio EPA, 2005, *Chapter 15 Use of Direct Push Technologies for Soil and Ground Water Sampling*, Division of Drinking and Ground Waters, February, www.epa.state.oh.us/port als/28/Documents/TGM-15.pdf.

Parker, Louise, et al., 2009, *Demonstration/Validation of the Snap Sampler Passive Ground Water Sampling Device for Sampling Inorganic Analytes at the Former Pease Air Force Base,* U.S. Army Corps of Engineers, ERDC/CRREL TR 09-12.

Parker, Louise and Nathan Mulherin, 2007, *Evaluation of the Snap Sampler for Sampling Ground Water Monitoring Wells for VOCs and Explosives.* U.S. Army Corps of Engineers, ERDC/CRREL TR-07-14.

QED, 2019, *Snap Sampler® Zero – Passive Sampling System, Standard Operating Procedure for the Snap Sampler Passive Groundwater Sampling Method*, November, www.snap sampler.com/local/uploads/content/files/Snap_Sampler_SOP.pdf.

Ramaroson, V., et al., 2018, *Tritium as Tracer of Groundwater Pollution Extension: Case Study of Andralanitra Landfill Site, Antananarivo, Madagascar*, Springer Link, Applied Water Science, Article 57, April, https://link.springer.com/article/10.1007/s13201-018-0695-9.

Ravansari, R, et al., 2020, *Portable X-ray fluorescence for environmental assessment of soils: Not just a point and shoot method*, Environmental International 134 (2020) 105250, Elsevier.

Rubin, Y., and S.S. Hubbard (2005), *Hydrogeophysics*, Water Science Technology Library, vol. 50, 523 pp., Springer, New York.

Solinst, 2019, *User's Manual Electronic Pump Control Unit, 109652 Rev. C, Model 464 Mk3*, Solinst, Canada. Available at: www.solinst.com/products/groundwater-samplers/464-pneumatic-pump-control-units/operating-instructions/electronic-control-unit-user-guide/.

Solinst, 2019, *Sampling Groundwater – Using Model 407 and 408 Pumps*, Georgetown, Ontario, Canada, September, www.soli nst.com.

Sterrett, R.J., 2007, *Groundwater and Wells,* 3rd ed., Johnson Screens, New Brighton, MN, ISBN 978-0978779306.

Sukop, M.C., 2000, *Estimation of Vertical Concentration Profiles from Existing Wells*, Vol. 38, No. 6, Ground Water, November–December.

Title 40 CFR, Part 50, Appendix A - *Reference Method for the Determination of Sulfur Dioxide in the Atmosphere (Pararosaniline Method).*
Title 40 CFR, Part 50, Appendix B - *Reference Method for the Determination of Suspended Particulate Matter in the Atmosphere (High Volume Method).*
Title 40 CFR, Part 50, Appendix C - *Measurement Principle and Calibration Procedure for the Measurement of Carbon Monoxide in the Atmosphere (Non-Dispersive Infrared Photometry).*
Title 40 CFR, Part 50, Appendix D - *Measurement Principle and Calibration Procedure for the Measurement of Ozone in the Atmosphere.*
Title 40 CFR, Part 50, Appendix F - *Measurement Principle and Calibration Procedure for the Measurement of Nitrogen Dioxide in the Atmosphere (Gas Phase Chemiluminescence).*
Title 40 CFR, Part 50, Appendix G - *Reference Method for the Determination of Lead in Suspended Particulate Matter Collected From Ambient Air.*
Title 40 CFR Part 50, Appendix J - *Reference Method for the Determination of Particulate Matter as PM10 in the Atmosphere.*
Title 40 CFR Part 50, Appendix L - *Reference Method for the Determination of Fine Particulate Matter as PM2.5 in the Atmosphere.*
US Army Corps of Engineers, 2014, *Demonstration of the AGI Universal Samplers (F.K.A. the GORE® Modules) for Passive Sampling of Groundwater*, Cold Regions Research and Engineering Laboratory, ERDC/CRREL TR-14-4, March.
USGS, 2020, *Passive Sampling of Groundwater Wells For Determination of Water Chemistry, Chapter 8 of Section D Water Quality, Book 1 Collection of Water Data by Direct Measurement*, https://pubs.usgs.gov/tm/01/d8/tm1d8.pdf.
USGS, 2013, *Monitoring-Well Installation, Slug Testing, and Groundwater Quality for Selected Sites in South Park, Park County, Colorado*, Open-File Report 2014-1231.
USGS, 2006, *Book 9 Handbooks for Water-Resources Investigations, National Field Manual for the Collection of Water-Quality Data, Chapter A4. Collection of Water Samples*, www.usgs.gov/mission-areas/water-resources.
USGS, 2004, *Use of Submersible Pressure Transducers in Water-Resources Investigations, Chapter A of Book 8, Instrumentation, Section A, Instruments for Measurement of Water Level, Techniques of Water Resources Investigations 8-A3*, Reston Virginia.
USGS, 2003, *Book 9 Handbooks for Water-Resources Investigations, National Field Manual for the Collection of Water-Quality Data, Chapter A2 Selection of Equipment for Water Sampling*, Version 2.0, March, pp. 16, 85.
USGS, 1995, *Methods of Conducting Air-Pressurized Slug Tests and Computation of Type Curves for Estimating Transmissivity and Storativity*, Open-File Report 95-424, Reston, Virginia.
USGS, 1986, *Techniques of Water-Resources Investigations of the United States Geological Survey*, Chapter A12, Fluorometric Procedures for Dye Tracing, Book 3.
Vroblesky, D.A., 2001a, *User's Guide for Polyethylene-Based Passive Diffusion Bag Samplers to Obtain Volatile Organic Compound Concentrations in Wells, Part 1: Deployment, Recovery, Data Interpretation, and Quality Control and Assurance.* U.S. Geological Survey Water-Resources Investigations Report 01-4060.
Vroblesky, D.A. ed., 2001b, *User's Guide for Polyethylene-Based Passive Diffusion Bag Samplers to obtain Volatile Organic Compound Concentrations in Wells, Part 2: Field Tests.* U.S. Geological Survey Water-Resources Investigations Report 01-4061.
Water Resources Research, 2012, *Utility of bromide and heat tracers for aquifer characterization affected by highly transient flow conditions*, Vol. 48.

5 Sample Preparation, Documentation, and Shipment

After the sample collection procedure is complete, sample containers must be preserved (if required), capped, custody-sealed, and transported along with appropriate documentation to the on-site or standard analytical laboratory for analysis. Great care should be taken when preparing samples for shipment because an error in this procedure has the potential to invalidate the samples and subsequent data.

5.1 SAMPLE PREPARATION

Before a sample bottle is filled, it must first be preserved as required by the laboratory that will be running the analyses and as specified in the Sampling and Analysis Plan (see Section 3.2.7). Sample preservation requirements vary, based on the sample matrix (e.g., soil, water, sludge) and the analyses being performed. Most analyses require some form of preservation that ranges from cooling to 4°C to preserving with acid. One should always request from the laboratory performing the analyses the specific preservation requirement for each of the analyses to be performed. For chemical analysis, the only preservation typically required for soil or sediment samples is cooling the sample to 4°C. For water samples, some analyses only require cooling to 4°C, whereas others also require a chemical preservative such as nitric acid (HNO_3), sulfuric acid (H_2SO_4), hydrochloric acid (HCl), or sodium hydroxide (NaOH). Soil samples for radiological analyses rarely require any sample preservation because the radiological composition and activity levels are not influenced by temperature or other factors, as is the chemical composition. Water or liquid waste samples for radiological analyses often require preservation with nitric acid (HNO_3) to prevent isotopes from adhering to the walls of the sample container.

When an acid or base preservative is required, enough preservative is added to the sample bottle either to lower the pH to < 2 or to raise the pH to > 10. The chemicals used to preserve a sample must be of analytical grade to avoid the potential for contaminating the sample. Cooling samples to 4°C is particularly important for those that are to be analyzed for volatile organic compounds because cooling slows the rate of chemical degradation.

To avoid any difficulties associated with adding chemical preservatives to sample containers in the field, most laboratories add these preservatives to sample bottles before being shipped to the field. This reduces the chances of improperly preserving sample bottles or of introducing field contaminants into a sample bottle while adding the preservative.

DOI: 10.1201/9781003284000-5

After a sample container has been filled, a Teflon®-lined cap or lid is screwed on tightly to prevent the container from leaking. The sample label is filled out, noting the sampling time and date, sample identification number, sampling location, sampling depth, analyses to be performed, sampler's initials, etc. (see Chapter 5, Section 5.2.5). A custody seal is then attached over the cap or lid just before placing the sample bottle into the sample cooler. The custody seal is used to detect any tampering with the sample before analysis.

5.2 DOCUMENTATION

Accurate documentation is essential for the success of a sampling program. It is only through documentation that a sample can be tied to a particular sampling time, date, location, and depth. Consequently, field logbooks must be kept by every member of the field team, and should be used to record information ranging from weather conditions to the time the driller stubbed his right toe. To assist the documentation effort, standardized forms are commonly used to outline the information that needs to be collected. Some of the more commonly used forms include:

- Borehole log forms
- Well completion forms
- Well development forms
- Well purging/sampling forms
- Water level measurement forms

Other documentation needs associated with sample identification and shipment include:

- Sample labels
- Chain-of-custody forms
- Custody seals
- Shipping airbills

In addition to these documentation requirements, a file must be kept to carefully track important information, such as

- Field variances
- Equipment shipping invoices
- Sample bottle lot numbers
- Documented purity specifications for preservatives, distilled water, and calibration standards
- Instrument serial numbers
- Copies of shipping paperwork
- Quality assurance nonconformance notices

5.2.1 Field Logbooks

Field logbooks are intended to provide sufficient data and observations to enable participants to reconstruct events that occurred during projects and to refresh the memory of the field personnel if called upon to give testimony during legal proceedings. In a legal proceeding, logbooks are admissible as evidence, and consequently, must be factual, detailed, and objective.

Field logbooks must be permanently bound, the pages numbered, and all entries written with permanent ink, signed, and dated. If an error is made in the logbook, corrections can only be made by the person who made the entry. A correction is made by crossing out the error with a single line, so as not to obliterate the original entry, and then entering the correct information. All corrections must be initialed and dated.

Observations or measurements that are taken in an area where the logbook may be contaminated can be recorded in a separate bound and numbered logbook before being transferred into the master field notebook. All logbooks must be kept on file as permanent records, even if they are illegible.

The first page of the logbook should be used as a table of contents to help locate pertinent data. As the logbook is being completed, the page numbers where important events can be found should be noted in this table of contents. The very next page should begin with a record of daily events. The first daily event entry should always be the date, followed by a detailed description of the weather conditions. All of the subsequent entries should begin with the time the entry was made. Any space remaining on the last line of the entry should be lined out to prevent additional information being added in the future. At the end of the day, any unused space between the last entry and the bottom of the page should be lined out, signed, and dated to prevent additional entries from being made at a later date.

To ensure a comprehensive record of all important events, each team member should keep a daily log. The field manager's logbook should record information at the project level, such as

- Time when team members, subcontractors, and the client arrive or leave the site
- Names and company affiliation of all people who visit the site
- Summary of all discussions and agreements made with team members, subcontractors, and the client
- Summary of all telephone conversations
- Detailed explanations of any deviations from the Sampling and Analysis Plan, noting who gave the authorization and what paperwork was completed to document the change
- Detailed description of any mechanical problem that occurred at the site, noting when and how it occurred, and how it is being addressed
- Detailed description of any accidents that occurred, noting who received the injury, how it occurred, how serious it was, how the person was treated, and who was notified
- Other general information, such as when and how equipment was decontaminated, what boreholes were drilled, and what samples were collected that day

The team member's logbook should record information more at the task level. Examples of the types of information that should be recorded in these logbooks include:

- Time when surveys began and ended on a particular site
- Details on the instruments used to collect analytical measurements

- Results from instrument calibration checks
- Details on remedial activities performed at the site
- Measurement data
- Level of personal protective equipment used at the site
- Sample collection times for all samples collected
- Total depth of any boreholes drilled
- Detailed description of materials used to build monitoring wells, including the type of casing material used; screen slot size; length of screen; screened interval; brand name, lot number, and size of sand used for the sand pack; brand name, lot number, and size of bentonite pellets used for the bentonite seal; brand name and lot number of bentonite powder and cement used for grout; well identification number
- Details on when, how, and where equipment was decontaminated, and what was done with the wastewater
- Description of any mechanical problems that occurred at the site, noting when and how it occurred, and how it was addressed
- Summary of all discussions and agreements made with other team members, subcontractors, and the client
- Summary of all telephone conversations
- Detailed description of any accidents that occur, noting who received the injury, how it occurred, how serious it was, how the person was treated, and who was notified

It is essential that field team members record as much information as possible in their logbooks because this generates a written record of the project. Years after the project is over, these notebooks will be the only means of reconstructing events that occurred. With each team member recording information, it is not uncommon for one member to record information that another member missed.

5.2.2 Photographic Logbook

A photographic logbook should be maintained to store historical photographs that are identified during the scoping process (Chapter 3, Section 3.2.2). This logbook should record as much information as possible about each photograph, such as the photograph number, date and time taken, subject, and name of the photographer.

One should also carefully file all electronic photographs taken during field investigations on a share drive so that they can be easily found and retrieved whenever needed. The electronic file name given to each photograph should be detailed enough to explain the contents of the photograph. Clear photographs of field activities can be very useful in reconciling any future uncertainties that may arise.

5.2.3 Field Sampling Forms

It is recommended that standardized field sampling/measurement forms be used to assist the sampler in a number of field activities. Some commonly used forms are presented in Figures 5.1 through 5.6. Forms are most often used to reduce the amount

Borehole #________

Page ____ of______

Borehole Log Form

Project ______________	Total Depth ______________	START FINISH
Location ______________	Borehole Diameter __________	Date ______ ______
Geologic Log by __________	Depth to Water ____________	Time ______ ______
Driller ______________	Rig ______________	How Left ______________
Geophysics by ___________	Bit(s) ______________	______________
Weather ______________	Drilling Fluid ____________	______________

Depth (ft)	Pene. Rate/ Blow Cts	Circulation Q (gpm)	OVA/ HNU	Sample		Geologic and Hydrologic Description		
				#	Interval	Lith. Symbol		% Core Recovery

FIGURE 5.1 Borehole log form. (From Byrnes, 1994, *Field Sampling Methods for Remedial Investigations*, Lewis Publishers, Boca Raton, FL.)

Well Completion Form

Location: ______ Elevation: Ground Level ______
Personnel: ______ Top of Casing ______

DRILLING SUMMARY:
Total Depth ______
Borehole Diameter ______

Driller ______

Rig ______
Bit(s) ______

Drilling Fluid ______

Surface Casing ______

WELL DESIGN:
Basis:
Geologic Log
Casing String(s): C = Casing S = Screen

Casing: C1 ______
C2 ______
C3 ______
C4 ______
Screen: S1 ______
S2 ______
S3 ______
S4 ______
Centralizers ______

Filter Material ______

Cement ______

Other ______

CONSTRUCTION TIME LOG:

Task	Start Date	Start Time	Finish Date	Finish Time
Drilling:				
Geophys. Logging:				
Casing:				
Filter Placement:				
Cementing:				
Development:				
Other:				

Comments:

Key:
Bentonite
Sand
Cement/Grout
Silt
Sand pack
Clay
Drill cuttings
Screen
Gravel

FIGURE 5.2 Well completion form. (From Byrnes, 1994, *Field Sampling Methods for Remedial Investigations,* Lewis Publishers, Boca Raton, FL.)

of documentation required in the field logbook. These are also effective in reminding the sampler of what information needs to be collected, and they make it obvious when the necessary information has not been collected.

When forms are used, they should be permanently bound in a notebook, the pages should be numbered, and all entries must be written with permanent ink. If an error is made in the forms notebook, corrections can only be made by the person who made

Well development form

Project name: ______________________________

Well Number and Location: ______________________________

Development Company and Crew: ______________________________

Water Levels Date/Time: Initial ______________ Final: ______________

Total Well Depth: Initial ______________ Final: ______________

Development Date and Time: Began: ______________ Ended: ______________

Development Method(s): ______________________________

____

Total quantity of water removed: ______________________________ gallons

Date/time and Pump Depth (ft)	Discharge Rate (gpm) and Measurement Method	Field Measurements				
		Temp (°C)	Specific Conductivity (µmhos/cm)	pH	Turbidity (NTUs)	Dissolved Oxygen (mg/L)

FIGURE 5.3 Well development form.

Well Purging and Sampling Form

Project Name: ______________________________ Well No: ______________________________

Dates(s): ___________________________________ Geologist: ______________________________

Purging Method: _________________________ Pumping Rate: _____________________________

Sampling Method: _______________________________ Pumping Rate: ______________________

DATA FROM IMMEDIATELY BEFORE AND AFTER DEVELOPMENT

Depth to Water Measured from TOC (ft): Before Purging: _________ { After Purging: ___________ / After Sampling: __________

Total Purging Time (min): _______________

Depth to Sediment in well (ft): Before Purging: _______________ After Purging: _______________

	Date/time	Cumulative Volume Removed (gals)	Water Temp °C	pH	Conductivity (μ mhos/cm)	Turbidity (NTUs)	Dissolved Oxygen (mg/L)
Before							
During							
During							
During							
During							
During							
During							
During							
During							
During							
After							

Comments

FIGURE 5.4 Well purging and sampling form.

Water Level Measurements

Measurement Team: ______________________________

Project Number and Location: ______________________________

Measuring Instrument: ______________________________

Well No:	Date	Time	Tape Reading		Depth to Water (ft)	Initials	Remarks
			Measure Pt.	Water Level			

Measuring Point: Point where measurement was taken. Top of casing (TOC); Top of protective Steel casing (TOSC); Land Surface (LS), etc.
Depth of Water: Measurements should be recorded to the nearest 0.01 ft.
Remarks: Any conditions that may influence the water level measurements.

FIGURE 5.5 Water level measurement form.

(From Byrnes, 1994, *Field Sampling Methods for Remedial Investigations*, Lewis Publishers, Boca Raton, FL.)

Soil Gas Measurement Form

Project name and number: ____________________

Site: ____________________

Weather Conditions: ____________________

Background Reading: ____________________

Date	Time	Measuring Device	Reading	Units	Initials	Comments

FIGURE 5.6 Soil gas measurement form.

(From Byrnes, 1994, *Field Sampling Methods for Remedial Investigations*, Lewis Publishers, Boca Raton, FL.)

the entry. A correction is made by crossing out the error with a single line, so as not to obliterate the original entry, and then entering the correct information. All corrections must be initialed and dated. The person who completed the form should sign and date the form at the bottom of the page. It is recommended that the field manager also sign the form to confirm that it is complete and accurate.

5.2.4 Identification and Shipping Documentation

The essential documents for sample identification and shipment include the sample label, custody seal, chain-of-custody form, shipping manifest, and shipping airbill. Together, these documents allow environmental samples to be shipped or transported to an analytical laboratory under custody. If custody seals are broken when the laboratory receives the samples, it must be assumed that the samples were tampered with during shipment. Consequently, the samples will likely need to be collected over again.

5.2.5 Sample Labels

The primary objective of the sample label is to link a sample bottle to a sample number, sampling date and time, and the analyses to be performed. The sample label in combination with the chain-of-custody form is used to inform the laboratory what the sample is to be analyzed for. At minimum, a sample label should contain the following information (Figure 5.7):

- Sample identification number
- Sampling time and date
- Analyses to be performed
- Preservatives used
- Sampler's initials
- Name of the company collecting the sample
- Name and address of the laboratory performing the analyses

To save time, and to avoid the potential for errors, all of the previous information should be added to the sample label before going into the field, with the exception of the sampling time and date, and sampler's initials. This information should be added to the label after capping the sample bottle, immediately following sample collection, and should reflect the time that sampling began, as opposed to the time that sampling was completed.

An effective sample numbering system is a key component of any field sampling program because it serves to tie the sample to its sampling location. The problem with using a simple numbering system such as 1, 2, 3, … is that the number tells you nothing regarding the location, depth, or sample media.

The number of digits used in a sample numbering scheme should be discussed with the laboratory performing the analyses, because the laboratory may limit the number of digits that can be used in a sample number. If only six digits are available, an effective numbering system can be developed with a little imagination. For

SAMPLE LABEL	
Client	Date
Lab No.	Sample ID
Initials	Time
Analyze for	
Preservatives	
H_2SO_4 HNO_3 NAOH Other:	
(Laboratory Name and Address)	

CUSTODY SEAL	CUSTODY SEAL
Date	Date
Signature	Signature

FIGURE 5.7 Example of a sample label and chain-of-custody seal.

example, assume that a project has a total of nine buildings where swipe, paint, dust, concrete, and shallow soil samples are to be collected. In this example, each of the six sample number digits could be used to represent the following:

First digit: Site number (1, 2, 3, …, 9)

1 = 103-R reactor building
2 = 106-D biological laboratory
3 = 110-A pump house
4 = 121-L treatment plant
5 = 158-A testing laboratory
6 = 185-B pump house
7 = 205-D storage building
8 = 242-S reactor building
9 = 251-D transfer station

Second digit: Media number (1, 2, 3, …, 5)

1 = swipe
2 = paint
3 = dust
4 = concrete
5 = shallow soil

Third, fourth, and fifth digits: Sampling location number

S99 = swipe (1, 2, 3, ..., 99)
P99 = paint (1, 2, 3, ..., 99)
D99 = dust (1, 2, 3, ..., 99)
C99 = corehole (1, 2, 3, ..., 99)
B99 = borehole (1, 2, 3, ..., 99)

Sixth digit: Sample number

1 = first sampling interval
2 = second sampling interval
3 = third sampling interval
4 = fourth sampling interval
5 = fifth sampling interval
6 = sixth sampling interval
7 = seventh sampling interval
8 = eighth sampling interval
9 = ninth sampling interval

The sample number "85B032" would therefore indicate that the sample was collected from the 242-S reactor building, it is a shallow soil sample, it was collected from borehole number 3, and it was collected from the second sampling interval. When using both numbers and letters in a sample number, one should try to minimize the use of letters such as S, I, and O, which can easily be misinterpreted as the numbers 5, 1, and 0.

Some laboratories will allow sample numbers as long as 10 or 12 characters in length. These additional characters should be taken advantage of, because the easier it is to interpret the sample identification number, the easier it will be to interpret the data reported by the laboratory.

5.2.6 Chain-of-Custody Forms and Seals

Chain of custody is the procedure used to document who has the responsibility for ensuring the proper handling of a sample from the time it is collected to the time the resulting analytical data are reported by the laboratory to the customer. After a sample bottle has been filled, preserved, and labeled, a custody seal is signed and dated, and then placed over the bottle cap to ensure that the sample is not tampered with (see Figure 5.7). The custody seal is a fragile piece of tape designed to break if the bottle cap is turned or tampered with.

The person who signs his or her name to the custody seal automatically assumes ownership of the sample until it is packaged and custody is signed over to the laboratory, carrier, or an overnight mail service. This transfer of the custody is recorded on a chain-of-custody form (Figure 5.8). The chain-of-custody form is also the document used by the laboratory to identify which analyses to perform on which

Chain of Custody Record

Laboratory Log Number ______

Client Name	Job Number or Purchase Order Number
Project Name	
Project Manager	Sampler(s)

Sample Number	Date Sampled	Time Sampled	Matrix Type	Sample Description	Number of containers	Analyses required							Hazardous sample Special handling required	Remarks

Signature	Company	Date	Time
Relinquished by			
Received by			
Relinquished by			
Received by			
Relinquished by			
Received by			

FIGURE 5.8 Example of a chain-of-custody form.

(From Byrnes, 1994, ***Field Sampling Methods for Remedial Investigations***, Lewis Publishers, Boca Raton, FL.)

samples. When the laboratory receives a sample shipment, it assumes custody of the samples by signing the chain-of-custody form. The sample bottles in the shipment are then counted, and the requested analyses on the sample bottles are compared against the requested analyses on the chain-of-custody form. If there are any inconsistencies, the laboratory should contact the project manager for clarification. These communications should all be carefully documented.

An overnight shipping airbill can be used to document the transfer of custody from the field to the overnight mail service. If this option is selected, the shipper's airbill number should be recorded on the chain-of-custody form, which can be taped inside the sample cooler. When the laboratory signs the shipper's delivery form and the chain-of-custody form inside the sample cooler, it has assumed custody of the sample.

Although most analytical laboratories dispose of chemically contaminated samples once the analyses have been completed, they often return radiologically contaminated samples to the customer, who must then assume responsibility for disposal. Laboratories must be licensed by the Nuclear Regulatory Commission or state to receive and handle radioactive materials. The license specifies the maximum quantity of radioactive material that the laboratory can receive and store.

5.2.7 Other Important Documentation

Careful documentation should also be maintained for all incoming shipments of materials and supplies used to support sampling activities. The most critical documents to keep on file include:

- Shipping invoices
- Sample bottle lot numbers
- Purity specification for preservatives, distilled water, and calibration liquids and gases
- Instrument serial numbers and calibration logs

Copies of shipping invoices can be used by the project manager to keep track of equipment costs, and can expedite the return of malfunctioning equipment. It is critical to keep track of sample bottle lot numbers and purity specifications for all chemicals used because improperly decontaminated bottles and low-quality preservation or decontamination of chemicals can contaminate samples. Instrument serial numbers are recorded to assist in tracking which instruments are working well and which must undergo repair.

REFERENCE

Byrnes, M.E., 1994, *Field Sampling Methods for Remedial Investigations*, Boca Raton, FL: Lewis Publishers.

6 Quality Control Sampling

Field and laboratory quality control samples are collected and analyzed for the purpose of verifying that field and laboratory procedures are supporting the collection of representative analytical data that can be used to facilitate environmental decision-making.

The most frequently collected field quality control samples include:

- Equipment rinsate blanks
- Trip blanks
- Field blanks
- Field split samples

An equipment rinsate blank is collected in the field by pouring organic-/analyte-free water over the top of the decontaminated sampling equipment (e.g., samplers, spoons, and bowls). The rinsate water is then captured (e.g., usually in a stainless steel bowl) and transferred into sample bottles for analytical testing. Equipment rinsate blanks should be run for all of the same analyses that environmental samples are to be tested for at the site. The purpose of collecting these equipment rinsate blanks is to evaluate the effectiveness of the equipment decontamination procedure. If contamination is detected in the equipment rinsate blank, it indicates that the equipment decontamination procedure has failed and the analytical results from samples collected during that day will likely need to be recollected. Typically, one equipment rinsate sample is collected for every day of the sampling.

A trip blank is typically prepared by an analytical laboratory by pouring organic-/analyte-free water into 40-mL volatile organic analysis (VOA) vials. These bottles are capped tightly at the laboratory and then shipped to the field. Whenever samples that are to be analyzed for volatile organic compounds are collected in the field, a trip blank must accompany the samples from the time of collection to the time it reaches the laboratory. One trip blank should be present in each sample cooler shipment when at least one sample in the cooler is required to be run for VOA. When the laboratory receives the sample cooler, the trip blank will be removed and run for VOA. The purpose of running the trip blank is to determine if the samples contained within the cooler were contaminated during shipment. If contamination is detected in the trip blank, the samples will likely need to be recollected.

A field blank is collected by pouring organic-/analyte-free water into sample bottles in the field at the site under investigation. Once the sample bottles are full, they are capped and then sent to the laboratory for analysis of the same parameters that environmental samples are to be tested for at the site. The purpose of collecting

DOI: 10.1201/9781003284000-6

field blanks is to evaluate the potential that exists for environmental samples to be contaminated by airborne (and other) contaminants during the sample collection process. For example, a field blank may determine that airborne organic vapors in the vicinity of a sampling site may potentially be contaminating the environmental groundwater samples being collected. The need to run field blank samples will vary from one site to another and will depend on contaminant concentrations (known or suspected) at the site. The number of field blanks to be run for a particular site should be negotiated with the lead regulatory agency.

A field split sample is collected by splitting one sample in the field into two samples. These two samples are then put into two separate sample bottles, which are then shipped to two separate laboratories for analysis of the same site-specific parameters, using the same analytical methods. The results from the analyses of these samples are used to estimate interlaboratory precision. The number of field split samples to be run for a particular site (if any) should be negotiated with the lead regulatory agency.

Laboratory quality control sampling includes the running of blank samples, matrix-spike samples, and matrix-spike duplicate samples. A blank sample is organic-/analyte-free water that is run through a laboratory instrument to verify that the instrument has not been contaminated by the running of earlier samples. The results from the running of blank samples should show no evidence of any compounds. If blank samples run in an instrument show that contamination is present, the environmental samples will need to be run (or rerun) on another instrument. Blank samples are most often analyzed at a 1-per-20 (5%) frequency, which is a default value that has been used in several U.S. Environmental Protection Agency (EPA) programs for many years.

Matrix-spike and matrix-spike duplicate samples are collected for the purpose of defining the accuracy and precision of analytical measurements. These samples are most often analyzed at a 1-per-20 (5%) frequency, which is a default value that has been used in several EPA programs for many years. For more details on how matrix-spike and matrix-spike duplicate samples are run, see the EPA's SW-846 guidance at the following web site: www.epa.gov/hw-sw846.

7 Data Verification and Validation

Once data packages supporting an environmental study have been received from the laboratory, the data must be verified and/or validated to ensure that it meets the data quality objectives identified for the study (see Section 3.2.5). The following guidance manuals should be used to support the data verification/validation process:

- EPA, 2017, *National Functional Guidelines for Inorganic Superfund Methods Data Review*, EPA-540-R-2017-001, January.
- EPA, 2017, *National Functional Guidelines for Organic Superfund Methods Data Review*, EPA-540-R-2017-002, January.
- EPA, 2002, *Guidance on Environmental Data Verification and Data Validation*, QA/G-8, November.

The following text focuses on the guidance provided in the EPA (2002) manual.

The EPA QA/G-8 guidance manual provides alternative approaches to assist users in verifying and validating environmental data. The term *data verification* refers to the process of evaluating the completeness, correctness, and conformance/compliance of a specific data set against methodological, procedural, or contractual requirements. On the other hand, the term *data validation* refers to an analyte- and sample-specific process that extends the evaluation of the data beyond data verification to determine the analytical quality of a specific data set (EPA 2002).

Data verification and data validation are typically sequential steps. For example, data verification is performed during or at the end of data collection activities, whereas data validation is conducted after this, most often by an individual who is independent of both the data collector and the data user. Data validation begins with the outputs from data verification (EPA 2002). Data verification and data validation occur prior to the formal Data Quality Assessment (DQA) process that is discussed in Chapter 8.

EPA guidance for data verification and data validation allows for a graded approach. This principle recognizes that there is no "one-size-fits-all" approach to quality given the wide variety of environmental programs (EPA 2002). The level of detail and stringency of data verification and data validation efforts should depend on the specific needs of the project and program in question. Because of this, one will likely only implement a subset of the techniques offered in the EPA QA/G-8 guidance manual. In general, research studies do not require the same degree of data verification or data validation as for samples used to make important environmental decisions such as determining if cleanup levels have been met. This is because the latter case has a much greater consequence if a decision error is made (e.g., receptors being exposed to contamination).

DOI: 10.1201/9781003284000-7

Because data verification is not as labor intensive as data validation, a project will typically perform data verification on 100% of the data sets for a project. Depending on what the data sets are to be used for, it is not uncommon for a project to only perform data validation on a percentage (e.g., 10%) of these data sets. This percentage will need to be agreed to by the lead regulatory agency.

7.1 DATA VERIFICATION

The goal of data verification is to ensure and document that the reported results reflect what was actually performed. To the extent possible, records are reviewed for completeness and factual content, and against project specifications. When deficiencies in the data are identified, those should be documented for the data user's review and, where possible, resolved by corrective action (EPA 2002).

Data verification may be performed by personnel involved with the collection of the data or by an external data verifier. Sampling protocols, analytical methods, and project-specific planning documents (e.g., Sampling and Analysis Plans) are examples of sources that can provide the requirements and specifications for the environmental data collection effort. Data verification evaluates how closely these documents and procedures were followed during data generation. Each person involved in data verification should understand data generation procedures and should know project documentation requirements. Therefore, in order for data verification to be most effective, these planning documents and procedures should be readily available to all of the people involved in the process (EPA 2002).

The first step in data verification is to identify the sources that provide the requirements and specifications for the environmental data collection effort. These sources often include:

- Project-specific planning documents (e.g., Sampling and Analysis Plans)
- Program-wide planning documents (e.g., Quality Management Plan)
- Standard operating procedure for field and laboratory methods
- Analytical methods (e.g., EPA SW-846)

As the data collection effort progresses from sample collection through sample analysis, the field and laboratory personnel produce a series of records that can be verified. These records may be verified at each sequential step or during the final record review process.

Table 7.1 identifies a number of common records generated by various types of sampling and analysis operations that can be evaluated as part of the data verification process (EPA 2002).

The project-specific data requirements are identified by reviewing the applicable documents identified in Table 7.1. A comparison is then made between the project-specific data requirements identified in the documents specified in Table 7.1 and the data actually collected in the field.

There are two general results or outputs from data verification. The first output is the verified data. Verified data are those data that have been checked for a variety

TABLE 7.1
Records Commonly Used as Inputs to Data Verification

Operation	Common Records
Sample collection	Daily field logs, drilling logs, sample collection logs, sampling chain-of-custody forms, shipper's copy of air bill, field survey forms
Sample receipt	Sampling chain-of-custody forms, receiver's copy of air bill, internal laboratory receipt forms, internal laboratory chain-of-custody forms, laboratory refrigerator or freezer logs
Sample preparation	Analytical services requests, internal laboratory receipt forms, internal laboratory chain-of-custody forms, laboratory refrigerator or freezer logs, preparation logs or bench notes, manufacturer's certificates for standards or solutions
Sample analysis	Analytical services requests, internal laboratory receipt forms, internal laboratory chain-of-custody forms, laboratory refrigerator or freezer logs, manufacturer's certificates for standards or solutions, instrument logs or bench notes, instrument readouts (raw data), calculation worksheets, quality control results
Records review	Internal laboratory checklists

Source: From EPA 2002, *Guidance on Environmental Data Verification and Data Validation,* QA/G-8, November.

of factors during the data verification process, including transcription errors, correct application of dilution factors, appropriate reporting of dry weight versus wet weight, correct application of conversion factors, etc. They may also include laboratory qualifiers if assigned. Any changes to the results as originally reported by the laboratory should either be accompanied by a note of explanation from the data verifier or the laboratory, or reflected in a revised laboratory data report (EPA 2002).

The second output from data verification is referred to as *data verification records.* A main part of these records may be a "certification statement" certifying that the data have been verified. The statement should be signed by responsible personnel either within the organization or as part of external data verification. Data verification records may also include a narrative that identifies technical noncompliance issues or shortcomings of the data produced during field or laboratory activities. If data verification identified any noncompliance issues, then the narrative should identify the records involved and indicate any corrective actions taken in response (EPA 2002).

7.2 DATA VALIDATION

The primary objectives of the data validation process are to evaluate whether the data quality goals established during the planning phase have been achieved, ensure that all project requirements are met, determine the impact on data quality of those project requirements that were not met, and document the results of data validation.

Data validation should be considered to be performed on field data as well as laboratory data sets. EPA (2002) recommends that data validation of field data sets should include:

- An evaluation of the field records for consistency
- Preparation of a summary of all samples collected
- A review of all quality control data
- Summarizing deviations and determining the impact on data quality
- Preparation of field data validation report

For laboratory data sets, EPA (2002) recommends that data validation should include the following steps:

- Assemble planning documents and data to be validated
- Review summary of data verification results to determine methodologically, procedurally, and contractually required quality control compliance/noncompliance
- Review verified, reported sample results collectively for the data set as a whole, including laboratory qualifiers
- Summarize data and QC deficiencies and evaluate the impact on overall data quality
- Assign data qualification codes as necessary
- Prepare analytical data validation report

Data validation is typically performed by persons independent of the activity being evaluated, with a strong background in analytical chemistry. The appropriate degree of independence is an issue that can be determined on a program-specific basis. At a minimum, it is preferable that the validator does not belong to the same organizational unit with immediate responsibility for producing the data set.

All verified data and data verification records, including all field records generated from the sample collection activities, should be available for data validation. Table 7.2

TABLE 7.2
Examples of Documents and Records Generated during Field Activities

Type of Document or Record	Purpose of Document or Record
Instrument calibration records	Maintains accurate record of instrument calibration
Field notebook or daily activity log	Maintains accurate record of field activities by providing written notes of all activities
Sample collection logs	Maintains accurate record of samples collected
Chain-of-custody	Maintains proof that samples were not tampered with and that samples were under appropriate possession at all times

Source: From EPA 2002, *Guidance on Environmental Data Verification and Data Validation,* QA/G-8, November.

presents a list of example field records that may be generated during field activities and the purpose of each document. These records should be considered to be evaluated as part of the data validation process.

When reviewing data packages from the analytical laboratory, the data validator should review all documentation needed to allow a determination to be made of the quality of the data. For example, the data validator should ensure that the correct sample preparation and analytical method was followed as required by the planning documents driving the sampling and analyses (e.g., Sampling and Analysis Plan). The data validator should ensure that quality control sampling was performed as required by the requested procedure and that the quality control results fall within the guidelines identified by the method. Data validators typically assign qualifiers to the data sets in order to identify potential deficiencies or concerns about the quality of the data. A focused data validation effort (more detailed evaluation) may be needed for a particular data record if a serious problem is identified.

The potential outputs from data validation include either the validated data or a data validation (or a focused data validation) report. Validated data should be the same as the verified data, with the addition of any data validation qualifiers that were assigned by the data validator. The following are common data validation qualifiers:

- J – Indicates that the constituent was analyzed for and was detected. The associated value is *estimated* due to a quality control deficiency identified during data validation. The data should be considered usable for decision-making purposes.
- R – Indicates that the constituent was analyzed for and detected. However, because of an identified quality control deficiency, the data should be considered *unusable* for decision-making purposes.
- U – Indicates that the constituent was analyzed for but was not detected. The data should be considered usable for decision-making purposes.
- UJ – Indicates that the constituent was analyzed for but was not detected. Because of a quality control deficiency identified during data validation, the value reported may not accurately reflect the minimum detection limit. The data should be considered usable for decision-making purposes.
- UR – Indicates that the constituent was analyzed for but was not detected. However, because of an identified quality control deficiency, the data should be considered *unusable* for decision-making purposes.

BIBLIOGRAPHY

EPA, 2017, *National Functional Guidelines for Inorganic Superfund Methods Data Review*, EPA-540-R-2017-001, January.

EPA, 2017, *National Functional Guidelines for Organic Superfund Methods Data Review*, EPA-540-R-2017-002, January.

EPA 2002, *Guidance on Environmental Data Verification and Data Validation,* QA/G-8, November.

8 Data Quality Assessment

Data Quality Assessment (DQA) is the statistical evaluation of environmental data to determine if they meet the planning objectives of the project, and thus are of the right type, quality, and quantity to support their intended use. The guidance provided in this chapter was derived from the Environmental Protection Agency (EPA) guidance manual QA/G-9R (EPA 2006a).

The five steps that comprise the data quality assessment process are as follows:

- DQA Step 1: Review the project's objectives and sampling design.
- DQA Step 2: Conduct a preliminary data review.
- DQA Step 3: Select the statistical method.
- DQA Step 4: Verify the assumptions of the statistical method.
- DQA Step 5: Draw conclusions from the data.

These steps are designed only to evaluate data that were collected using a *statistically* based sampling design.

8.1 DQA STEP 1: REVIEW THE PROJECT'S OBJECTIVES AND SAMPLING DESIGN

DQA begins by reviewing the Data Quality Objectives (DQO) defined in the Sampling and Analysis Plan (Chapter 3, Section 3.2.7). Reviewing this document provides the context for understanding the purpose of the data collection effort and establishes the qualitative and quantitative basis for assessing the quality of the data set for the intended use. The sampling design presented in the Sampling and Analysis Plan provides important information about how to interpret the data. By studying the sampling design, the reviewer can gain an understanding of the assumptions under which the design was developed, as well as the relationship between these assumptions and the study objective. By reviewing the methods by which the samples were collected, measured, and reported, the reviewer prepares for the preliminary data review and subsequent steps of DQA (EPA 2006a).

In the process of reviewing the DQO, one should pay particular attention to decision rules identified in DQO Step 5 (Chapter 3, Section 3.2.5.5), the formulation of the null hypothesis and statistical hypothesis tests and identification of error tolerances in DQO Step 6 (Chapter 3, Section 3.2.5.6), and the sample design in DQO Step 7 (Chapter 3, Section 3.2.5.7).

DOI: 10.1201/9781003284000-8

8.2 DQA STEP 2: CONDUCT A PRELIMINARY DATA REVIEW

When sufficient documentation is present, the first activity in DQA Step 2 is to review any relevant quality assurance reports that describe the data collection and reporting process as it was actually implemented. These quality assurance reports provide valuable information about potential problems or anomalies in the data set. Specific items that may be helpful include:

- Data verification and validation reports that document the sample collection, handling, analysis, data reduction, and reporting procedures used
- Quality control reports from laboratories or field stations that document measurement system performance (EPA 2006a)

These quality assurance reports are useful when investigating data anomalies that may affect critical assumptions made to ensure the validity of the statistical tests.

Next, one should calculate basic quantitative statistical characteristics of the data (e.g., mean, median, mode, percentiles, range, and standard deviation). It is often useful to prepare a table of descriptive statistics for each population. Following this one should graph the data. This visual display is used to identify patterns and trends in the data that might go unnoticed when using purely numerical methods.

As no single graphical representation will provide a complete picture of the data set, the reviewer should choose different graphical techniques to illuminate different features of the data. If the sampling plan relied on any critical assumptions, consider whether a particular type of graph might shed light on the validity of that assumption. Usually, graphs should be applied to each group of data separately or each data set should be represented by a different symbol (EPA 2006a). Examples of commonly used graphical plots that should be considered when performing this data evaluation include:

- Histogram plots
- Box-and-whiskers plot
- Quantile plot
- Normal probability plot
- Scatter plot
- Time plot
- Posting plot

Appendix B of EPA (2006a) provides details on what these plots entail. The two outputs a reviewer should have documented at the conclusion of DQA Step 2 include:

- Basic statistical quantities
- Graphs showing different aspects of the data

8.3 DQA STEP 3: SELECT THE STATISTICAL METHOD

This step involves the selection of a statistical method that is to be used to draw conclusions from the data. If a particular statistical procedure has been specified in the planning process, the reviewer should use the results of the preliminary data review to determine if it is appropriate for the data collected. If not, then the reviewer should document what the anomaly appears to be and select a different method. Chapter 3 of *Data Quality Assessment: Statistical Methods for Practitioners* (EPA QA/G-9S) (EPA 2006b) provides alternatives for several statistical procedures. If a particular procedure has not been specified, then the reviewer should select one based upon the reviewer's objectives, the preliminary data review, and the key assumptions necessary for analyzing the data. Table 8.1 presents a summary of a few of the more common statistical methods recommended in EPA (2006a) along with some associated data requirements and assumptions.

All statistical tests make assumptions about the data. For instance, the so-called parametric tests assume some distributional form (e.g., a one-sample *t*-test assumes that the sample mean has an approximate normal distribution). The alternative, nonparametric tests make much weaker assumptions about the distributional form of the data. However, both parametric and nonparametric tests assume that the data are statistically independent or that there are no trends in the data. While examining the data, the reviewer should always list the underlying assumptions of the statistical test. Common assumptions include distributional form of the data, independence, dispersion characteristics, approximate homogeneity, and the basis for randomization in the data collection design. For example, the one-sample *t*-test needs a random sample, independence of the data, approximately normally distributed sample mean, no outliers, and a few "nondetects" (EPA 2006a).

The full statement of a statistical hypothesis has two major parts: the null hypothesis and the alternative hypothesis. It is important to take care in defining the null and alternative hypotheses because the null hypothesis will be considered true unless the data demonstratively shows proof for the alternative. In layman's terms, this is equivalent of an accused person appearing in civil court; the accused is presumed to be innocent unless shown by the evidence to be guilty because of a preponderance of evidence. Note the parallel "presumed innocent" and "null hypothesis considered true," "evidence" and "data," and "preponderance of evidence" and "demonstratively shows." It is often useful to choose the null and alternative hypotheses in light of the consequences of making an incorrect determination between them. The true condition that occurs with the more severe decision error is often defined as the null hypothesis, thus this kind of decision error is hard to make. The statistical hypothesis framework would rather allow a false acceptance than a false rejection. As with the accused and the assumption of innocence, the judicial system makes it difficult to convict an innocent person (the evidence must be very strong in favor of conviction) and, therefore, allows some truly guilty to go free (the evidence was not strong enough). The judicial system would rather allow a guilty person to go free than have an innocent person found guilty (EPA 2006a).

All hypothesis tests have a similar structure and follow five general steps, which include:

- Set up the null hypothesis
- Set up the alternative hypothesis
- Choose a test statistic

TABLE 8.1
Common Hypothesis Tests Recommended by EPA

Type of Comparison	Statistical Hypothesis Test	Random Sample	Independence	Approximately Normal Distribution	No Outliers	No (or few) Nondetects	Other Assumptions
Compare a mean to a fixed number – for example, to determine whether the mean contaminant level is greater than 10 ppm	One-sample *t*-test	•	•	•	•		
	Wilcoxon signed Rank test	•	•	—	—	—	Not many data values are identical. Symmetric
	Chen test	•	•	—	—	•	Data come from a right-skewed distribution (such as a lognormal distribution)
Compare a median to a fixed number – for example, to determine whether the median is greater than 8 ppm	Wilcoxon signed Rank test	•	•	—	—	—	Not many data values are identical. Symmetric
	Sign test	•	•	—	—	—	Not many sample values are equal to the fixed level (reduces efficiency)
Compare a proportion or percentile to a fixed number – for example, to determine if 95% of all companies emitting sulfur dioxide into the air emit below a fixed discharge level	One-sample proportion test	•	•	—	—	—	Not many sample values are equal to the fixed level (reduces efficiency)

Compare two means – for example, to compare the mean contaminant level at a remediated Superfund site to a background site or to compare the mean of two different drinking water wells	Student's two-sample t-test	•	•	•	•	—	Approximately the same underlying variance for both populations
	Satterthwaite's two-sample t-test	•	•	•	•	—	
Compare two proportions or percentiles – for example, to compare the proportion of children with elevated blood lead in one area to the proportion of children with elevated blood lead in another area	Two-sample test for proportions	•	•	—	—	—	

Source: This table summarizes some of the more commonly used hypothesis tests identified in Appendix C of EPA 2006a, *Data Quality Assessment: A Reviewer's Guide*, QA/G-9R, February.

- Select the critical value or p-value
- Draw a conclusion from the test

The two important outputs that the reviewer should have documented from DQA Step 3 include:

- The chosen statistical method
- A list of the assumptions underlying the statistical method

8.4 DQA STEP 4: VERIFY THE ASSUMPTIONS OF THE STATISTICAL METHOD

In this step, the reviewer should assess the validity of the statistical test chosen in DQA Step 3 by examining its underlying assumptions. This step is necessary because the validity of the selected method depends on the validity of key assumptions underlying the test. The data generated will be examined by graphical techniques and statistical methods to determine if there have been serious deviations from the assumptions. Minor deviations from assumptions are usually not critical as the robustness of the statistical technique used is sufficient to compensate for such deviations (EPA 2006a).

If the data do not show serious deviations from the key assumptions of the statistical method, then the DQA process continues to Step 5 (Drawing conclusions from the data). However, it is possible that one or more of the assumptions may be called into question, and this could result in a reevaluation of one of the previous steps. If a different statistical method is required to be selected, refer to EPA (2006b) for a comprehensive list of options.

The two outputs a reviewer should have documented at the conclusion of DQA Step 4 include:

- Documentation of the method used to verify each assumption together with the results from these investigations
- A description of any corrective actions that were taken (EPA 2006a)

8.5 DQA STEP 5: DRAW CONCLUSIONS FROM THE DATA

In this step, the reviewer performs the statistical hypothesis test or computes the confidence/tolerance interval, and draws conclusions that address the project objectives. All of the activities conducted up to this point should ensure that the calculations performed on the data set and the conclusions drawn here (in Step 5) address the reviewer's needs in a scientifically defensible manner (EPA 2006a).

REFERENCES

EPA 2006a, *Data Quality Assessment: A Reviewer's Guide,* QA/G-9R, February.

EPA 2006b, *Data Quality Assessment: Statistical Methods for Practitioners,* EPA QA/G-9S, February.

9 Equipment Decontamination

When performing environmental investigations, all sampling equipment must be treated as if it is contaminated, and therefore should be thoroughly decontaminated between sampling points. Decontamination procedures for chemically and radiologically contaminated equipment are presented in Sections 9.1 and 9.2, respectively.

Decontamination is defined as the process of removing, neutralizing, washing, and rinsing surfaces of equipment and personal protective clothing to minimize the potential for contaminant migration. It is performed to ensure the collection of representative environmental samples. The only way to eliminate decontamination is by using disposable equipment.

It is critical to test the effectiveness of any decontamination procedure so that the credibility of environmental samples cannot be questioned. At chemically and radiologically contaminated sites, this is performed through the collection and analysis of equipment rinsate blanks, as described in Chapter 6. If the results from the analysis of equipment rinsate samples show contamination, then the decontamination procedure was not effective. In this case, analytical data for any samples collected with the contaminated equipment should not be used. At radiologically contaminated sites, it is not uncommon to also scan the decontaminated equipment using handheld gross activity measurement instruments or collect swipe samples (Chapter 4, Section 4.2.9.1) for radiological counting. If the results from the scanning measurements and swipe sampling exceed radiological release limits, additional decontamination steps must be taken to remove the contamination from the equipment. If the results are below the release limit, the equipment may be released for use in collecting additional samples.

Wastewater generated by the decontamination process must be contained and treated in accordance with the site-specific Waste Management Plan that has been approved by the lead regulatory agency (see Chapter 11).

9.1 CHEMICAL DECONTAMINATION PROCEDURE

The following sections describe chemical decontamination procedures for large equipment used to support environmental sampling as well as for the environmental sampling equipment itself.

9.1.1 Large Equipment Decontamination Procedure

The following procedure should be used to decontaminate all types of large equipment that support sampling activities. Such equipment may include, but is not limited to,

DOI: 10.1201/9781003284000-9

drilling pipe, drilling casing, drilling augers, drill bits, drill rod, well casing, and well screens.

1. Remove soil adhering to the equipment by scraping, brushing (wire brush), or wiping.
2. Use a pressure washer to steam-clean the equipment thoroughly with potable water and a nonphosphatic laboratory-grade detergent. The pressure washer should produce a minimum water pressure of 80 psi and a minimum temperature of 180°F.
3. Use a pressure washer to rinse the equipment with tap water.
4. Allow the equipment to air-dry.
5. Wrap the equipment in plastic sheeting to keep it clean before use.

The wastewater and solid waste material generated from this decontamination procedure must be containerized, handled, and disposed of following the requirements of a site-specific Waste Management Plan (see Chapter 11). This procedure should preferably be performed at a permanent (Figure 9.1) decontamination pad. However, if a permanent pad is not available, a temporary pad can be used (Figure 9.2).

9.1.2 Sampling Equipment Decontamination Procedure

The following procedures should be considered to decontaminate all types of equipment used to perform chemical sampling activities. These equipment include, but are not limited to:

- Soil gas sampling equipment
- Deep soil sampling equipment
- Shallow soil sampling equipment
- Sediment sampling equipment
- Surface water/groundwater sampling equipment
- Concrete/asphalt sampling equipment
- Paint sampling equipment
- Air sampling equipment
- Support equipment coming in contact with the sample (e.g., bowls and spoons)

The first three decontamination procedures presented below are recommended by the Environmental Protection Agency (EPA Region 4 2020, EPA 2002) as methods that could be used to decontaminate sampling equipment employed to collect samples containing trace organic and inorganic constituents.

9.1.2.1 Organic and Inorganic Decontamination Procedure (EPA Region 4 2020)

1. Clean with tap water and Luminox® detergent using a brush (if necessary) to remove particulate matter and surface films

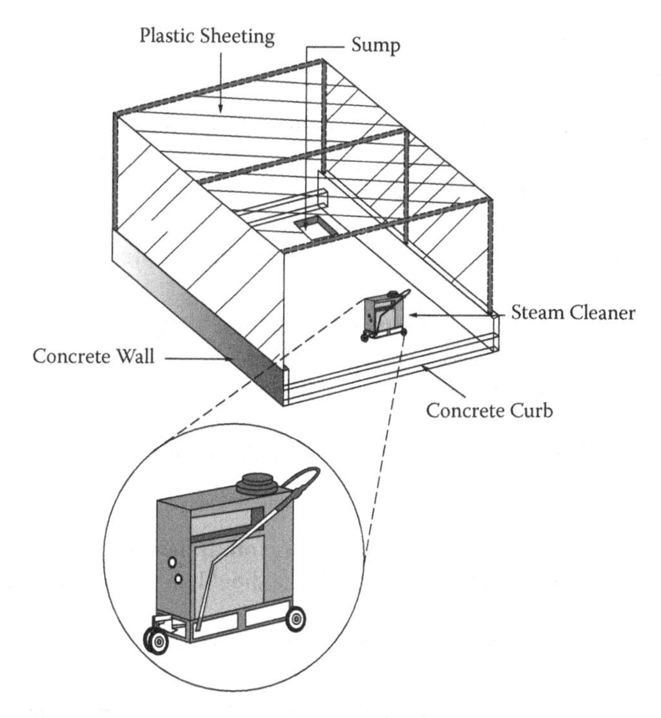

FIGURE 9.1 Permanent decontamination pad with sump and steam cleaner.

(From Byrnes, 1994, *Field Sampling Methods for Remedial Investigations*, Lewis Publishers, Boca Raton, FL.)

FIGURE 9.2 Temporary decontamination pad.

(From Byrnes, 1994, *Field Sampling Methods for Remedial Investigations*, Lewis Publishers, Boca Raton, FL.)

2. Rinse thoroughly with tap water
3. Rinse thoroughly with organic-free water and place on a clean foil-wrapped surface to air-dry
4. Wrap the dry equipment with aluminum foil. If the equipment is to be stored overnight before it is wrapped in foil, it should be covered and secured with clean unused plastic sheeting

9.1.2.2 Alternative Solvent Rinse Organic and Inorganic Decontamination Procedure (EPA Region 4 2020)

1. Clean with tap water and Luminox® detergent using a brush (if necessary) to remove particulate matter and surface films.
2. Rinse thoroughly with tap water.
3. Rinse thoroughly with deionized water.
4. Rinse with an appropriate solvent (generally isopropanol).
5. Rinse with organic-free water and place on a clean foil-wrapped surface to air-dry.
6. Wrap the dry equipment with aluminum foil. If the equipment is to be stored overnight before it is wrapped in foil, it should be covered and secured with clean unused plastic sheeting.

9.1.2.3 Organic and Inorganic Decontamination Procedure (EPA 2002)

1. Clean the sampling equipment with tap water and phosphate-free laboratory detergent (e.g., Liquinox®), using a brush if necessary to remove particulate matter and surface films.
2. Rinse thoroughly with tap water from any municipal water treatment system.
3. Rinse thoroughly with deionized/organic-free water. (Deionized/organic-free water is tap water that has been treated with a standard deionizing resin column and activated carbon, and must meet the criteria of deionized [analyte-free] water. It should not contain detectable pesticides, herbicides, or extractable organic compounds, or volatile organic compounds above minimum detectable levels.)
4. Rinse thoroughly with pesticide-grade isopropanol. Do not rinse polyvinyl chloride or plastic items with isopropanol. (For equipment highly contaminated with organics [such as oily waste], a laboratory-grade hexane may be a more suitable alternative to isopropanol.)
5. Rinse thoroughly with deionized/organic-free water and allow the equipment to dry completely.
6. Remove the equipment from the decontamination area. Equipment stored overnight should be wrapped in aluminum foil and covered with clean, unused plastic.

Three additional chemical sampling equipment decontamination procedures can be considered. The first procedure may be used when samples are to be analyzed for inorganic parameters only, the second when samples are to be analyzed for organic parameters only, and the third if samples are to be analyzed for both organic and inorganic parameters.

Inorganic Decontamination Procedure

1. Remove soil adhering to the equipment by scraping, brushing (wire brush), or wiping.
2. Wash thoroughly with a strong nonphosphate detergent/soap (e.g., Liquinox®) wash.
3. Rinse thoroughly with tap water.
4. Rinse with American Society for Testing Materials (ASTM) Type II (or equivalent) water.
5. Rinse with dilute hydrochloric or nitric acid solution.
6. Rinse with ASTM Type II (or equivalent) water.
7. Allow the equipment to air-dry.
8. Wrap the equipment in aluminum foil (shiny side out) to keep it clean before use.

Organic Decontamination Procedure

1. Remove soil adhering to the equipment by scraping, brushing (wire brush), or wiping.
2. Wash thoroughly with a strong nonphosphate detergent/soap (e.g., Liquinox®) wash.
3. Rinse thoroughly with tap water.
4. Rinse with ASTM Type II (or equivalent) water.
5. Rinse with pesticide-grade acetone (or methanol).
6. Rinse with pesticide-grade hexane.
7. Allow the equipment to air-dry.
8. Wrap the equipment in aluminum foil (shiny side out) to keep it clean before use.

Combined Inorganic/Organic Decontamination Procedure

1. Remove soil adhering to the equipment by scraping, brushing (wire brush), or wiping.
2. Wash thoroughly with a strong nonphosphate detergent/soap (e.g., Liquinox®) wash.
3. Rinse thoroughly with tap water.
4. Rinse with ASTM Type II (or equivalent) water.
5. Rinse with dilute hydrochloric or nitric acid solution.
6. Rinse with ASTM Type II (or equivalent) water.
7. Rinse with pesticide-grade acetone (or methanol).
8. Rinse with pesticide-grade hexane.
9. Allow the equipment to air-dry.
10. Wrap the equipment in aluminum foil (shiny side out) to keep it clean before use.

FIGURE 9.3 Decontamination line.

The dilute acid solution, acetone, hexane, isopropanol, and distilled/deionized water are most easily handled when they are contained within Teflon squirt bottles. Squirt bottles made of substances other than Teflon should not be used because they have the potential to contaminate the solutions. When using these solutions, the drippings should be caught in a bucket or tub and then transferred into a waste drum (Figure 9.3).

A decontamination line should be set up as shown in Figure 9.3. Tub 1 and tub 2 are to be used for the soap wash and clean water rinse, respectively. Tub 3 is used to collect acid- and solvent-rinse solutions. A table is employed to set the decontaminated equipment on, to allow it to air-dry. The decontamination line should never be set up downwind of any sampling operations because contaminants carried in the air could affect the equipment. Therefore, the decontamination line is preferably set up crosswind of sampling operations.

Decontaminating pumps is more difficult than decontaminating other sampling equipment because many of them are not easy to disassemble. The procedure to use for these tools is to place the pump in a large decontamination tub full of nonphosphate detergent and tap water. The pump is turned on, which forces the wash water through the pump and discharge line. The same procedure is repeated in a tub full of clean tap water. The outer surfaces of the pump and the pressure and discharge lines are then decontaminated using the sampling decontamination methods described earlier.

All of the water generated from this decontamination procedure must be containerized and then analytically tested to determine the appropriate method for disposal. Discharge permits must be acquired from the city or county before discharging the decontaminated water into any sanitary or storm sewer system.

9.2 RADIOLOGICAL DECONTAMINATION PROCEDURES

The following four methods are available to support the decontamination of radiologically contaminated sampling and drilling equipment, and are presented in the order of increasing levels of contamination. These procedures were derived from *Sampling and Surveying Radiological Environments* (Byrnes 2001). One should begin with the first decontamination method if that is practical. If the first is not effective in removing the contamination, then proceed to the next method.

9.2.1 Tape Method

1. Apply masking tape or duct tape to the surface of the equipment.
2. Remove the tape.
3. Repeat steps 1 and 2 until the entire surface of the equipment has been covered.
4. Scan the surfaces of the equipment for gross alpha and gross beta/gamma activity to determine if the equipment meets radiological release limits.
5. Collect swipe samples from the areas showing the highest activity (based on the scanning measurements) to determine if the equipment meets radiological release limits for removable gross alpha and gross beta/gamma activity.
6. If radiological release limits have been met, release the equipment for reuse. Wrap the decontaminated sampling equipment in aluminum foil to keep it clean before use. Drilling equipment may be wrapped in plastic sheeting to keep it clean before use. If radiological release limits have not been met, proceed to the methods described in Section 9.2.2 and perform the next level of decontamination.
7. Document the results from the scanning surveys and swipe samples.

9.2.2 Manual Method

1. Remove soil adhering to the equipment by scraping, brushing (wire brush), or wiping.
2. Wash thoroughly with a strong nonphosphate detergent/soap wash.
3. Rinse the equipment with tap water.
4. Rinse the equipment with ASTM Type II (or equivalent) water. (Note that this step is required only when decontaminating sampling equipment.)
5. Allow the equipment to air-dry.
6. Scan the surfaces of the equipment for gross alpha and gross beta/gamma activity to determine if the equipment meets radiological release limits.
7. Collect swipe samples from the areas showing the highest activity (based on the scanning measurements) to determine if the equipment meets radiological release limits.
8. If radiological release limits have been met, release the equipment for reuse. Wrap the decontaminated sampling equipment in aluminum foil to keep it clean before use. Drilling equipment may be wrapped in plastic sheeting to keep it clean before use. If radiological release limits have not been met, proceed to the methods described in Section 9.2.3.
9. Document the results from the scanning surveys and swipe samples.

9.2.3 HEPA Vacuum Method

1. Remove soil adhering to the equipment by scraping, brushing (wire brush), or wiping.
2. Use a high-efficiency particulate air (HEPA) vacuum system to vacuum surfaces of the equipment.
3. Rinse the equipment with tap water.
4. Rinse the equipment with ASTM Type II (or equivalent) water. (Note that this step is required only when decontaminating sampling equipment.)
5. Allow the equipment to air-dry.
6. Scan the surfaces of the equipment for gross alpha and gross beta/gamma activity to determine if the equipment meets radiological release limits.
7. Collect swipe samples from the areas showing the highest activity (based on the scanning measurements) to determine if the equipment meets radiological release limits.
8. If radiological release limits have been met, release the equipment for reuse. Wrap the decontaminated sampling equipment in aluminum foil to keep it clean before use. Drilling equipment may be wrapped in plastic sheeting to keep it clean before use. If radiological release limits have not been met, proceed to the methods described in Section 9.2.4.
9. Document the results from the scanning surveys and swipe samples.

9.2.4 High-Pressure-Wash Method

1. Remove soil adhering to the equipment by scraping, brushing (wire brush), or wiping.
2. Use a pressure washer to steam-clean the equipment thoroughly with potable water and a nonphosphatic laboratory-grade detergent. The pressure washer should produce a minimum water pressure of 80 psi and a minimum temperature of 180°F.
3. Use a pressure washer to rinse the equipment with tap water.
4. Rinse the equipment with ASTM Type II (or equivalent) water. (Note that this step is only required when decontaminating sampling equipment.)
5. Allow the equipment to air-dry.
6. Scan the surfaces of the equipment for gross alpha and gross beta/gamma activity to determine if the equipment meets radiological release limits.
7. Collect swipe samples from the areas showing the highest activity (based on the scanning measurements) to determine if the equipment meets radiological release limits.
8. If radiological release limits have been met, release the equipment for reuse. Wrap the decontaminated sampling equipment in aluminum foil to keep it clean before use. Drilling equipment may be wrapped in plastic sheeting to keep it clean before use. If radiological release limits have not been met for sampling equipment, dispose of the equipment properly in a radiological permitted landfill. If radiological release limits have not been met for large equipment, it may be considered for reuse only if it does not come in direct

contact with future environmental samples to be collected and only if the activity levels are low enough to not be a health hazard to workers. This is because the radiological contamination is "fixed" on the equipment. However, this radiologically contaminated equipment must not leave the radiologically controlled site where it is currently located.

9. Document the results from the scanning surveys and swipe samples.

BIBLIOGRAPHY

Byrnes, M.E., 2001, *Sampling and Surveying Radiological Environments*, Boca Raton, FL: Lewis Publishers.

Byrnes, M.E., 1994, *Field Sampling Methods for Remedial Investigations*, Boca Raton, FL: Lewis Publishers.

EPA Region 4, 2020, *Field Equipment Cleaning and Decontamination*, LSASDPROC-205-R4, June.

EPA, 2002, *RCRA Waste Sampling Draft Technical Guidance,* EPA 530-D-02-002.

EPA, 1987, *A Compendium of Superfund Field Operations Methods,* EPA/540/P-87/001a.

EPA, 1989, *RCRA Facility Investigation (RFI) Guidance,* PB89-200299.

10 Health and Safety

As a result of the passing of the Resource Conservation and Recovery Act (RCRA) in 1976 and the Comprehensive Environmental Response, Compensation, and Liability Act (CERCLA) in 1980, numerous jobs were created in the hazardous waste industry. In the process of creating these jobs, another dilemma had been created. Laws to protect the workers' health and safety were lacking, thereby putting this group at the mercy of their employer's safety policies. This void was addressed in 1986 when the President signed the Superfund Amendments and Reauthorization Act (SARA; Chapter 2, Section 2.1.1) into law, which addresses the need to protect employees from exposure to hazardous wastes. Specific safety requirements are outlined under 29 CFR 1910.120.

The development of a Project Health and Safety (H&S) Plan is the single most important element in protecting the health and safety of the field worker. It is only through thorough planning that potential hazards may be anticipated, and steps taken to minimize the risk of harm to the workers. The primary objective of the H&S Plan is to identify, evaluate, and control safety and health hazards, and provide emergency response. The specific elements that the plan must address include:

- A hazard risk analysis
- Employee training requirements
- Personal protective equipment (PPE)
- Medical surveillance requirements
- Frequency and type of air monitoring
- Site Control Plan
- Decontamination
- Emergency response
- Confined-space entry procedure
- Spill containment procedure

Before writing the H&S Plan, an off-site data-gathering effort should be conducted to identify all suspected conditions that may pose inhalation or skin absorption hazards that are immediately dangerous to life or health (IDLH), or other conditions that may cause death or serious injury. An on-site survey should then be performed using appropriate protective equipment. After the site has been determined as safe for the initiation of field activities, a monitoring program should be implemented to continue to ensure the worker's safety.

After the H&S Plan has been written, it is essential that all field personnel, including subcontractors, be required to read it. The field manager should ensure that workers have read and understood the H&S Plan by having them signed a training record.

DOI: 10.1201/9781003284000-10

The following sections will discuss the various aspects of training requirements, medical surveillance, hazards overview, PPE, air monitoring, site control, decontamination, handling drums, and other equipment.

10.1 TRAINING REQUIREMENTS

According to 29 CFR 1910.120, all employees exposed to hazardous substances, health hazards, or safety hazards must be thoroughly trained in:

- Safety, health, and other hazards present on the site
- Proper use of PPE
- Work practices that can minimize risks from hazards
- Safe use of engineering controls and equipment on the site
- Medical surveillance requirements

At the time a job is assigned, an employee must receive a minimum of 40 hours of initial instruction off-site, and a minimum of 3 days of actual field experience under the direct supervision of a trained, experienced supervisor. Workers who may be exposed to unique hazards must be provided with additional training. For example, employees who are responsible for responding to hazardous emergencies must be trained in what emergencies to anticipate, and how to respond to each situation.

Personnel managers who are responsible for supervising employees engaged in hazardous waste operations must meet all the training requirements outlined above, and have at least eight additional hours of specialized training on managing such operations. This training must be received at the time the job is assigned.

Employees and managers are required to receive 8 hours of refresher training once a year. This training should provide a refresher summary of the most important topics discussed during the 40-hour training, as well as some new material.

10.2 MEDICAL SURVEILLANCE

Field personnel who work in areas where there is a potential of exposure to hazardous substances at or above regulatory limits should have a medical examination before entering the field and at least once every 12 months after this time. An additional medical examination should be performed if a worker begins to show signs or symptoms of overexposure, or is leaving the company.

The objectives of the initial medical examination are to:

- Determine an individual's fitness to perform the work
- Determine the worker's ability to wear protective equipment
- Provide baseline data for comparison with future medical data

The fitness of a worker should be determined by a physician by assessing the person's medical history and by performing a physical examination. The physical examination should be conducted by a licensed physician, and should involve an assessment of at least:

- Height, weight, temperature, pulse, respiration, and blood pressure
- Head, nose, and throat
- Eyes, including vision tests that measure refraction, depth perception, and color vision
- Ears, including audiometric tests, and ear drum for perforation
- Chest, including heart and lungs
- Peripheral vascular system
- Abdomen and rectum
- Spine and other components of the musculoskeletal system
- Genitourinary system
- Skin
- Nervous system

Tests that should be performed in conjunction with the physical examination include:

- Blood
- Urine
- Chest X-ray

The type of protective equipment that the worker can wear without endangering his/her health must be determined by the physician. Workers who have severe lung disease, heart disease, or back problems should never be put in a position where they need to wear heavy protective equipment such as a self-contained breathing apparatus (SCBA). According to 29 CFR 1910.134, a special assessment of a worker's capacity to perform work while wearing a respirator must be performed and the results must be documented before beginning the work.

Medical screening is used not only to determine one's fitness level or ability to wear protective equipment, but also to establish baseline data. Baseline data is the information that is later used to evaluate whether any exposures have occurred in the field. After the baseline examination, annual medical monitoring should be performed. The content of the annual examination is determined by the physician, and may vary from partial to full medical examination. In the case of potential overexposure, a full examination should always be performed.

Termination examinations are typically performed when an employee leaves a company. The purpose of the examination is to document the health of the individual at that point in time. If the individual should suffer an exposure several years later while performing work for another company, the termination examination records would provide important data that the previous company could defend itself with, should there be a legal issue.

10.3 HAZARD OVERVIEW

Hazardous waste sites pose a multitude of health and safety concerns, which could result in serious injury or death. Some of the hazards of greatest concern include:

- Electrical hazards
- Chemical exposure

- Biological exposure
- Radiological exposure
- Oxygen deficiency
- Fire and explosion
- Heat stress
- Noise
- Heavy equipment accidents

Hazardous waste sites are particularly dangerous because the identity of substances present are often unknown, the typical disorderly nature of these sites makes them unpredictable, and the need to wear bulky protective clothing usually adds to the hazard.

Overhead power lines, downed electrical wires, and buried cables all pose a danger of shock or electrocution. To minimize the hazards posed by working in this environment, it is important to:

- Wear special clothing designed to protect against electrical hazards as described by 29 CFR 1910.137
- Use low-voltage equipment with ground-fault interrupters and watertight, corrosion-resistant connecting cables
- Monitor weather conditions, and suspend work during electrical storms
- Properly ground capacitors that may retain a charge before handling

When working in chemical and radiological environments, exposure can be controlled by using engineering controls, wearing appropriate PPE, using health and safety monitoring equipment, and minimizing the length of exposure. Before beginning work in these environments, an H&S Plan must be written that addresses all the safety concerns, including how to respond in case of an emergency. All employees should be trained in these procedures before beginning work.

Oxygen deficiency hazards can be overcome by avoiding work in confined spaces whenever possible, and by requiring the use of oxygen meters. Fire and explosion hazards can be minimized by controlling the use of open flames, using nonsparking tools, and using an explosivity meter in potentially explosive environments. Fire escape routes should be identified before beginning work.

When work is being performed in hot and humid environments, heat stress is a common hazard. To avoid this condition, the Health and Safety Officer is responsible for seeing that workers take regular breaks, recharge their body fluids, and wear appropriate protective equipment.

Whenever noise is a potential hazard, employees should be required to wear appropriate hearing protection. When working around heavy equipment, accidents can be avoided by wearing PPE such as hard hats, safety glasses, and steel-toed boots; using two-way radios to improve communication; and by working at a steady, unrushed pace.

10.4 ENGINEERING CONTROLS

Engineering controls should be used to help reduce the concentration of contaminated air in a working environment. These types of controls can significantly reduce contamination levels in the breathing zone, thus reducing the respiratory protection required to be worn at the job site.

10.5 AIR MONITORING

The health and safety of the site worker may be jeopardized by airborne contaminants. This danger can be controlled by using screening instruments to monitor air quality. Most air-screening instruments used for H&S are designed to measure gases and vapors in the air, and provide immediate monitoring results. These instruments are used by field samplers to provide early warning signs of a hazardous environment. Some of the advantages of these direct-reading instruments include the ability to provide immediate information regarding explosive, flammable, and oxygen-deficient environments, and provide information on a specific gas or a general screen of gases.

Some of the most common direct-reading instruments include the combustible gas indicator (CGI), flame ionization detector (FID), photoionization detector (PID), oxygen meters, and a multitude of other instruments discussed earlier in Chapter 4, Section 4.1.8.

When working in radiological environments, the inhalation of radioactive dust particles can be avoided by wearing air-filtering respirators. As there are no radiological instruments that can provide immediate air quality information, respirators should always be worn when working in these environments.

10.6 RADIOLOGICAL SCREENING INSTRUMENTS

Radiological screening instruments should always be used to monitor field personnel before leaving a radiological environment. The purpose of the screening is to determine if a worker has been contaminated, and to prevent the spread of contamination from the site.

To perform the screen, the probe of the alpha, beta, or gamma meter is passed very slowly over the worker's body, focusing on those areas that have the highest probability of being contaminated. This screen typically focuses on the soles of the worker's shoes, and on his/her hands, arms, shoulders, and knees.

If elevated activity levels are identified on a piece of clothing, it must be removed and properly disposed of. If contamination is detected on a person's skin, the contaminated area should be carefully washed with soap and water, and then rescreened to determine whether the contamination has been removed. To prevent the skin pores from contracting or expanding, the temperature of the wash water should be approximately equal to body temperature. If the contamination was not removed, the procedure should be repeated. A soft brush can be used, but care should be taken not to imbed the contamination deeper into the skin.

10.7 SITE CONTROL

Site control is a requirement of 29 CFR 1910.120 because it is needed to prevent the spread of contamination and to protect outsiders from the site hazards. Site control can most easily be obtained by breaking the site into an Exclusion Zone, a Contamination Reduction Zone, and a Support Zone (Figure 10.1). A Hot Line and a Contamination Control Line should be used to control the spread of contamination beyond the boundary of the Exclusion Zone and Contamination Reduction Zone, respectively. These lines should be marked very clearly using signs, placards, or hazard tape.

The Exclusion Zone is the area where the contamination is located at the beginning of the field program, and every effort should be taken to keep the contamination from spreading from this area. This zone can be subdivided based on the types of contaminants, or degree of hazards. Subdividing the zone allows more flexibility in safety requirements, operations, and decontamination procedures.

The Contamination Reduction Zone is the area used to decontaminate workers exiting the Exclusion Zone. The contamination level should decrease in this zone as one gets closer to the Support Zone. To reduce the spread of contamination within this zone, a Contamination Reduction Corridor should be used. This corridor should contain all of the contamination reduction wash tubs, and should be the only place

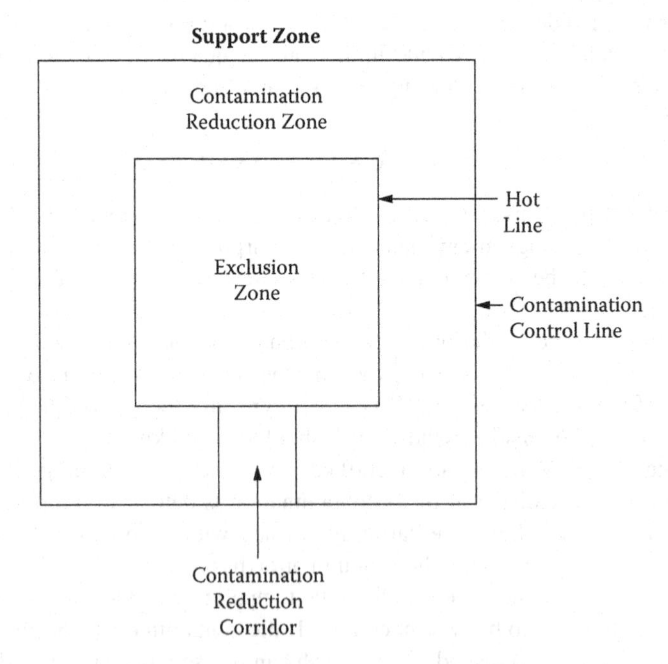

FIGURE 10.1 Zones used to control the spread of contamination.

(From Byrnes 1994, *Field Sampling Methods for Remedial Investigations*, Lewis Publishers, Boca Raton, FL.)

where workers can enter and exit the Exclusion Zone. This corridor should have at least two lines of decontamination tubs: one for personnel and the other for heavy equipment. This area should also be used to store emergency response equipment, fire extinguisher, PPE, sampling equipment, sampling packaging materials, and a worker temporary rest area.

The Support Zone is the uncontaminated area where workers should not be exposed to any hazardous conditions. Any functions that cannot be performed in a hazardous area are performed here. Personnel in this zone are responsible for alerting the proper agencies in the event of an emergency. All emergency telephone numbers, evacuation route maps, and vehicle keys should be kept in this zone.

10.8 PERSONAL PROTECTIVE EQUIPMENT

The purpose of PPE is to shield or isolate individuals from chemical, physical, or biological hazards that may be encountered while working in a contaminated environment. Various types of PPE are effective in protecting the respiratory system, skin, eyes, face, hands, feet, head, body, and auditory system.

It is mandatory that hazardous waste operations, including site characterization, comply with the Occupational Safety and Health Administration (OSHA) outlined in 29 CFR 1910.120 and 29 CFR 1910.132 through 1910.137. Under 29 CFR 1910.120, OSHA requires that all employees working in potentially hazardous environments be trained in how to safely use PPE before beginning their jobs. This requirement has resulted in a dramatic reduction in the number of injuries that occur each year.

The Environmental Protection Agency (EPA) has established four levels of worker protection (Table 10.1). These range from Level A protection, which is designed to protect workers in extremely hazardous and potentially oxygen-deficient environments, to Level D, which is designed for very low-hazard environments. Within each of these levels, there are recommended types of PPE designed to protect the worker. Table 10.1 presents recommended PPE for each level of protection. The level of protection selected should also take the worker's location and/or job function into consideration. For example, the worker in the Exclusion Zone may require Level A protection, whereas Level B or C protection may be adequate for support staff in the Contamination Reduction Zone (Table 10.1).

When choosing chemical protective clothing, it is necessary to keep in mind that no one type of clothing material can protect against all chemicals. Therefore, the clothing should be selected using historical information about the chemicals used at the facility. Similar information for body suits can be obtained from vendors.

When using PPE, each worker must assume responsibility for monitoring the effectiveness of the clothing and equipment. Problems with the PPE may be realized through:

- Degradation
- Perception of odors
- Skin irritation
- Unusual residues on PPE

TABLE 10.1
Personal Protective Equipment for Worker Protection Levels A–D

Level A:
(1) Respirator (supplied air)
Positive pressure/pressure demand
SCBA
Supplied air (with emergency egress unit)
(2) Fully encapsulating chemical-resistant suit
(3) Gloves (chemical resistant)
(4) Boots (chemical resistant), steel toe, and shank

Level B:
(1) Respirator (supplied air)
Positive pressure/pressure demand
SCBA (may be external backpack)
Supplied air line
(2) Chemical-resistant clothing
Coverall, splash suits (one- or two-piece)
(3) Gloves (chemical resistant)
(4) Boots (outer; chemical resistant), steel toe, and shank

Level C:
(1) Air-purifying respirator (NIOSH-approved)
(2) Chemical-resistant clothing, splash suit (one- or two-piece)
(3) Gloves (outer; chemical resistant)
(4) Boots (outer; chemical resistant), steel toe, and shank

Level D:
(1) Coveralls
(2) Boots/shoes (leather or chemical resistant)
(3) Miscellaneous (i.e., head, eyewear, hearing protection, etc.)

- Discomfort
- Resistance to breathing
- Interference with vision or communication
- Personal responses such as rapid pulse, nausea, and chest pain

Before using a respirator, or SCBA, the equipment should always be inspected to ensure that the rubber or elastomer parts are pliable, and that there are no cracks, pinholes, or disfigurations in the facepiece, headband, and valves.

A worker must be fit tested before using a respirator to assure an effective seal. Some physical conditions that can prevent a secure respirator facepiece seal include:

- Facial hair
- Facial scars
- Hollow temples

- Prominent cheekbones
- Dentures or missing teeth

Whenever these or other conditions prevent a good face seal, it is required that the individual not be allowed to wear a respirator.

A respirator can be fit tested by several methods. The most effective test involves a person wearing the respirator to enter a test atmosphere containing a test agent in the form of an aerosol, vapor, or gas. An analytical instrument is then used to quantitatively measure the concentration of the gas both inside and outside the respirator. Another method involves exposing the worker to an irritant smoke, or odorous vapor. The worker will not smell the vapor through the respirator if there is a good seal. Finally, a negative and positive pressure test can be performed. To perform the negative pressure test, the worker cups the respirator inlets with his/her hands and gently inhales for about 10 seconds. Any inward rush of air indicates a poor fit. To perform the positive pressure test, the worker gently exhales while covering the exhalation valve of the respirator to ensure that a positive pressure is built up. Failure to build a positive pressure indicates a poor fit.

When reusable chemical protective suits are used, they must be inspected regularly for cracks, punctures, and discolorations on the inside of the suit. One problem with reusable suits is that chemicals that have begun to permeate the clothing may not be removed during decontamination. Therefore, they may continue to diffuse through the material toward the inside surface and become a hazard for the next person who uses the suit. When determining if clothing can be reused, one should evaluate permeation rates, toxicity of contaminants, care taken when clothing was decontaminated, and whether decontamination is degrading the material.

REFERENCES

29 CFR 1910.120, www.osha.gov.

29 CFR 1910.132 through 1910.137, www.osha.gov.

Byrnes, M.E., 1994, *Field Sampling Methods for Remedial Investigations*, Boca Raton, FL: Lewis Publishers.

11 Management of Investigation-Derived Waste

When performing remedial investigations, a number of different types of investigation-derived waste (IDW) will be generated. This includes a variety of wastes generated from field investigations such as drilling, sampling, decontamination, and other activities associated with defining the nature and extent of contamination. A few examples of waste materials generated by these types of activities include:

- Drill cuttings
- Well development and purge water
- Waste soil, sediment, and water
- Used personal protective equipment (e.g., gloves, chemical-resistant clothing, etc.)
- Decontamination liquids
- Used sampling equipment
- Paper towels
- Sample bottles

IDW does not include wastes that are generated from the removal or displacement of environmental media, or debris as a result of remediation activities. Rather, these wastes are referred to as *remediation wastes* and are regulated separately.

After being generated, IDW must be handled and disposed of in accordance with all applicable federal and state regulatory requirements. Due to the complexity of these regulations, it is critical that each environmental sampling program has a dedicated waste management specialist on staff who is familiar with all of the applicable federal and state waste management and waste disposal regulations. This specialist should oversee the management of IDW and develop all waste profiles and waste designations for the project. Misclassifying a waste material can have very serious financial and legal consequences, and as a result it should not be treated lightly. Because the management and disposal of IDW can be quite expensive, project managers should strive to minimize the generation of this waste as much as possible.

Generally speaking, IDW is a waste that is generated in the process of investigating or examining a potentially contaminated site. IDW includes Comprehensive Environmental Response, Compensation, and Liability Act (CERCLA) waste (see Chapter 2, Section 2.1.1), Resource Conservation and Recovery Act (RCRA) solid waste (see Chapter 2, Section 2.1.2.1), RCRA hazardous waste (see Chapter 2, Section

DOI: 10.1201/9781003284000-11

2.1.2.2), and environmental media or debris. CERCLA waste is defined by 40 CFR 300.440(a)(1), which is located within the National Oil and Hazardous Substances Pollution Contingency Plan (NCP).

CERCLA IDW that does not exhibit a characteristic of a hazardous waste (e.g., ignitability, corrosivity, reactivity, toxicity) may be disposed of in a CERCLA-permitted disposal facility. IDW that occurs in the form of an RCRA hazardous waste may either exhibit a characteristic of a hazardous waste (e.g., ignitability, corrosivity, reactivity, toxicity), or the waste may carry a listed hazardous waste code (e.g., F, P, K, U). IDW also includes media (e.g., soil, sediment, water) and other debris that is not hazardous, but is contaminated with hazardous constituents. It is important to note that not all IDWs are hazardous wastes. In fact, the generator can often dispose of some nonhazardous IDWs as any other solid waste.

Although the characterization of IDW most often requires some form of sampling and analysis, this is not always the case. There may be circumstances in which preexisting sampling data may be used to make a hazardous waste determination. Factors such as the age of the sampling data, potential for changed conditions, and whether the sampling data used are representative of the IDW in question must be considered.

Section 2.1.2.2 in Chapter 2 of this book provides some examples of materials that have been excluded from the definition of an RCRA hazardous waste, along with guidance on topics such as:

- How a listed waste code may be removed from a waste through a "contained-out" determination
- What defines a "mixed waste"
- Land disposal restrictions

If polychlorinated biphenyls (PCBs) are determined to be present in the IDW, then Toxic Substance Control Act (TSCA) regulations (40 CFR 761) will apply. As the topic of waste management is beyond the expertise of the author, the readers are advised to seek guidance from a professional waste management specialist before disposing of any IDW. A few references that the readers may find of value as they relate to the management of IDW follow:

- EPA, 2014, *RCRA Orientation Manual*, EPA 530-F-11-003, October
- 40 CFR 260–282
- 40 CFR 300
- 40 CFR 761

Appendix: General Reference Tables

TABLE A.1
Unit Conversion Table

Conversion Table

Measurement Conversions
ENGLISH TO METRIC

Known	Multiplier	Product
LENGTH		
inches	2.54×10^4	micron (= 10,000 Angstrom units)
inches	25.4	millimeters
feet	30.48	centimeters
yards	0.9144	meters
miles (statute)	1.6093	kilometers
AREA		
square inches	6.4516	square centimeters
square feet	0.0929	square meters
square yards	0.8361	square meters
square miles (1 square mile = 640 acres)	2.5900	square kilometers
VOLUME		
cubic inches	16.3871	cubic centimeters
cubic feet	0.02832	cubic meters
cubic yards	0.7646	cubic meters
cubic miles	4.1684	cubic kilometers
quarts (U.S. liquid)	0.9463	liters (= 1000 cm^3)
gallons (U.S. liquid) (= 0.8327 Imperial gal)	3.7854	liters
barrels (petroleum –1 bbl = 42 gal)	158.9828	liters
acre-feet (= 43,560 ft^3 = 3.259×10^5 gal)	1233.5019	cubic meters
MASS		
ounces	28.3495	grams
pounds	0.4536	kilograms
short tons (2000 lb)	0.9072	megagrams = (metric tons)
long tons (2240 lb)	1.0160	megagrams
carats (gems)	0.2000	grams
VOLUME PER UNIT TIME		
cubic feet per second (= 448.83 gal/min)	0.02832	cubic meters per second
cubic feet per second	28.3161	cubic decimeters per second (= liters per second)
cubic feet per minute (= 7.48 gal/min)	0.47195	liters per second
gallons per minute	0.06309	liters per second
barrels per day (petroleum – 1 bbl = 42 gal)	0.00184	liters per second

(Dietrich 1982)

Conversion Table (cont.)

METRIC TO ENGLISH

Known	Multiplier	Product
LENGTH		
micron (= 10,000 Angstrom units)	3.9370×10^{-5}	inches
millimeters	0.03937	inches
centimeters	0.0328	feet
meters	1.0936	yards
kilometers	0.6214	miles (statute)
AREA		
square centimeters	0.1550	square inches
square meters	10.7639	square feet
square meters	1.1960	square yards
square kilometers	0.3861	square miles (1 square mile = 640 acres)
VOLUME		
cubic centimeters	0.06102	cubic inches
cubic meters	35.3146	cubic feet
cubic meters	1.3079	cubic yards
cubic kilometers	0.2399	cubic miles
liters (= 1000 cm^3)	1.0567	quarts (U.S. liquid)
liters	0.2642	gallons (U.S. liquid)
liters	0.006290	barrels (1 bbl = 42 gal)
cubic meters	0.0008107	acre-feet (= 43,560 ft^3 = 3.259×10^5 gal)
MASS		
grams	5.0000	carats (gems)
grams	0.03527	ounces
kilograms	2.2046	pounds
megagrams (= metric tons)	1.1023	short tons (2000 lb)
megagrams	0.9842	long tons (2240 lb)
VOLUME PER UNIT TIME		
cubic meters per second	35.3107	cubic feet per second (= 448.83 gal/min)
cubic decimeters per second (liters per second)	0.03532	cubic feet per second
liters per second	2.1188	cubic feet per minute
liters per second	15.8503	gallons per minute
liters per second	543.478	barrels per day (petroleum – 1 bbl = 42 gal)

(Dietrich 1982)

Conversion Table (cont.)

Volume

1 cubic ft	=	7.4805 U.S. gallons	=	6.2321 imperial gallons	=	28.37 liters
1 U.S. gallon	=	0.13368 cubic ft	=	0.83271 imperial gallons	=	3.7825 liters
1 imperial gallon	=	0.16046 cubic ft	=	1.2009 US. gallons	=	4.5437 liters
1 liter	=	0.03531 cubic ft	=	0.26417 US. gallons	=	0.22009 imperial gallons
1 cubic ft	=	0.028317 cubic meter	=	0.000022957 acre-ft		
1 cubic meter	=	35.3145 cubic ft	=	0.00081071 acre-ft		
1 acre-ft	=	43,560 cubic ft	=	1,233.5 cubic meters		
1 cubic mile	=	3.3792 million acre-ft				
1 cfs-day	=	86,400 cubic ft	=	1 cubic ft per second for 24 hr		

Volume Conversion Factors

Initial Unit	Coefficient (multiplier) to obtain:					
	Cfs-day	Mil. cu. ft	Mil. gal.	Acre-ft	In. per sq mi.	Mil. cu. meters
Cfs-days	–	0.086400	0.64632	1.9835	0.037190	0.0024466
Mil. cu. ft	11.574	–	7.4805	22.957	0.43044	0.028317
Mil. gal.	1.5472	0.13368	–	3.0689	0.057542	0.0037854
Acre-ft	0.50417	0.043560	0.32585	–	0.018750	0.0012335
In. per sq. mi.	26.889	2.3232	17.379	53.333	–	0.065785
Mil. cu. meters	408.73	35.314	264.17	810.70	15.201	–

Velocity				
1 mile per hr	=	1.467 ft per sec		
1 mile per hr	=	88 ft per min		
1 ft per sec	=	0.682 mile per hr		
1 ft per min	=	0.0114 mile per hr		
1 ft per sec	=	0.3048 meter per sec		
1 meter per sec	=	3.281 ft per sec		
Pressure (0°C = 32°F)				
1 ft of head, fresh water	=	0.433 lb per sq in, pressure		
1 lb per sq in, pressure	=	2.31 ft of head, fresh water		
1 meter of head, fresh water	=	1.42 lb per sq in, pressure		
1 lb per sq in, pressure	=	0.704 meter of head		
1 atmosphere (m.s.i.)	=	33.907 ft of water		
Weight				
1 cubic ft of fresh water	=	62.4 lb	=	28.3 kg
1 cubic ft of sea water	=	64.1 lb	=	29.1 kg
1 cubic ft of fresh water	=	1000 kg	=	1 metric ton

(Dietrich 1982)

Conversion Table (cont.)

Rates of Flow

1 cubic ft per sec	=	448.83 U.S. gallons per min	=	646,317 U.S. gallons per day	=	0.028317 cu meter per sec
1 cubic ft per min	=	7.4805 US. gallons per min	=	10,722 US. gallons per day	=	0.00047195 cu meter per sec
1 US. gallon per min	=	0.002228 cubic ft per sec	=	0.13368 cubic ft per min	=	1440 US. gallons per day = 0.000063090 cu meter per sec
1 US. gallon per day	=	0.000093 cubic ft per min	=	0.0006944 US. gallons per min		
1 cubic ft per sec	=	1.9835 acre-ft per day	=	723.98 acre-ft per year		
1 acre-ft per day	=	0.50417 cubic ft per sec	=	365 acre-ft per year	=	0.014276 cu meter per sec
1 acre-ft per year	=	0.00138 cubic ft per sec	=	0.00274 acre-ft per day		
1 inch per hr on 1 acre	=	1 cubic ft per sec (approx.)				
1 inch per hr on 1 sq mi	=	645.33 cubic ft per sec				

Volume Conversion Factors

Initial Unit	Coefficient (multiplier) to obtain:					
	Cu ft per sec	Gal per min	Mil gal per day	Acre-ft per day	Inches per day per sq mi	Cu meters per sec
Cu ft per sec (cfs)	–	448.83	0.64632	1.9835	0.037190	0.028317
Gal per min (gpm)	0.0022280	–	0.001440	0.0044192	0.00008286	0.000063090
Mil gal per day (mgd)	1.5472	694.44	–	3.0689	0.057542	0.043813
Acre-ft per day	0.50417	226.29	0.32585	–	0.01850	0.014276
Inches per day per sq mi	26.889	12,069	17.379	53.333	–	0.76140
Cu meters per sec	35.314	15,850	22.834	70.045	1.3134	–

(Dietrich 1982)

TABLE A.2
Unified Soil Classification System

Unified Soil Classification System				
MAJOR DIVISIONS			GROUP SYMBOLS	TYPICAL NAMES
COARSE-GRAINED SOILS More than half of material is larger than no. 200 sieve size.	GRAVELS More than half of coarse fraction is larger than no. 4 sieve size.	Clean gravels	GW	Well-graded gravels, gravel-sand mixtures, little or no fines
			GP	Poorly graded gravels, gravel-sand mixtures, little or no fines
		Gravels with fines	GM	Silty gravel, gravel-sand-silt mixtures
			GC	Clayey gravels, gravel-sand-clay mixtures
	SANDS More than half of coarse fraction is smaller than no. 4 sieve size	Clean sands	SW	Well-graded sands, gravelly sands, little or no fines
			SP	Poorly graded sands, gravelly sands, little or no fines
		Sands with fines	SM	Silty sands, sand-silt mixtures.
			SC	Clayey sands, sand-clay mixtures

Unified Soil Classification System

FINE-GRAINED SOILS More than half of material is smaller than no. 200 sieve size.	SILTS AND CLAYS	Low liquid limit	ML	Inorganic silts and very fine sands, rock flour, silty or clayey line sands, or clayey silts, with slight plasticity
			CL	Inorganic clays of low to medium plasticity, gravelly clays, sandy clays, silty clays, lean clays
			OL	Organic silts and organic silty clays of low plasticity
		High liquid limit.	MH	Inorganic silts, micaceous or diatomaceous fine sandy or silty soils, elastic silts
			CH	Inorganic clays of high plasticity, fat clays.
			OH	Organic clays or medium to high plasticity, organic silts
Highly organic soils			Pt	Peat and other highly organic silts

NOTES:

1. Boundary Classification: Soils possessing characteristics of two groups are designated by combinations of group symbols. For example, GW-GC, well-graded gravel-sand mixture with clay binder.
2. All sieve sizes on this chart are U.S. Standard.

(Dietrich 1982)

TABLE A.3
Classification of Soils

Concept and Classification of Soils

DEFINITION: Soil is a natural, historical body with an internal organization reflected in the profile and its horizons, consisting of weathered rock materials and organic matter, with the former usually predominant, and formed as a continuum at the land surface largely within the rooting zones of plants.

HYPOTHETICAL SOIL PROFILE:
with notations for master horizons

pre-1980 nomenclature	current nomenclature	
01	Oi	Loose leaves and organic debris, largely undecomposed..
02	0e	Organic debris, partially decomposed
A1	Ah	A dark-colored horizon of mixed mineral and organic matter and with much biological activity
A2	E	A light-colored horizon of maximum eluviation; prominent in some soils but absent in others
A3	EB	Transitional to B but more like A (or E) than B; may be absent
B1	BE	Transitional to A (or E) but more like B than A (or E); may be absent
B2	B	Maximum accumulation of silicate clay minerals or of sesquioxides and organic matter; maximum expression of blocky or prismatic structure; or both
B3	BC	Transitional to C but more like B than C; may be absent
C	C	Weathered parent material: occasionally absent; formation of horizons may follow weathering so closely that the A or B horizon rests on consolidated rock
R	R	Layer of consolidated rock beneath the soil

(Dietrich 1982)

REFERENCE

Dietrich, R.V., Dutro, J.T., Jr., and Foose, R.M., 1982, American Geologic Institute Data Sheets.

Index

A

B

C

H

N

O

P

Z

Printed in the United States
by Baker & Taylor Publisher Services